Writing in the Sciences

Exploring Conventions
of Scientific Discourse

Writing
in the Sciences

Exploring Conventions
of Scientific Discourse

ANN M. PENROSE
North Carolina State University

STEVEN B. KATZ
North Carolina State University

St. Martin's Press • New York

Editor-in-chief: Steve Debow
Manager, Publishing services: Emily Berleth
Publishing services associate: Meryl Gross
Production supervisor: Kurt Nelson
Project management: Omega Publishing Services, Inc.
Text design: Gene Crofts
Cover design: Evelyn Horovicz
Cover photos: top: Uniphoto/NASA; *bottom:* Uniphoto

Library of Congress Catalog Card Number: 95-073180

Manufactured in the United States of America

1   0   9   8
f   e   d   c   b   a

For information, write:
St. Martin's Press, Inc.
175 Fifth Avenue
New York, NY 10010

ISBN: 0-312-11971-2

Acknowledgments

Figure 3.1. Source: *Scientific Style And Format: The CBE Manual for Authors, Editors, and Publishers* by Edward J. Huth. Table 3.1, p. 590, "Sections of a Research Report," 1994. Reprinted with the permission of Cambridge University Press.

Figures 3.3 and 3.4. Source: *Physics in Medicine and Biology,* 1989, Vol. 34, No. 7, 795–805. Courtesy of IOP Publishing Limited, Bristol, England.

Figure 3.8. Source: Brody and Pelton, *Canadian Journal of Zoology.* Copyright © 1988 by NRC Research Press. Reprinted by permission.

Figure 5.1. Source: Reprinted with the permission of the American Society for Microbiology.

Figure 7.4. Source: Terence R. Monmaney. Excerpt from "Marshall's Hunch," originally published in *The New Yorker,* September 20, 1993, pp. 64–72. Copyright © 1993. Reprinted by permission of the author.

Acknowledgments and copyrights are continued at the back of the book on page 309, which constitutes an extension of the copyright page.

Contents

• •

· · · · · · · · · · · · · ·

Part Two   Sample Research Cases

9   Research on the "Ulcer Bug"

MARSHALL AND WARREN AND COLLEAGUES

Preface

....

Writing in the Sciences can be used in scientific writing courses, in technical writing courses with large numbers of science majors, or by instructors in the sciences seeking to integrate writing into their undergraduate courses. It can also be used as a supplemental text in specialized or graduate courses and by students and professionals seeking to polish the communication skills they use every day. This book grows out of our own experience teaching a multigenre, multidisciplinary course in scientific writing in our English department over the past eight years. Our goal was to design a text that treats the major genres of scientific communication in a specific yet flexible way that would be applicable to a variety of disciplines. Our approach and methods are based on research in rhetoric, composition, linguistics, psychology, sociology, and on the philosophy of science and science education recently adopted by the National Academy of Sciences.

The National Academy of Sciences (1995) formally recognizes communication not only as necessary for the conduct of research but also as fundamental to the scientific enterprise itself. The *Science Citation Index* lists over 3000 peer-reviewed publications, and more than 40,000 scientists each year apply for research funding from the National Science Foundation and National Institutes of Health (Seiken 1992). Given these figures, it is not surprising our students found that scientists spend an average of five hours a day writing, reading, and reviewing scientific texts.

Our focus is not only on the distinctive features of the major genres in science and research fields but also on how and why such texts are created by scientists. In keeping with the goal of treating genres as they are used in various disciplines, we also approach the task of teaching students in the sciences how to write in their chosen fields in a descriptive rather than a prescriptive way. That is, we present general rhetorical concepts and analytical tools that students will need to recognize the conventions actually used by scientists in their fields. This approach has several advantages.

On the practical level, teaching students how to investigate their own disciplines enables us to address writing in all sciences by looking at generalizable conventions across fields, rather than trying to investigate and prescribe the conventions of each particular field or, more likely, of just a few fields.

Teaching students how to investigate their own fields also enables instructor and student to investigate variation across fields if they so desire. We have found that exceptions to general conventions can sometimes and in some fields be the rule, and that these exceptions are often illuminating when seen against and understood in terms of the processes and products of scientific writing generally. For example, in theoretical and observational research fields such as geology, astrophysics, and some wildlife sciences, the standard research report form is quite different from that of the experimental sciences. Comparing these forms reveals a great deal about the questions, methods, and operating assumptions of these respective fields. This comparison of disciplines is offered as a pedagogical device for teasing out distinctive features of the discourse of a discipline.

Students will find these investigative skills applicable not only in their undergraduate and graduate communication courses, but also throughout their careers as professional researchers and scientists.

Writing in the Sciences contains professional examples of each type of discourse introduced. Part One consists of chapters on the major genres and social dimensions of scientific communication. In each chapter we describe common conventions and explore variations across fields, discuss how and why such documents are created by drawing on studies of scientists' composing processes and contexts, and discuss how these documents are read, "used," and evaluated in the scientific community. These chapters draw extensively on three "research cases" presented in Part Two. These cases, from medicine (Chapter 9), botany (Chapter 10), and astrophysics (Chapter 11), comprise sample reports, proposals, and other texts drawn from contemporary lines of research in these fields. Instructors may choose to use these documents as we do, as illustrations for students working through Chapters 1 through 8; or these later chapters may serve as the starting point for a course that works inductively back through the text.

The book provides maximum flexibility for structuring a course. Chapters may be used in different orders depending on the focus of the course and the needs of students. Graphics are treated in several chapters, where they are discussed in the context of specific genres. We have included a separate chapter on literature review skills, which students find particularly challenging; this material can be used on its own, or in conjunction with the chapters on writing research reports and writing proposals. We also have included a chapter on ethics in scientific communication to provide a critical context for the study of scientific conventions and genres.

In short, this book is designed to help instructors of scientific writing guide students from a broad range of disciplines in researching the conventions of their fields. Our ultimate goal is to enhance students' understanding of the rhetorical processes of science, in order that they may learn to communicate persuasively and responsibly as scientists.

An Instructor's Guide, containing commentary on selected exercises, additional examples, and sample syllabi, is available. To request a copy, contact Scientific American/St. Martin's Press Faculty Services at 1-800-446-8923 or <faculty services@sasmp.com>.

We were fortunate to have many fine minds helping us with this project. We are extremely grateful to our colleagues Michael Carter and Carolyn Miller; the value of their careful readings was surpassed only by the value of their collegial support. We would like to thank our research assistant, Patricia Watson, for her tireless effort in running down examples and information and for her ideas about the field of science and the writing of this book. We'd also like to acknowledge other English Department colleagues who contributed materials, advice, and moral support, including Cynthia Haller, Carl Herndl, Carol Hermann, Robanna Knott, Brad Mehlenbacher, and Kelley Sassano.

We owe a special debt to two scientists at North Carolina State University, JoAnn Burkholder from the Department of Botany, and Stephen Reynolds from the Department of Physics, who provided the sample texts on which much of this book is built. We are indebted as well to Carl Blackman of the Environmental Protection Agency, who graciously offered reflections on his own development as a professional scientist and communicator. The willingness of these scientists to share their work, time, and expertise with us illustrates what we believe to be the best features of science: its collaborative nature and social spirit.

Many readers contributed comments and reactions during the development of *Writing in the Sciences*. We gratefully acknowledge their participation and candid commentary and thank them for their assistance: Christy Friend, University of Texas, Austin; Kyle S. Glover, Oklahoma State University; Alan G. Gross, University of Minnesota; Gary Layne Hatch, Brigham Young University; Susan Peck MacDonald, University of California, San Diego; Marshall Myers, Texas Tech University; John R. Nelson, Jr., University of Massachusetts, Amherst; Christopher Sawyer-Lauçanno, Massachusetts Institute of Technology; and Dale Sullivan, Northern Illinois University.

We are grateful for the guidance and support of the professionals at St. Martin's Press who have skillfully steered the project through publication, including Steve Debow, Emily Berleth, Sandy Schechter, Evelyn Horovicz, and former editorial director Karen Allanson. Special thanks also to Rich Wright at Omega Publishing Services for his bookmaking expertise.

We would also like to take this opportunity to express our gratitude to David Brooks and Alison Katz for their encouragement and support throughout the writing process.

Finally, we are indebted to our students at North Carolina State. Over the past eight years, students in English 333 "Communication for Science and Research" have played no small part in the writing of this book. As in science, they have been not only observers but also participants in this experiment.

Ann M. Penrose
Steven B. Katz
Raleigh, North Carolina

1

Science as a
Social Enterprise

We begin this introduction with the case of a young scientist who made a revolutionary discovery. In September, 1983, Barry Marshall, an unknown internist at Australia's Royal Perth Hospital, presented his and pathologist J. R. Warren's findings at the Second International Workshop on Campylobacter Infections in Brussels in 1983. They had already published separate, technical "letters" together in the *Lancet* (Warren and Marshall 1983). In these letters they reported the presence of bacteria (later classified as *Helicobacter pylori [H. pylori]*) in the stomach lining of patients with gastritis. Following the presentation of their joint paper at this international conference, Marshall stood before doctors, researchers, and specialists in microbiology, gastroenterology, and infectious diseases. In response to a question from one of the experts, Marshall declared that he thought that this bacterium was the cause of all stomach ulcer disease and that chronic ulcer recurrence could be eradicated in most if not all patients with a treatment of common antibiotics and bismuth such as Pepto Bismol (Chazin 1993; Monmaney 1993).

Marshall and Warren's theory that ulcers were the result of bacterial infection in the stomach lining was greeted with both intense interest and skepticism by experts in the field. After all, the claim that *anything* could even live in the intensely acidic environment of the stomach was as unbelievable as it was revolutionary. If the claim were true, the implications for ulcer sufferers—and for the major drug companies manufacturing acid blocker treatments—would be substantial. While they were polite and respectful in the professional journals, experts in the field later told the popular press

1

that when Marshall first presented the theory, they thought he was "brash" (SerVaas 1994, 62), "a madman" (Chazin 1993, 122), "a medical heretic" (Monmaney 1993, 65), "a crazy guy saying crazy things" (Monmaney 1993, 66). The experts were intrigued, however, as evidenced by letters and editorials in scientific journals devoted to gastroenterology and internal medicine. Eventually, a series of research reports in the *Lancet, Gastroenterology,* and the *Annals of Internal Medicine* provided support for the theory (see Chapter 9 for examples).

By 1993, more than 1500 studies around the world had lent support to Marshall's theory (Monmaney 1993). In a major shift in policy in February 1994, a panel of the National Institutes of Health released a consensus statement in which it accepted that there is a relationship between *H. pylori* and ulcer disease and recommended that antibiotics be used in the treatment of stomach ulcers where the bacterium is present (NIH 1994). In April 1996, the FDA approved the first drug specifically designed for treating ulcers based on Marshall's findings. Today, Marshall's therapy of antibiotics in conjunction with bismuth is increasingly being prescribed by physicians not only as a treatment for chronic duodenal and gastric ulcers but also as a possible cure.

Dr. Marshall, later tenured at the University of Virginia School of Medicine and now Director of the Helicobacter Foundation, started a revolution in the field of gastroenterology (Carey 1992). Yet, it took more than a decade for the theory to be accepted. And most of the development and support has come from other physicians and scientists. While Marshall continued to publish technical letters, abstracts, reports, and some retrospective studies, that decade was a period of trial and tribulation for him. Following his presentation at the Brussels conference, Marshall had great difficulty getting his clinical research published and felt his message about ulcer treatment was not being heeded (Chazin 1993).

Why was there such negative reaction to Marshall's presentation in Brussels? Despite all the positive editorials by scientists expressing interest in the theory and the subsequent research and confirmation of the claim, why did Marshall have difficulty getting his research published? Why wasn't his theory immediately embraced, as he, the press, and the public thought it should have been? A close examination of the answers to these questions can provide insight into how knowledge is created and shaped in science and into the role communication plays in that process.

Certainly, part of the answer has to do with the somewhat shaky nature of Marshall's methods—with his science itself. In 1982, J. Robin Warren, a pathologist at the Royal Perth Hospital, had observed the presence of the bacterium in the stomach and showed Marshall the slides. Together they studied a hundred patients suffering from peptic ulcers and found that in 87 percent of the cases *H. pylori* was present. This was the study they presented in Brussels and published in the *Lancet* in 1984. But by his own accounts—and those of experts such as David Graham, chief of gastroenterology at Houston's Veterans Affairs Medical Center and one of Marshall's most outspoken opponents—Marshall was "not the greatest researcher of all time" (Chazin 1993, 123). One scientific editorial pointed to the 1984 Marshall and Warren study in the *Lancet* as "well planned" (*Lancet* 1984, 1337), but in 1988 Marshall's large-scale study was rejected by the *New England Journal of Medicine* as "inconclusive" (Chazin 1993, 123). Other editorials in scientific journals criticized the validity of his methods and conclusions (see Lam 1989), and Marshall admitted to the press that he was more interested in curing patients than in developing adequate methods and conducting large clinical experiments needed to support his claim; he had hoped that clinical success with patients would be enough to convince his colleagues (Chazin 1993). Thus, Marshall was faced with the problem that to some degree all scientists, not just the lucky ones who wind up doing important research, face in their career. How do you get other scientists to listen to you?

To gain the acceptance he thought his theory deserved, Marshall did something that most scientists would—and should—never do. Unable to convince his colleagues that *H. pylori* caused stomach ulcers, in 1984 Marshall created a potent mixture containing the bacteria and drank it, inducing a case of acute gastritis in himself (Marshall et al. 1985). Marshall's methods were not only unorthodox but radical and potentially dangerous. Although Marshall's experimenting on himself led to further research, the incident made Marshall's critics even more skeptical of his professionalism and less accepting of his theory (Morris 1991; Carey 1992).

Another reason Marshall's theory wasn't readily accepted has to do with the prevailing assumptions about, and treatment of, gastrointestinal diseases. Warren and Marshall were not the first to observe bacteria in connection with gastric inflammation. Medical researchers as far back as the late 1800s had reported and even published images of the bacteria living in the stomach lining (Blaser 1987). But these findings were either dismissed as contaminants introduced during biopsy or as unrelated agents existing near ulcers; because of the presence of hydrochloric acid, gastroenterologists assumed that the stomach was "a sterile organ" (Warren and Marshall 1983, 1273). As Monmaney (1993) reported, the causes of ulcers have been and still are to a large extent attributed to weak stomach linings, and/or to an increase in stomach acids caused by emotional trauma, tension, nervousness, or modern life itself: "Marshall's theory challenged widely held and seemingly unassailable notions about the cause of ulcers. No physical ailment has ever been more closely tied to psychological turbulence" (64).

The contrast between the initial response to Marshall's hypothesis and the subsequent success of his theory suggests several important points about how knowledge is shaped in science and the importance of communication in that process. We will highlight these briefly here by way of introduction and then will delve into them in more depth throughout this book as you begin to explore them in your own field.

- Scientific experimentation and knowledge are governed by tacit beliefs and assumptions about what is factual, valid, and acceptable; these beliefs and assumptions are "social" in nature.
- Communication is central to the growth of scientific knowledge in each discipline, and thus to the advancement of science itself.
- Persuasion is an integral part of scientific communication; it includes the use of sound arguments and an appropriate style of presentation, as well as acceptable theories, methods, and data.
- As a social enterprise, scientific fields are also to some degree governed by explicit conventions and rules about how and what to communicate in science, conventions and rules that professional researchers and scientists expect each other to follow; failure to follow these can result in a failure to communicate and thus can hamper the advancement of scientific knowledge.

• Collaboration and cooperation are essential to the development of scientific theories, research, and knowledge. As you will learn in your exploration of communication in science, collaboration and cooperation are central to research and to the actual writing of research papers and proposals. Scientific knowledge is built and shared through collaboration and cooperation.

1.2 The social nature of science

As you will see when you explore your own field, science is a social enterprise. In one sense, this means that science is a part of the larger society in which it is situated. Science is shaped by the values of the dominant culture in which scientists participate and live, sharing many of its assumptions, goals, biases, and problems (NAS 1995). Conversely, science also exerts a powerful influence on society. Think about decisions you've made recently about such practical things as medical treatment, diet, energy and fuel consumption, weather. Indeed, at a very deep level, our very way of thinking about the world is rooted in current scientific practices and beliefs.

Our focus here, however, is on the social nature of the activity within scientific communities rather than on the broader social context in which these communities are embedded. Scientists in a discipline constitute a *community* in which knowledge is built, is validated, has meaning. Robin Warren needed Barry Marshall's clinical knowledge to understand how the bacterium he was observing was related to gastric symptoms (Chazin 1993). Marshall needed Warren in order to understand the biology of the bacterium he was observing in his patients. Both Marshall and Warren needed other scientists and the NIH to further test, validate, accept, and extend their work in theory and practice. We saw earlier that this can be a slow process. Like any society, scientific communities operate by a system of assumptions and beliefs which govern the perception and understanding of phenomena, the methods used, the research that is conducted, and the kinds of conclusions that can be drawn.

EXERCISE 1.1   To get an idea of how assumptions and beliefs operate in science, try the following experiment. In the next five minutes, connect all the dots below with only four straight lines and without lifting your pencil from the page. If you have done this problem before, do what Einstein did: a *Gedanken* (thought) experiment: connect all the dots with only one line; there are several ways of doing this.

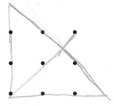

Exercise 1.1 presents the classic "nine-dot problem" (Adams 1976). For those of you who are having trouble solving this problem, don't worry; most people do. The frustration you are feeling is the creative tension, which scientists, like everybody else, experience when they're trying to solve a problem and can't find the solution. The tension makes you want to keep working on the problem and is one of the driving forces of science. Often the difficulty is that the solution cannot be found within your current cognitive framework; your assumptions about the problem, the experimental and cultural context in which you are working, and your expectations about the solution all influence your perception. Thus, according to Kuhn (1970), finding the solution involves something like a "gestalt" switch, where you see the problem in a new context, a context that allows you to find the solution.

Notice that solving the nine-dot problem involves going outside the conceptual space you perceived and believed to be a square. The perception that dots form lines is, of course, one of the basic principles of geometry, which defines a *line* as the track made by a moving point, and so in some sense this perception is based on a social belief or convention. But the belief that the solution must fall within the square prevents the perception of the easy solution. To solve the nine-dot problem, you must "violate" or ignore conventional beliefs and assumptions. This is what leading gastroenterologists had to do to accept Marshall's theory that stomach ulcers are caused by bacteria rather than stress.

The important point here is that assumptions and beliefs influence scientists' perception of phenomena. In *The Structure of Scientific Revolutions,* Thomas Kuhn (1970) calls these sets of assumptions in science *paradigms*. Paradigms are learned tacitly, through observation and imitation of scientists and practitioners in the field (Polanyi 1958); and they are learned explicitly through education, textbooks, and practice (Kuhn 1970). Thus some of the assumptions in paradigms are only implicit or subconscious, and some are explicit in the form of rules and accompanying examples. As students of science, you are currently engaged in learning the paradigms of your fields. As we will explore in this book, paradigms include conventions, assumptions, and rules about communication in science as well.

Paradigms constrain thinking, as illustrated in the nine-dot problem. Yet, at the same time paradigms provide the support and context for discovery. Without the background of the square created by the pattern of the dots, there would be no clear way to connect the dots, nothing to break out of, no problem to solve. In the case of *H. pylori,* the discoveries that had been made—and the assumptions, examples, and knowledge about bacteria and about ulcers that already existed in the field—provided both the scaffolding and the backdrop against which to discover the connection between the bacterium and gastritis. Although Warren, Marshall's coworker, stated that the discovery of the ulcer bacteria "was something that came out of the blue," he also admitted he "happened to be there at the right time, because of the improvements in gastroenterology in the seventies" (Monmaney 1993, 68). We observe here that the discovery by an individual scientist that takes place outside the dominant paradigm nevertheless depends on that paradigm for the perception of it. Without paradigms, we would notice nothing at all. As a set of assumptions, rules, and examples, the prevailing paradigm in a field helps define the problems in that field, specify methods that are allowed to solve the problem, and predict what results can be expected.

This is not to deny the role of the individual scientist's perception and judgment. Some assumptions in science are personal, subjective, and even aesthetic (the scientific values of simplicity and elegance, for example, are aesthetic). The National Academy of Science calls attention to this dimension of science:

> Researchers continually have to make difficult decisions about how to do their work and how to present that work to others. Scientists have a large body of knowledge that they can use in making these decisions. Yet much of this knowledge is not the product of scientific investigation, but instead involves value-laden judgments, personal desires, and even a researcher's personality and style. (1989, 1)

In general practice, this subjective dimension is balanced by social paradigms. Paradigms as social mechanisms act as a check on personal judgment and individual error, and so they make science as we know it possible (NAS 1995). In fact, what is personal, private, and subjective must be validated socially by the community of scientists to count as "objective" scientific knowledge. Thus, while paradigms that operate in a particular discipline tend to influence perception and constrain scientific research and thought, they also serve to correct and enhance it.

Sometimes, as in the case of Drs. Marshall and Warren, the individual scientist or group of scientists opposing a dominant paradigm in their field turn out to be right. When other scientists begin to shift their beliefs and assumptions and work on the problems created by the new theory, the field undergoes a paradigm shift—what Kuhn (1970) calls a "scientific revolution." Scientific revolutions, such as that of the shift from Ptolemaic to Copernican astronomy, or from the belief that the earth was flat to the belief that earth is round, are very well-known paradigm shifts. Other scientific debates in contemporary society that may involve conflicting paradigms include the discussion of whether evolution proceeds gradually or by leaps (gradualism versus punctuated equilibrium); whether the universe is expanding and contracting or is a steady-state universe; the cause of dinosaur extinction (continental drift versus meteor storm and/or global climate change). Can you think of any debates that may involve conflicting paradigms in your field?

In these special instances of revolutionary or extraordinary science, as in normal science (Kuhn 1970; Toulmin et al. 1984), scientific knowledge must be validated by the community of scientists. As we will discuss next, communication is essential to that process and thus is central to science itself.

1.3 The centrality of communication in science

The centrality of communication in science? To the casual observer, the phrase may seem somewhat nonsensical if not patently false. After all, what matters in science is the science itself—hypotheses, research methods, experimental apparatus, results. Certainly, these are fundamental to good science. Yet, without the communication of those hypotheses, research methods, experimental apparatus, and results to other scientists, no science would be possible. Scientists in and across fields would not be able to share or build knowledge on the results of other scientists. Science would become a private, redundant, and ultimately futile endeavor.

Some external factors exert enormous pressure on scientists *not* to share. Scientists who work for industry or on security projects for the government, for instance, often find there are restrictions on what they can talk about. But scientists generally agree that secrecy is bad for science (NAS 1989, 1995; see Huizenga 1992). One of the fundamental principles of science is free and open communication. The National Academy of Science puts it bluntly: "If scientists were prevented from communicating with each other, scientific progress would grind to a halt" (1989, 10).

Barry Marshall's sharing of his theory prompted a rethinking of the field of gastritis, ulcer treatment, and the curing of patients. His presentation in Brussels did "arouse much interest" (*Lancet* 1984, 1336), and the desire and necessity to check out his theory propelled scientists in the fields of microbiology, gastroenterology, and internal medicine into a flurry of research. Hundreds of articles investigating the existence, nature, classification, detection, and treatment of *H. pylori* were produced, eventually lending support to Marshall's contention.

For the National Academy of Science, communication is the engine that drives the "social mechanism" of science. We have already noted that the communication of hypotheses, research methods, experimental apparatus, and results—in journals, at conferences, over E-mail—is essential for the growth of science. *Sharing ideas is essential to the evolution of every scientific field.* In this sense, it is communication that binds any discipline into a community, that makes science social. Language is the basis of any society. Without language, there could be no communication, no cooperation, no concerted research effort that we note in the investigation of *H. pylori* or other scientific endeavors.

Indeed, so important is communication in science that *the credit for a scientific discovery is awarded not to the scientist who discovers a phenomenon, but to the scientist who publishes first!* (Actually, the date the paper is received by the journal is what determines historical priority.) This is another way in which communication is central in the conduct of science. The practice of attributing originality to the scientist whose paper first reaches the offices of the journal, along with the promise of quick turnaround time for publication and support by the institution and/or journal in disputes concerning the ownership of ideas and discoveries, was begun by the Royal Society in the eighteenth century. By protecting the rights of the author, the Royal Society hoped to ensure open communication and the sharing of ideas in science by alleviating the (real) fear of scientists that their ideas or results would be stolen by other scientists; the National Academy of Sciences cites the example of Isaac Newton who wrote in Latin anagrams so his findings could be on record but not publicly available (NAS 1989). Marshall and Warren were not the first to observe *H. pylori*. But they were the first to recognize and write about its role in gastritis, and their joint letters to the *Lancet* thus mark the historical point of discovery of that phenomenon (Warren and Marshall 1983).

Since historical priority is awarded to the scientist(s) whose manuscript reaches publication first, scientists who do original work and want that work recognized and used by the field must also publish quickly. Marshall presumably felt this need, for he, like many other scientists, submitted initial reports of his important discovery in the form of brief technical letters or preliminary notes rather than as fully elaborated research reports (the Watson and Crick letter in *Nature* about the double-helix structure of DNA is a famous example). Speed of publication is necessary not only because of the rapid advance of the field but also because of scientists' need to ensure their claim of originality.

Thus, while scientific journals protect the science that is submitted and published, the necessary speed of publication in most sciences also creates a competition in science: "researchers who refrain from publishing risk losing credit to someone else who publishes first" (NAS 1989, 9). It is a matter of historical fact that eminent scientists such as Darwin, Watson and Crick, and others were pressured *to write up* and publish their results to beat the competition. Darwin didn't want to write at all, and only did so when he heard that Alfred Russell Wallace was about to publish a theory of evolution (Campbell 1975); Watson and Crick were hotly competing with another lab to be the first to announce the structure of DNA (Watson 1968; Halloran 1984). Learning to write quickly and well is important in science.

There are other, less obvious but absolutely crucial ways in which communication has come to play a key role in science. The processes of writing and submitting papers, of giving presentations, and of writing grant proposals in a real sense *define the nature and activity of the field and the state of knowledge within it.* The acceptance or rejection of conference abstracts, presentations, papers, grant proposals, and the like by conference organizers, journals, funding agencies, and peers becomes a vehicle not only for the dissemination but also for the control of scientific research. As the National Academy of Science states, "At each stage, researchers must submit their work to be examined by others with the hope that it will be accepted. This process of public, systematic skepticism is critical in science" (1989, 10).

Peer review is the primary mechanism through which such gatekeeping is accomplished. In addition to the dissemination of scientific research, it is the function of research journals to ensure "quality control" by deciding what is acceptable to publish in the field. These decisions are typically accomplished via a "peer review" system in which journal editors send the manuscripts they receive from researchers to other researchers working the same field for their expert opinion and evaluation. After soliciting evaluations on a given paper from several such experts, an editor is better able to make a decision about whether the paper merits publication in the journal or not. Through this process, journal editors and reviewers influence what scientists read and, to some extent, what scientists work on (Bazerman 1983; NAS 1989). (Marshall's three-year clinical study was not published because it was inconclusive (Chazin 1993); but his smaller studies and technical notes were allowed into print.) Similarly, research funds are typically allocated using a peer review process (Seiken 1992). Funding agencies such as the National Science Foundation, the Department of Energy, and the National Institutes of Health evaluate proposals or requests for funding by assigning them for peer review and then, based on recommendations, decide which studies to fund, thus largely determining what kind of research can proceed (Myers 1985; Seiken 1992).

Scientists take this system of checks and balances seriously and believe that violating or circumventing it makes for bad science. Scientists become concerned when scientific results are released to the public prematurely or independently of scientific publication. As the National Academy of Science states, "Bypassing the standard routes of validation can short-circuit the self-correcting mechanisms of science. Scientists who release their results directly to the public—for example, through a press conference called to announce a discovery—risk adverse reactions later if their results are shown to be mistaken or are misinterpreted by the media or the public" (1989, 10). It is because of this gatekeeping function of the peer review process, the need and desire to control the flow of information and ensure "quality control," that E-mail and fax, which have no quality control mechanism in place, also have become a concern for professional scientific organizations (NAS 1995).

These gatekeeping mechanisms and the entire consensus process by which theories and results are verified and accepted as objective knowledge in the scientific community depend on the fair, accurate assessment of research. Because of this need, *papers must be written in a way that makes the science accessible, testable, and acceptable.* As philosopher of science Karl Popper (1959) proposed, a scientific hypothesis or theory must be susceptible to falsification to be proved true; that is, a hypothesis or theory must be wholly testable (as opposed to tested) before it can be accepted by the scientific community as a valid hypothesis or theory. One way in which findings are tested is through replication, in which later scientists repeat or build on the methods and results of an earlier study. The practice of replication as standard procedure was recommended at the beginning of modern science by Francis Bacon to address the untrustworthiness of the senses and mind in interpreting what we see, playing tricks with our perception and understanding of reality (Bacon 1605).

In addition, *the development of knowledge in science depends on the willingness and ability of scientists to share information* after *publication.* Research reports must provide enough information for readers to evaluate the plausibility and rigor of the researchers' theory, methods, and results; but scientists can't and don't include every detail of their experiment in their reports (Berkenkotter and Huckin 1995). Researchers must be willing to make further details available to others working in the field. This is especially important when you consider that scientists cannot be present at each other's observations and lab experiments. Rather, scientists must rely on how observations, lab experiments, and results are presented in writing. Thus, as Popper (1959) pointed out, the statements contained in a research report come to embody and represent the science itself. Hypotheses, theories, experiments, and results are primarily presented, obtained, and critiqued through publication.

• •

EXERCISE 1.2   1. Carefully read the summary contained in Figure 1.1. Describe the sequence of events. Now think about the discussion in this chapter. Based on the information given here, why do you think Pons and Fleischmann did what they did? Explain the reaction of the scientific community. What principles of science were involved? What principles of scientific communication did Pons and Fleischmann violate? Were they justified in doing so? Can you think of a time when scientists would be justified in doing so?

2. Now compare the story of Pons and Fleischmann with that of Marshall and Warren. What are the similarities between these two cases? What are the differences? Do you think the differences between these two cases have anything to do with the eventual acceptance of Marshall and Warren's theory, or with the rejection of Pons and Fleischmann's? Explain. What should Pons and Fleischmann (or Marshall) have done differently? Why?

• •

On March 23, 1989, at a Salt Lake City press conference called by the University of Utah, two electrochemists, Dr. B. Stanley Pons (Univerity of Utah) and Dr. Martin Fleischmann (University of Southhampton, England) announced to the world that they had achieved cold fusion. They claimed that their electrolysis experiment produced four times the amount of energy required to run the experiment—not by the tremendous heating and smashing and splitting of atoms (fission), but by bringing together positively charged atomic (deuterium) nuclei at normal room temperatures (fusion). The benefits of their method of achieving cold fusion would be that deuterium is available in seawater, produces much less dangerous radioactivity, and thus that the process would not require nuclear reactor facilities. Pons and Fleischmann, both chemists, thought they had made a major breakthrough in nuclear physics, where research into the possibility of cold fusion had been going on for years without much hope of success; they thought their breakthrough would benefit the entire world (Crease and Samios 1989; Maddox 1989). A flurry of experiments followed, with major research labs around the world (MIT, Cal Tech, Harwell in Britain, for example) diverting attention and money to cold fusion projects.

On March 24, the day after Pons and Fleischmann's press conference, *Nature* received a paper by a team of physicists, led by Stephen E. Jones, working on cold fusion at Brigham Young University; this paper made much more modest claims about cold fusion. Pons and Fleischmann apparently had a prearranged agreement with Jones, made at a March 6 meeting between the scientists and the presidents of their respective universities, to submit their papers simultaneously to *Nature* on March 24 (Huizenga 1992). However, on March 11, unknown to Jones, Pons and Fleischmann submitted a paper to the *Journal of Electroanalytical Chemistry* (Fleischmann and Pons 1989); the revised version was received March 22, the day before the press conference on March 23, and appeared in the *Journal of Electroanalytical Chemistry* on April 10, 1989. The paper had been faxed around the world so many times that only the words "Confidential—Do Not Copy" were legible (Huizenga 1992, 24). This paper was later followed by the publication of extensive corrections, called "errata," including the omission of the third author, Marvin Hawkins. Pons and Fleischmann never submitted a paper to *Nature* (Huizenga 1992), as widely believed and reported in the press.

On April 26, 1989, the University of Utah asked for $5 million from the Utah state legislature, and Pons and Fleischmann appeared before the US Congress to ask for an additional $25 million immediately (and $125 million later) for a Cold Fusion Institute to continue their research. The paper by Jones et al. was published in *Nature* on April 27. At the American Physical Society meeting in Baltimore on May 1 and 2, other groups reported negative results from cold fusion experiments, and at the American Electrochemical Society meeting in Los Angeles on May 8, Fleischmann reported flaws in some of Pons and his original results. On May 18, the first full-fledged critique of Fleischmann and Pons's paper appeared in *Nature*. Petrasso et al. (1989a) criticized the research on the grounds that it lacked adequate controls and that the equipment may have been miscalibrated, and attributed the reports of energy production and other by-products to those

(continued)

FIGURE 1.1 **Chronology of communication events in cold fusion.**

errors. Even though it was "only" a "preliminary" or "technical note," many scientists thought that Fleischmann and Pons's paper should have been revised again before being published. As Huizenga comments:

> When the paper was finally available for examination by an anxious scientific community, most readers were shocked by the blatant errors, curious lack of important experimental detail and other obvious deficiencies and inconsistencies. David Bailey, a physicist at the University of Toronto, said the paper was "unbelievably sloppy." He was quoted as saying, "If you got a paper like that from an undergraduate, you would give it an F." (1992, 24)

Scientists also complained that there wasn't enough information in the paper for scientists to replicate the experiment. Pons and Fleischmann refused to answer criticisms directly or provide crucial details of their experiment (Petrasso et al. 1989b; Huizenga 1992). As reported by the *New York Times:*

> Drs. Pons and Fleischmann offered little help to the unfortunates struggling to repeat their work; they declined to provide details of their techniques and refused to send samples of their equipment to laboratories for analysis. . . . When someone claimed that it was not possible to produce cold fusion, the two Utah (sic) scientists would add more instructions. As Robert Park, head of the Washington office of the American Physical Society remarked, "Anytime someone did the experiment with no results they would say, 'You didn't do the experiment right,' and offer up another tidbit." (Crease and Samios 1989, 3D)

When asked for more information, Pons and Fleischmann claimed "that they preferred to press on with more urgent work rather than stop to handle the reviewers' criticisms" (Crease and Samios 1989, 3D).

As researchers failed in their attempts to test or reproduce Fleischmann and Pons's results, many of the big laboratories terminated their expensive cold fusion experiments. By early July, a special advisory panel to the Department of Energy, co-chaired by John Huizenga, had recommended against awarding special funds for cold fusion research. Undeterred, Pons and Fleischmann and other supporters continue to stand by their discovery and to work on cold fusion. According to other scientists, however, subsequent research, including a paper published by Pons and Fleischmann as late as 1993 (Pons and Fleischmann 1993), has added little to what is already known (Amato 1993; Dagani 1993). The field is now split between "believers" and "nonbelievers" who continue to try to convince each other of their positions (Dagani 1993; Greenland 1994).

Among other sources, we are indebted to Huizenga (1992) for the basis of the chronology here.

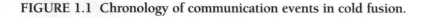

FIGURE 1.1 Chronology of communication events in cold fusion.

. . . .

1.4 The role of persuasion in scientific communication

We have already discussed the importance of sharing information in science. But facts do not speak for themselves. Rather, facts are interpreted and presented as evidence in scientific arguments contained, for example, in research reports, conference presentations, or grant proposals. In the two cases we have examined, that of Marshall and Warren and that of Pons and Fleischmann, the initial failure to gain acceptance for a theory can be directly attributable to a failure to convince colleagues of the validity of the work. The problem is not only a matter of methods and data; it is also a matter of the accessibility, quality, and presentation of evidence—the persuasiveness and style of the argument made.

Neither Marshall and Warren nor Pons and Fleischmann did a particularly good job of persuading their colleagues. In the end, Marshall's critics were persuaded by the results of other researchers, made possible in part because Marshall and Warren provided enough information to make their theory testable; Pons and Fleischmann did not. While Marshall and Warren's theory has revolutionized the field of gastroenterology, Pons and Fleischmann's theory is still hotly disputed, as is their professional credibility.[1]

Persuasion is central to scientific communication. Persuasion tends to be a dirty word in our culture, and a tricky subject in science, which traditionally prides itself on objectivity. But in addition to acceptance by editors and reviewers of journals and granting institutions, the work of scientists ultimately must be accepted by the scientific community at large. As mentioned previously, science consists of those findings that have survived the scrutiny of the community and the test of time. Individual findings take on the status of scientific knowledge as they are accepted by more and more members of the field. Thus, the process of building scientific knowledge is best described not via individual facts, but through the achievement of consensus about what count as facts.

And this consensus is created through scientific argument. In the cases of Marshall and Warren and of Pons and Fleischmann, we get a glimpse of the importance of persuasion, argumentation, and debate in the construction of scientific knowledge. In later chapters you will explore this dimension of science in communication in your own field. To briefly illustrate the role of argumentation in science here, let's return to the case of Barry Marshall. In the exchange of technical letters in the *Lancet* that followed Warren and Marshall's, scientists focused not only on methods and data, but on proving or disproving the validity of Marshall's argument. Debate raged not so much about data, but about reasoning from the data. Commenting on a 1989 paper by the Marshall team, Walter Peterson (1989, 509) pointed to "a number of problems *with this paper* that compel me to urge that *its recommendations* not be accepted" (emphasis ours). Once published, the letters themselves became the object of critique.

---

[1] As recently as 1996, a judge ruled against Pons and Fleischmann in a libel suit they had brought against an Italian journalist who had reviewed their work in a book on scientific fraud. In his decision, the judge ruled that the journalist's review was justified, citing "important opposition from the scientific community, not just against the theory of the research and the way the experiments were conducted, but also the way the data were divulged and the conclusions reached about the future direction of research" (Nature 1996).

EXERCISE 1.3   In Chapter 9 we have reprinted five letters commenting on developments following Warren and Marshall's *H. pylori* announcement: Veldhuyzen van Zanten et al (1988); Lam (1989); Marshall, Warren, and Goodwin (1989); Loffeld, Stobberingh, and Arends (1989); Bell (1990). In each letter, what is being debated? Is it the facts of the case or the writers' argument? Note all the places where the *reasoning* of the scientists—and thus the persuasiveness of their argument—is being questioned. What is being questioned? Their logic? Their evidence? Their terminology? Their beliefs? Their judgment? Look for instances of each of these categories and create additional categories if necessary.

Persuasion is created not only by the logic of arguments but also by presentation and style. Earlier, we said Marshall had trouble getting people to listen to him because of the way he answered questions at the conference in Brussels. Certainly, a part of the problem was Marshall's position as a young internist speaking before seasoned experts in gastroenterology, an outsider working against the dominant assumptions in the field. Marshall himself admitted to the press that the odds were stacked against him. But critiques by his colleagues in the popular press indicated that they were skeptical not only because of his youth and casual appearance but also because of how he presented himself. Beyond one's past "reputation," there is the persona one projects through language—what Aristotle called *ethos:* the persuasive character of the speaker or writer created in and through language. Though the contents of his conference paper may have been appropriately qualified and cautious, Marshall struck listeners as brash and reckless because of his presentation style and the way he answered questions:

> Unschooled at such presentations and filled with boyish eagerness, he refused to respond to questions in the measured, cautious manner of most researchers. Asked whether he thought the bacteria were responsible for some ulcer disease, Marshall replied, "No, I think they're responsible for *all* ulcer disease." Such blanket statements, backed only by small studies and anecdotal case histories, alarmed many researchers. (Chazin 1993, 121–22)

Pons and Fleischmann too, were criticized for overstating their claims. Physicist Stephen Jones, on the other hand, made more modest claims in the proper scientific publication *Nature,* and so he was more believable. As the *New York Times* reported, Jones's "colleagues took him seriously not because he was one of their own, nor even because he showed up at all the important meetings to defend his work. Rather, it was because his work betrayed an awareness of potential pitfalls" (Crease and Samios 1989, 3D). As we will see in subsequent chapters, that awareness is reflected not only in what is said, but how it is said. Sociological research has shown that the kinds of arguments and styles employed in formal scientific communication often differ from those in informal settings. Much gets said in the lab that would not be said in more formal forums such as the research report or grant proposal. In formal communication, scientists employ a style that subordinates their personal preferences and professional allegiances

(Gilbert and Mulkay 1984). Regardless of the validity of his claims, Marshall's enthusiasm seemed inappropriate in this formal context. Even Walter Peterson, one of Marshall's most staunch opponents, says, "We scientists should have looked beyond Barry's evangelical patina and not dismissed him out of hand" (Chazin 1993, 124). But the question is: Can scientists look beyond style of argument, appearance, and delivery when this is the way science is presented?

The reaction of scientists to Marshall's and to Pons and Fleischmann's presentations of their research illustrate the central role of argument and style in scientific communication. To be persuasive, scientists must make the claims of their research believable in the context of the previous research and the existing paradigm of the field; and they must present these arguments in professional forums and styles that are acceptable in the scientific community.

1.5 Scientific communication and convention

As illustrated by the preceding discussion, the forums and styles a scientist chooses can make a difference in how well the results of his or her research is heard and understood. The conferences you attend and participate in, the publications you submit your work to, the funding agencies to which you apply, even the institutions that you work for— all can make a difference in how well your research is received, and whether it is used by other scientists.

As you will discover in working through subsequent chapters, different types or genres of writing follow different conventions and rules, both implicit and explicit. Learning what these conventions and rules are and how to use them to demonstrate the nature, reliability, and significance of your research to other scientists in your field is one of the things that distinguishes a student scientist from a professional scientist. Learning to be a professional scientist means learning both the science and the conventions of communication in your discipline.

While we can't go deeply into these conventions and rules in this introduction, we can talk about them in more general terms. For example, although scientific research is systematic, the process of doing science is really a much more creative and sometimes haphazard process than is commonly thought (the National Academy of Science published *On Being a Scientist* in part to demonstrate that fact). Yet, when science is written up and published in a typical research report, it is not *presented* to the scientific community as a personal narrative or story of "what happened" (including missteps and mistakes) in the actual order that it happened. Although the report may include a chronological description of methods, this description is embedded in a broader argument in which a claim or hypothesis is supported or refuted. Thus several scholars have pointed out that the traditional report does not accurately represent the processes of scientific research (Medawar 1964; Gross 1985). In writing a report, the actual process of scientific imagination and discovery is reconceptualized along the lines of the argument needed to justify the science to the scientific community (Carnap 1950; Holton 1973).

The four-part structure of the conventional research report (introduction, methods, results, discussion) requires the writer to begin not with the first step in the experiment but with an argument for the significance of the hypothesis. Personal narrative does play a role in science; it is often employed to communicate science to general audiences, particularly in general-interest magazines and television programs, and often on E-mail, where discussions are more informal. But in a contemporary research report in most fields, the narrative form probably would be considered inappropriate in convincing other scientists of the validity of research, and might actually undermine that attempt.

Throughout this book you will be learning the communication conventions accepted by your field. We also will talk about some of the ethical considerations of this socialization process in Chapter 8. It is important for students contemplating becoming professional scientists to know the conventions of their field, to understand the underlying assumptions and attitudes that give rise to those conventions, and to understand how to work within them. As you will explore in more detail in subsequent chapters, these conventions include not only the structural features of major genres, such as research reports and grant proposals, but also the styles in which these are presented. For now, we would point out that all these genres and styles are social conventions that create the channels through which and influence the ways in which scientists interact and collaborate with one another.

1.6 The role of collaboration in scientific communication

If science is a community, and communication in science is social, it should be no surprise that the process of writing itself is also social in nature. In the sciences, writing tends to be more collaborative than in other disciplines because of the necessarily cooperative nature of science, and because of the pronounced role communication plays in it. (In our citing of scientific research in this book, note the number of times "et al.," designating multiple authorship, has been used; even the shorter technical letters to the editor are often written collectively, or represent a group of scientists). The discovery of the role of *H. pylori* in stomach ulcers clearly has a collaborative history. Marshall drew on earlier studies to support his claim that bacteria in the stomach already had been noted but overlooked; Marshall and Warren wrote joint letters and a research report together; finally, other scientists validated, confirmed, and extended Marshall and Warren's research. The creation of scientific knowledge is truly a collaborative process.

Collaboration is necessary for the development of scientific theory and the conduct of research. This is especially true in contemporary society, where "big science" involves teams of experts from varied fields, including administrators, politicians, and other nonscientists. For example, imagine the range of experts involved in the construction, deployment, repair, maintenance, and use of the Hubble Space Telescope. In Figure 1.2 you will find a part of a transcript of two biochemists working on a proposal for the National Institutes of Health (Mehlenbacher 1992). This excerpt, drawn from the research of one of our colleagues investigating the social process of scientists writing grant proposals, illustrates scientific collaboration in action.

Raymond: If we remove the histidines there's gonna be a gap.

Larry: We really don't need to remove it. In the literature the most important suggestion is to substitute histidine. It's structurally similar, so it's not going to affect whether you know anything.

Raymond: Okay, I kind of felt that, so it's really a precision problem here. You're not really deleting it, you're substituting it.

Larry: Substituting it, yes.

Raymond: In the results it's deletion. All right.

Larry: Then we make a note.

Raymond: Will anybody get confused about that?

Larry: We'll have to make that, it's not clear.

Raymond: So we should both decide to stick with one word. Or else we should use the word people use to make sure that the meaning is defined. That's in my section, too, so why don't we decide on that? You wanna use deleted or substituted.

Larry: Substituted. Deleted is wrong.

Raymond: Okay, I'll make the change in my stuff.

FIGURE 1.2 Excerpt from transcript of collaborative writing session.

This brief exchange illustrates several important features of the conduct of scientific research: that professional scientists help each other with both the thinking and the writing of science; that the actual writing of a report or proposal requires us to carefully examine and often revise our ideas and interpretations; that our thinking is shaped by our awareness of what our readers will need and understand and how they will react to our thinking. In the process of anticipating readers' needs, collaborators may rethink not only their text but also their research plan and hypotheses (Latour and Woolgar 1979). Communication is fundamental in science. Indeed, sociologists Latour and Woolgar have characterized the "entire laboratory activity [as] a process of inscription, gradually turning the materials under study into the words and symbols that appear in an article or other scientific communication" (Bazerman 1983, 165).

ACTIVITIES AND ASSIGNMENTS

1. To expand your ability to shift paradigms, reclassify, reorganize, and solve problems, represent the information in Figure 1.3 in at least three different ways. This exercise does not necessarily entail any great artistic skill or knowledge in graphics, though you will want to think visually here.

Directions: Make the following information more accessible and easier to read by classifying and reorganizing it. There are several different ways of laying this information out.

When time is limited, travel by Rocket, unless cost is also limited, in which case go by spaceship. When only cost is limited, an astrobus should be used for journeys of less than 10 orbs, and a satellite for longer journeys. Cosmocars are recommended when there are no constraints on time or cost, unless the distance to be traveled exceeds 10 orbs. For journeys longer than 10 orbs, when time and cost are not important, journeys should be made by superstar.

Source: From Wright and Reid, "Written Information: Some Alternatives to Prose for Expressing the Outcomes of Contingencies," *Journal of Applied Psychology* 57 (1973): 160–66.

FIGURE 1.3 Breaking paradigms: Representing information in different ways.

2. Working in a group of students from your field, brainstorm about a theory that has been accepted by your discipline. Prepare to discuss what you know about its history: previous theories to explain the phenomenon, the personalities of the scientists involved, the debates surrounding the newer theory, and the evidence that made it acceptable or unacceptable in your field. What does your discussion tell you about how knowledge is constructed in your field?

3. To demonstrate the value of collaboration as a means of generating ideas and discovering insights, do the NASA Simulation Exercise in Figure 1.4 (p. 18).

4. Warren and Marshall's joint letters to the *Lancet* announcing their *H. pylori* discovery are reprinted at the start of Chapter 9. Write a 1–2 page analysis in which you compare and contrast Warren's letter with Marshall's. How are the letters related? Which arguments are similar and which are different? Do you detect any difference in emphases and purpose between them? Why do you think Warren and Marshall decided to submit these letters together?

Your spaceship has just crash-landed on the lighted side of the moon. You were sched-uled to rendezvous with a mother ship 200 miles away on the lighted surface of the moon, but the rough landing has ruined your ship and destroyed all the equipment on board, except for the 15 items below.

Your crew's survival depends on reaching the mother ship, so you must choose the most critical items to take on the 200-mile trip. Working alone, your first task is to rank the 15 items in terms of their importance to your crew in reaching the rendezvous point. In the column labelled "Your Rank" place a number 1 by the most important item, num-ber 2 by the second most important, and so on through number 15, the least important.

When you have finished, your instructor will give you additional directions for working in groups, and then for calculating errors based on the NASA rank.

	Your Rank	Group Rank	NASA Rank	Your Error	Group Error
Box of matches					
Food concentrate					
50 feet of nylon rope					
Parachute silk					
Solar powered portable heating unit					
Two .45 caliber pistols					
One case of dehyrdated Pet milk					
Two 100-pound tanks of oxygen					
Stellar map (of the moon's constellations)					
Self-inflating life raft					
Magnetic compass					
5 gallons of water					
Signal flares					
First-aid kit containing injection needles					
Solar-powered FM receiver-transmitter					

FIGURE 1.4 NASA simulation exercise. Adapted from Hall (1971).

2

•••

Forums for Communication in Science

•••••••••••••••

2.1 The socialization process: Entering a new community

As we join a particular research community, we gradually acquire knowledge of the ways in which that community develops and communicates knowledge. This happens naturally, and usually without conscious effort, as we read papers written by other researchers, attend conferences, and talk informally with our new colleagues in the classroom, the field, the development lab, or other research workplace. We are gradually *socialized* into the communication patterns the community has adopted.

You have probably already started thinking as a biologist or soil scientist or wildlife researcher. If you've taken a course or two in your area of science or have pursued a personal interest in the field outside your formal studies, the socialization process has begun. Not only have you begun developing *content knowledge*—the principles, concepts, and terminology that members of the field take for granted; but you've also begun to acquire *procedural knowledge*—knowledge of how to do things in this area of science: how to solve problems, how to test hypotheses, how to use the basic methods of your field, and how to communicate your concerns, questions, and findings to others in the community. The more familiar you become with the ways of thinking, speaking, and writing in your field, the easier it is for you to quickly understand written texts, to grasp important concepts and recognize issues that matter to the field, and to contribute to written and spoken conversations—that is, to participate in the development of new knowledge.

Your content and procedural knowledge develop together, though not necessarily at the same rate, and are mutually reinforcing. You cannot write like a physicist if you don't know anything about physics. And you can't acquire knowledge of physics if you can't read the texts written by physicists, apply the experimental procedures and theorems that physicists have developed, and communicate with others about your questions and your developing understandings. Thus, the process of socialization involves learning the subject matter of your particular branch of science and also learning how to think and communicate as a scientist.

Theoretically, the more you know about the community you want to join and about the socialization process itself, the easier your initiation into the community will be. Thus, you can simply let the socialization process take its course, or you can actively work to "socialize yourself" into the community by seeking out the community's distinctive patterns of communication and interaction, rather than waiting to stumble upon them. The discussion and activities in this chapter are intended to help you recognize the communication patterns that have developed in the scientific community you are joining.

2.2 Research journals and their readers

An important first step for any new researcher is to become familiar with the primary journals in the field, for these professional publications represent the principal medium through which individual scientists share their theories and experimental results with others in the scientific community. As Chapter 1 illustrates, the progress of science rests upon the exchange of knowledge among scientists. Indeed, Winsor (1993) describes scientific knowledge as "constructed . . . not just in labs or at field sites, but in arguments that scientists conduct through the medium of scientific papers" (128). In journal articles, researchers do much more than describe methods and report results; they present and defend a particular interpretation of those results in light of the field's developing knowledge. It is through such arguments that scientific knowledge advances, as theories are presented, challenged, refuted, and refined. These papers thus represent the heart of the scientific enterprise, and much professional effort is devoted to writing them.

The easiest way to identify the essential journals in your field is to notice those around you. Start by paying attention to what other researchers are reading—your professors, lab and teaching assistants, other students. When you are assigned research articles to read in your classes, take time to notice where and when the papers were published, what institutions the researchers represent, and what granting agencies provided funding for the research. Identifying the professional associations or institutions that sponsor the journals will help you understand who reads them. Most journals are quite explicit about their audiences and goals, publishing editorial goals and guidelines in every issue or in one special issue each year. For example, the American Institute of Physics publishes or cosponsors dozens of journals, each intended for a specific segment of the physics community. See Figure 2.1 for a sampling of these journals and their intended audiences. Notice as you read the editorial policies for these journals that the target audiences vary along a number of dimensions, not just in their areas of specialization.

American Journal of Physics

Published by AIP for the American Association of Physics Teachers. Devoted to meeting the needs and interests of college and university physics teachers and students by focusing on the instructional and cultural aspects of physics. Contains feature articles that describe novel approaches to laboratory and classroom instruction and other areas of physics pedagogy. Information about the journal can be obtained online at http://www.amherst.edu/~ajp.

The Astrophysical Journal

Published by the University of Chicago Press for the American Astronomical Society. Published three times each month in both electronic and paper formats, this publication contains articles on all aspects of astrophysics, astronomy, and related sciences.

Journal of Applied Physics

Publishes the results of applied physics research with important applications in other fields and in engineering and industry. *Journal of Applied Physics* offers significant new experimental and theoretical results with novel applications. A key component of this journal is *Applied Physics Reviews,* which features review articles covering important current topics of experimental or theoretical research in applied physics and applications to physics from other branches of science and engineering.

Physical Review A: Atomic, Molecular, and Optical Physics
Physical Review B: Condensed Matter
Physical Review C: Nuclear Physics
Physical Review D: Particles and Fields
Physical Review E: Statistical Physics, Plasmas, Fluids, and Related Disciplinary Topics

Published by AIP for the American Physical Society, these journals publish original research in the specified areas.

Physical Review Letters

Published by The American Physical Society. Contains short, original communications in active and rapidly developing areas of physics. The reports are written so that their significance can be appreciated by physicists working outside the area while remaining stimulating to physicists in the same field. Information about American Physical Society journals can be obtained online at http://publish.aps.org.

(continued)

FIGURE 2.1 Statements of editorial policy from selected journals published by the American Institute of Physics (AIP) and affiliated societies.

Physics Today

A semipopular publication that contains articles and news of interest to the reader with only a general interest in physical science, as well as the professional physicist. Provides original articles and timely news on ground-breaking research in physics and related sciences. Offers fresh perspectives on contemporary technical innovations, basic and applied research, the history of science, education, and science policy.

FIGURE 2.1 Statements of editorial policy from selected journals published by the American Institute of Physics (AIP) and affiliated societies.

We saw in Chapter 1 that the study of science is quite dynamic: as fields develop and diversify, and as the interests of different fields begin to converge, new areas of research become active and important. Consequently, these new communities, often comprising researchers from different fields, establish specialized journals devoted to their emerging interests. The field of marine ecology, for example, includes researchers in such diverse fields as microbiology, botany, oceanography, conservation, and resource management. The journal *Marine Ecology Progress Series* was created to meet the needs of these researchers, who, despite their different disciplinary backgrounds, have a common set of research interests. The purpose and scope of this journal is clearly announced by the editors inside the front cover of each issue. This statement is reproduced in Figure 2.2.

A journal like *MEPS* unites researchers from a variety of fields who have a common interest. Most researchers thus belong to several research communities and subcommunities, each of which is linked by one or more specialized journals. As our research interests and needs develop, we seek out those journals that are most pertinent to our research agenda, often reading in more than one field.

To stay active in a research field or interdisciplinary area, then, we need to be aware of the purposes and intended audiences of the journals published in that area. This knowledge helps us decide which journals we will want to find time to read (of the thousands in publication); and, equally important, it helps us decide where we should submit our own work for publication. Choosing an appropriate journal is a critical step in the publication of a scientific paper. As we will discuss in Chapter 3, "appropriateness for the journal" is the primary criterion used by reviewers in evaluating papers for potential publication. Further, the ultimate influence of a paper in the scientific community may be significantly determined by the journal in which it appears. In science as in other domains, the amount of attention an announcement receives is to some extent determined by the context in which it appears—that is, by where, when, how, and to whom the announcement is made (Miller 1992). The choice of a journal ultimately determines who will have access to our work.

As a case in point, Chapter 10 includes reports of a research project headed by Dr. JoAnn Burkholder, an aquatic botanist at North Carolina State University. In 1988, Burkholder's team discovered a new genus of algae, a "predatory" dinoflagellate that kills fish by releasing a lethal neurotoxin into the water. In a series of studies over the

MARINE ECOLOGY PROGRESS SERIES

HISTORY: Marine Ecology Progress Series (MEPS) was established in response to requests from numerous marine ecologists from all over the world. When founded, MEPS was based on 'Marine Ecology' – the first comprehensive, integrated treatise on life in oceans and coastal waters – conceived, organized and edited by Professor Otto Kinne, and published by John Wiley & Sons.

AIM: MEPS serves as a world-wide forum for all aspects of marine ecology, fundamental and applied. The journal covers: microbiology, botany, zoology, ecosystem research, biological oceanography, ecological aspects of fisheries and aquaculture, pollution, environmental protection, conservation, resource management.
Ecological research has become of paramount importance for man's future. The information presented here should, therefore, encourage critical application of ecological knowledge for the benefit of mankind and, in fact, of life on earth. Marine Ecology Progress Series has received recognition as the leader in its field. The journal strives for

- complete coverage of the field of marine ecology
- the highest possible quality of scientific contributions
- quick publication
- a high technical standard of presentation.

SCOPE: MEPS is international and interdisciplinary. It presents rigorously refereed and carefully selected research articles, reviews and notes, as well as comments, concerned with:

Environmental Factors: Tolerances and responses of marine organisms (microorganisms, autotrophic plants, and animals) to variations in abiotic and biotic components of their environment; radioecology.

Physiological Mechanisms: Synthesis and transversion of organic material (mechanisms of auto- and heterotrophy); thermo-, ion-, osmo- and volume-regulation; stress resistance; non-genetic and genetic adaptation; population genetics; orientation in space and time; migrations; behavior; molecular basis of ecological processes.

Cultivation: Maintenance and rearing of, as well as experimentation with, marine organisms under environmental and nutritional conditions which are, to a considerable degree, controlled; analysis of the physiological and ecological potential of individuals, populations and species; determination of nutritional requirements; ecological aspects of aquaculture; water-quality management; culture technology.

Dynamics: Production, transformation, and decomposition of organic matter; flow patterns of energy and matter as these pass through organisms, populations and ecosystems; biodiversity; variability of ecosystem components; trophic interrelations; population dynamics; plankton ecology; benthos ecology; estuarine and coastal ecology; wadden-sea ecology; coral reef ecology; deep-sea ecology; open-ocean ecology; polar ecology; theoretical ecology; ecological methodology; ecological modelling and computer simulation.

Ocean Management: Human-caused impacts: their role as modifiers and deformers of living systems, their biological consequences and their management and control; inventory of living resources in coastal areas, estuaries and open oceans; ecological aspects of fisheries; pollution of marine areas and organisms; protection of life in the seas; management of populations, species and ecosystems; management of coastal zones and sea areas; biotechnology.

FIGURE 2.2 **Statement of editorial policy for** *Marine Ecology Progress Series.* **MEPS 1996; 134.**

next few years, Burkholder and her colleagues determined that the alga was responsible for massive fish kills in estuaries along the southeastern coast of the United States and was very likely associated with fish kills in other regions as well. The striking nature of the finding (this was the first known instance of predatory behavior among such organisms), along with its potential international ramifications, warranted submission of the results to the British journal *Nature,* which has a broad, interdisciplinary, and international readership. Published in *Nature* in 1992, this first major paper from the project announced the discovery of the new organism, described its behavior, and briefly listed related fish kills that had been observed. In contrast, subsequent papers focused more narrowly on the results of individual studies of the organism in the lab and in particular estuarine locations. For these later reports, the researchers targeted more specialized audiences within the marine ecology community, publishing in such forums as the *Journal of Plankton Research* and the *Marine Ecology Progress Series* described above.

The publications from the Burkholder project also illustrate another important feature of scientific research journals—the variety of *genres,* or types of articles, that they publish. The team's original announcement in *Nature* (included in Chapter 10) was presented as a research "letter," an important and increasingly common genre in scientific writing, which is designed to enable researchers to make quick announcements of important findings to the relevant scientific community. According to *Nature's* Guide to Authors, reproduced in Figure 2.3, *letters* are "short reports of outstanding novel findings whose implications are general and important enough to be of interest to those outside the field."

GUIDE TO AUTHORS

PLEASE follow these guidelines so that your manuscript may be handled expeditiously.

Nature is an international journal covering all the sciences. Contributors should therefore bear in mind those readers who work in other fields and those for whom English is a second language, and write clearly and simply, avoiding unnecessary technical terminology. Space in the journal is limited, making competition for publication severe. Brevity is highly valued. One printed page of *Nature*, without display items, contains about 1,300 words.

Manuscripts are selected for publication according to editorial assessment of their suitability and reports from independent referees. They can be sent to London or Washington and should be addressed to the Editor. Manuscripts may be dealt with in either office, depending on the subject matter, and will where necessary be sent between offices by overnight courier. High priority cannot be given to pre-submission enquiries; in urgent cases they can be made in the form of a one-page fax. All manuscripts are acknowledged on receipt but fewer than half are sent for review. Those that are not reviewed are returned as rapidly as possible so that they may be submitted elsewhere without delay. Contributors may suggest reviewers; limited requests for the exclusion of specific reviewers are usually heeded. Manuscripts are usually sent to two or three reviewers, chosen for their expertise rather than their geographical location. Manuscripts accepted for publication are processed from the London office.

Nature requests authors to deposit sequence and crystallographic data in the databases that exist for this purpose.

Once a manuscript is accepted for publication, contributors will receive proofs in about 4 weeks. *Nature's* staff will edit manuscripts with a view to brevity and clarity, so contributors should check proofs carefully. Manuscripts are generally published 2–3 weeks after receipt of corrected proofs. *Nature* does not exact page charges. Contributors receive a reprint order form with their proofs; reprint orders are processed after the manuscript is published and payment received.

Categories of paper

Review Articles survey recent developments in a field. Most are commissioned but suggestions are welcome in the form of a one-page synopsis addressed to the Reviews Coordinator. Length is negotiable in advance.

Progress articles review particularly topical developments for a nonspecialist readership. They do not exceed 4 pages in length. Suggestions may be made to the Reviews Coordinator in the form of a brief synopsis.

Articles are research reports whose conclusions are of general interest and which are sufficiently rounded to be a substantial advance in understanding. They should not have more than 3,000 words of text (not including figure legends) or more than six display items and should not occupy more than five pages of *Nature*.

Articles start with a heading of 50–80 words written to advertise their content in general terms, to which editors will pay particular attention. The heading does not usually contain numbers, abbreviations or measurements. The introduction to the study is contained in the first two or three paragraphs of the article, which also briefly summarize its results and implications. Articles have fewer than 50 references and may contain a few short subheadings.

Letters to Nature are short reports of outstanding novel findings whose implications are general and important enough to be of interest to those outside the field. Letters should have 1,000 or fewer words of text and four or fewer display items. The first paragraph describes, in not more than 150 words and without the use of abbreviations, the background, rationale and chief conclusions of the study for the particular benefit of nonspecialist readers. Letters do not have subheadings and contain fewer than 30 references.

Commentary articles deal with issues in, or arising from, research that are also of interest to readers outside research.

News and Views articles inform nonspecialist readers about new scientific advances, sometimes in the form of a conference report.

FIGURE 2.3 Guide to authors from the editors of *Nature. Nature* 1995; 378:6552.

In contrast to full-length reports which often run to 6000 words or more, letters and research notes are generally limited to 1000 or 1500 words, enabling researchers to write them more quickly and making it possible for reviewers to read and respond to them quickly, with the goal of getting important information into print as soon as possible. Despite the obvious appeal of this genre for writers, the more traditional full-length report—with its extensive discussion of goals, methods, results, and implications—remains the most common type of journal article published in the sciences. This genre is discussed at length in Chapter 3.

In addition to the primary scholarship presented in letters and full-length reports, many journals, including *Nature* (see Figure 2.3), publish secondary scholarship, or review articles, in which the authors survey recent research on a particular topic of interest to readers of the journal. Often, a journal editor will ask a known expert to write a review article on a current topic, but some journals accept unsolicited reviews as well. These reviews of previously published results generally provide a quick history of recent research, comparing and contrasting results from a range of studies in an effort to help readers understand where research consensus is forming and what issues are still open for exploration. This genre will be discussed in more detail in Chapter 4.

· · · · · · · · · · · · · ·
2.3 Research conferences and professional associations

Like research journals, conferences vary in size, scope, and audience. Most are sponsored by professional organizations whose membership and areas of interest vary widely. For example, botany student Alexander Krings reports that researchers in plant community ecology may belong to a small, highly specialized group such as the International Association for Vegetation Science (about 1200 members) and/or to the larger Ecological Society of America (7000 members), whose members come from a wide array of ecology subfields. (Krings's profile of this community is presented in Figure 2.4.)

Conferences represent another formal mechanism which, like the letter genre discussed earlier, has developed to speed up the exchange of scientific knowledge. Most groups meet annually, allowing researchers to present recent results directly to others in their field even before studies appear in print. The conference thus provides opportunities for immediate feedback from other researchers, often on work in progress.

Local, regional, and national conferences play different roles in different fields, and learning about these roles is an important part of the socialization of a new researcher. An easy way to find out about important conferences in your research field is to take notice of conference announcements posted in your department and to pay attention to the details when your professors, lab directors, or teaching assistants go out of town to present papers. Ask where they went, whom they spoke to, what kind of work they presented. Borrow their conference programs, and browse them to get a sense of the kinds of topics that were covered at the meeting and the kinds of sessions that were scheduled. Conference announcements and programs are increasingly available through the web sites of professional associations. In addition to panels of formal

The Plant Community Ecology Research Community

Alexander Krings

Plant community ecology, or phytosociology, is a study of interactions, of structure, and of composition. The field is broad with roots in the past. Its history in the Western world, some say, began with early plant geographers such as Alexander von Humboldt (1769–1859). Humboldt traveled extensively in the New World, prolifically describing the natural world around him. Indeed he is credited with coining the word "association" that is still in use among ecologists today. After a period of continued development of the science in the nineteenth century, it finally grew apart from plant geography in the early part of this century. Today, the field is largely dominated by professionals and academicians studying the causes of community structure as well as the relationships and dynamics of communities. The methods used in the field have, since Day 1, been largely observational. As the root drive of the science is to understand and describe natural plant communities, most efforts have concentrated on strict observation of, and data gathering from, existing communities. There is, however, a growing trend toward controlled experiments in certain subfields, such as those dealing with physiological efficiency research, where observational data no longer prove reliable or accurate.

As with any scientific field of study, the channels of communication within plant community ecology are broad and diverse. They may involve the casual conversation with a colleague in the hallway or the professional presentation of research results as part of an international conference. As the channels of communication are so diverse, they may be grouped easily according to the different levels on which they function—local/intimate, local/formal, and regional/international.

The local/intimate category seems by far the most time consuming. Forms of communication that fall into this category include casual hallway conversations with colleagues, meetings with graduate students about their respective research proposals/projects, participating on various committees, etc. This is the most intensive level for the individual researcher, as this level builds the groundwork for the science. Basic questions such as sampling design, matters of statistics, and objectives of studies are addressed on this level. It is this level that sets the stage for the data gathering, the experiments, and the general additions to the body of scientific knowledge. Without the intimate, one-on-one interactions with colleagues, students, and other professionals, there would be no foundation to the science.

The local/formal category closely follows the local/intimate category in terms of time consumption. Forms of communication on this level include the preparation and delivery of lectures, attendance at thesis defenses, and the attendance or delivery of seminars. Frequently, despite the "formality" accorded to the preceding events, there is ample opportunity for one-on-one questioning and discussion. In cases such as seminars, new

(continued)

FIGURE 2.4 Profile of the field of plant community ecology. Student sample.

[Handwritten margin annotations: "Introduction" "focus of his field" "history" "Categories to help him organize his observation" "Specific examples"]

ideas can have their first chance to be examined while at the same time being distributed to various professionals in a research community. Lectures in themselves play a critical role in the furthering of the science in that they prepare students for the questions that they are expected to ask in their field. Lectures also serve as gateway communication channels between students eager to enter a field and already established professionals.

The third category of communication channels is the regional/international level. Regional, as used here, more aptly refers to interinstitutional/interagency communication. This level encompasses the justification for various projects, future research needs, peer review of present research, and necessary funding sources. There are various subchannels of communication at this level. Practicing professionals communicate through journals as well as through professional organizations and their conferences.

Of all the various journals in ecology, few deal exclusively with plant community ecology. Of those that do, the most notable examples are the *Journal of Vegetation Science, Vegetatio,* the *British Journal of Ecology,* and *Castanea.* A good example of the breadth of the field and the magnitude of its undertakings can be seen in the diversity of the publishing centers. The first three journals mentioned are European creations, whereas the last is a regional, Appalachian-based endeavor. Consequently, the information published in the differing journals may be quite distinct and tailored to a specific audience. The large European journals usually report on worldwide research focusing on an academic audience, whereas *Castanea* concentrates on more regional issues and research, targeting both an academic and a professional audience. In general, the Council of Biology Editors style manual is used regarding general matters of style for journal articles, but most researchers simply follow the guidelines issued at least once a year in the journals in which they want to publish.

The publication and review of journal articles are probably the most common communication channels that involve the members of the field as a whole. Other inclusive channels at the regional level include conferences, where practitioners have the opportunity of meeting the "faces behind the publications" and a chance, as Dr. Tom Wentworth (North Carolina State University–Botany) puts it, "to catch up on what is new in the field without having to read the entire spectrum of publications available." Most often it is the professional organizations and agencies that organize conferences. In the field of plant community ecology, individuals may belong to broad-based organizations such as the Ecological Society of America (estimated 7000 members) or more field-specific organizations such as the International Association for Vegetation Science (estimated 1200 members). Depending on individual interest, or even regional interest, societies such as the Torrey Botanical Club, the Association of Southeastern Biologists, or the American Association for the Advancement of Science fit almost everyone's need. Of crucial importance as well is the fact that such a varying array of journals and organizations can cover many aspects and details that no one journal could cover by itself. This fact is crucial because it furthers the already existing hierarchy of levels of communication and adds to the stability of the field as a whole.

(continued)

FIGURE 2.4 Profile of the field of plant community ecology. Student sample.

Also within the regional level of communication channels in plant community ecology lies the ever important source of funding. A good deal of time is spent by professionals, especially researchers, procuring adequate funding to carry out their scientific investigations. Grant proposals may be written and rewritten, often being tailored to specific funding sources and their interests, before final success. Often individual researchers' interests may be modified by the funding agencies' interests. It seems better to at least carry out part of your original inquiry with some modifications than not to be able to carry out any investigation at all. Funding sources in plant community ecology include, but are not limited to, the National Science Foundation, the US Forest Service, the US Fisheries and Wildlife Service, the US Department of Agriculture, the National Biological Survey, and the National Park Service.

As the forms of communication on the regional and international level largely overlap, they have been included together for the purpose of discussion, the main difference being the larger scope in terms of institutional backgrounds, geographical distances, and interests over which the communication takes place. Although peer review and the publishing of journal articles, as discussed before, are probably the primary channels of communication internationally, conferences also play a key role. In fact, international conferences are probably the primary means of direct, fact-to-face communication on this level. Although an international perspective is often intriguing, the drawback is a certain lack of depth to the discussion. Primarily, international forms of communication address the field as a whole, its context in society and science, its needs, and its position in shaping international trends. Region-specific details are discussed not so much for their individual validity but for their relation to the whole. Nonetheless, international conferences provide an important service in that they broaden the horizons of individual researchers, allowing them not only to utilize shared concepts, but to continue working with the idea of being part of a greater community of researchers that transcends all national boundaries.

FIGURE 2.4 Profile of the field of plant community ecology. Student sample.

oral presentations, you'll discover that most organizations also use other meeting formats, such as poster sessions, invited symposia, and workshops. Chapter 5 will discuss the preparation of conference papers and posters.

2.4 Research proposals and their audiences

Research proposals represent another critical channel of communication in science and research. As illustrated in the profile in Figure 2.4, the quest for research funding is a central activity in science. Just as journal editors and conference organizers exert control over what information is made available to the community, so funding agencies influence what kinds of research are undertaken in the first place. We saw in Chapter 1 that in denying funding to Pons and Fleischmann for their work on cold fusion, the Department of Energy dealt a major blow to their research in cold fusion as well as to the credibility of their research team. Because these all-important decisions are based on reviewers' assessments of the quality and persuasiveness of the proposal document, many scientists consider proposal writing the most important writing they do.

As a new member of a research community, you'll want to find out which public and private sources provide funding for work in your field. Notice where your professors or employers apply for funding, and pay attention to the funding agencies acknowledged in journal articles you read. Like the professional associations which sponsor journals, funding agencies have specific interests; most will accept proposals only on specified topics and may limit their funding to projects with certain types of applications (such as industrial, educational, or environmental).

For example, consider two proposals written by the Burkholder team for their research on toxic algae. The first proposal, for basic research on the algae's place in the estuarine food web, was submitted to—and funded by—the National Science Foundation. A second proposal (included in Chapter 10) described a more applied project, the development of gene probes to help detect the organism in water samples. This project was funded by the National Sea Grant College Program, which had issued a call for proposals in the area of marine biotechnology.

Large federal agencies like the National Science Foundation encompass a range of specific interests, making it important to target an appropriate division *within* the agency as well. Burkholder and coinvestigator Alan Lewitus sent their proposal to NSF's Biological Oceanography division. Another large federal agency, NASA (the National Aeronautics and Space Administration), supports a range of research projects in astrophysics and other fields through its many subdivisions. Stephen Reynolds, an astrophysicist whose reports of research on supernova remnants are included in Chapter 11, has submitted proposals to several different NASA programs, including one to the ROSAT Guest Observer Program in 1993, proposing a set of observations to be carried out by the ROSAT satellite; and another to NASA's Astrophysics Theory Program in 1994, in which Reynolds and two colleagues proposed to develop a new theoretical model of supernova dynamics to be used in interpreting data gathered from the ROSAT and other sources (this proposal is included in Chapter 11).

In sum, before sitting down to write a proposal, researchers select a target agency carefully and become thoroughly familiar with its funding history and preferences. Funding agencies publish guidelines for proposal writers as well as formal *requests for proposals (RFPs)*. Much can be learned about a funding agency's interests and purposes by reading these materials. We will take a closer look at how to do this in Chapter 6.

• • • • • • • • • • • • •

2.5 Other communication channels within scientific communities

In this chapter, and in the book as a whole, we have concentrated on the most common formal channels for communication in the research sciences: journals, conferences, and grant proposals. In Chapter 7, we will also examine the issues involved in adapting scientific information for general audiences, but we will not include more specialized types of scientific communication in this text. Scientists who conduct their research in applied settings—for example, as employees of pharmaceutical companies, environmental consulting firms, or regulatory agencies such as the EPA or FDA—will need to become familiar with other types of internal and external reporting in addition to the research genres described in this text. The general principles presented in this book will be useful on these occasions for writing, but we should note that these specialized communications are also governed by more specific conventions that can be recognized and learned. In addition to style sheets and manuals developed "in house," technical writing books designed for engineers and business managers often offer useful advice for scientists working in industrial or governmental contexts.

As we conclude this overview of formal communication channels, it is important to recognize that most of the day-to-day work of science takes place through much less formal channels—conversations in the lab or field, informal memos within a research group, departmental presentations and colloquia, and casual communication via telephone, fax, E-mail, on-line discussion groups, World Wide Web sites, and other technological channels.[1] Communication media are developing rapidly in a number of directions, and reliance on these technologies varies widely across and within scientific disciplines. As access to technologies increases within a given field, new communication channels will become navigable. Their purposes and audiences will gradually become defined, as will the other conventions governing their use. Students entering your research field a decade from now will no doubt find an even more complex set of communication channels to negotiate.

ACTIVITIES AND ASSIGNMENTS

The following activities are designed to help you begin to investigate the communication channels in your research community. Your instructor may ask you to work alone or with others in your field.

1. Write a one-page introduction to the "business" of your field for outsiders who are unfamiliar with that branch of science. What do botanists (or soil scientists or organic chemists) do? Where do they work? What do they study? What kinds of questions do they ask? What kinds of methods do they use? Why is their research important? Who uses the results of this research? Who is affected by its results?

2. Interview a member of your research community to learn the kinds of writing, reading, speaking, and listening he or she does in professional life. A good way to elicit this kind of information is to ask your interviewee to describe a typical day or couple of days

[1] The *Directory of Scholarly Electronic Conferences* (Kovacs et al 1994), for example, lists hundreds of LIST-SERV discussion groups that exist in most disciplines in the biological and physical sciences, as well as other fields, all over the world. This directory can be accessed on line from many library information systems, or a printed copy can be obtained from the Association for Research Libraries; contact Ann Okerson (ann@cni.org) for more information.

in the lab or office: Whom did he or she talk to? What meetings did he or she attend? What reading or writing activities did she or he engage in (e.g., recording data, writing notes, drafting part of a report or proposal, revising an article, reading or skimming journal articles or abstracts, reviewing manuscripts or proposals)? Your interview subject should also be able to help you identify the major research journals, professional conferences, and funding agencies in your field. If you are working with a group, decide together on an appropriate interview subject, develop a set of questions, and arrange for one or more members of the group to conduct the interview. Feel free to interview more than one person, particularly if your group includes students from different subfields. Your instructor may ask you to summarize your findings for the class in an informal oral presentation, or to present them in writing, as described in Activity 3.

3. Use the research you've conducted in Activities 1 and 2 as the basis for a written profile of communication patterns in your research community, as illustrated in Figure 2.4. Your goal is to describe the distinctive practices that have developed in your community. Include the names of the most important journals, professional associations, and funding agencies as examples, but focus more broadly on how, why, and with whom scientists in your field communicate. Include both written and spoken channels of communication. Tell your readers what kinds of knowledge or information are exchanged via the various channels, to whom that information is directed or from whom it is received, and what form it must take in each case.

   Your profile should be written for readers who are not familiar with your particular field of study. Begin the paper with a brief introduction to the field (as described in Activity 1) to give readers some idea of the type of research that is carried out in this field. Then go on to describe the channels of communication that have developed to facilitate this research.

4. In groups, exchange your written profiles with other students in the class. (If profiles were presented orally, use your notes from the oral presentations.) Ideally, each member of the group should represent a different research field or subfield. After all members of the group have read the set of profiles, develop a list of similarities and differences across fields. Prepare a brief oral presentation for the class in which you compare and contrast the conventional ways of communicating in this set of research fields. For example, consider whether scientists in these fields ask similar or different types of questions, whether their objects of study are similar or different in significant ways, and whether and how their methods differ. Determine whether differences in research goals or methods have led to different patterns of communication in these communities. For example, you might examine the importance of conferences versus publications in different fields, the amount of interaction researchers in each area have with outside audiences, the amount of collaborative work in each field and the mechanisms for supporting that work, and so forth.

5. Go to the library, and find the major research journals you've identified. Working from the editors' statement of purpose or guidelines for authors (as well as information gathered in your interview), annotate your list with a brief summary of the purpose and audience of each journal, as in Figure 2.1.

6. Select one research journal in your field for closer study. Write a three- or four-page rhetorical analysis in which you discuss the following basic elements of communication:

a. *Author.* Who publishes this journal? Whose research is reported there?

b. *Audience.* Who are the intended readers? What types or levels of expertise do they have? Are they primarily researchers, policy makers, resource managers, educators, or other experts?

c. *Subject.* What topics or kinds of topics does the journal cover? How general or specialized are these topics?

d. *Text.* Describe the distinctive features of the text which helped you make inferences about author and audience. Do the articles have a formal or informal tone? Does the journal include more than one type of article (e.g., research reviews, letters, technical notes)? If so, how do these texts differ in style, tone, format, and purpose? (*Hint:* Don't overlook obvious features of the journal that might provide important information about its primary purpose and audience: the journal's title, institutions represented on the editorial board, announcements and advertisements, types of books reviewed, etc.)

7. Write a letter to a real or a hypothetical research collaborator in which you explain why you'd like to submit the results of your collaborative project to a particular research journal. Assume the two of you had narrowed your choices to two journals in the field. You've done some more investigating of the two journals, and you're ready to argue that one rather than the other will be the better place to report your findings. Your argument should be based on a clear comparison of the purpose, scope, and target audiences of the two journals.[2]

---

[2] We are indebted to Christina Haas for this assignment idea.

3

Reading and Writing Research Reports

3.1 Argumentation in science

Despite the proliferation of other forms of communication in the sciences, the published research report remains the most common medium used by individual scientists to communicate their findings to the research community at large—and thus to contribute to the developing knowledge of the field. Given the critical role of the research report in the development and exchange of scientific knowledge, it is important to understand its distinctive features and the common strategies scientists use in writing and reading these documents. Research reports in scientific journals are variously referred to in different fields as *journal articles, papers,* or *reports.* In this text, we use these terms interchangeably to designate the class of texts reporting original results or theoretical developments in professional research journals. Our purpose in this chapter is to describe some of the general conventions research reports follow and to help you analyze the specific conventions governing reports in your field.

The communication of research results is far from straightforward. It is traditional to think of scientific reports as purely factual or explanatory, but as we saw in Chapter 1, the report serves an important interpretive and persuasive function. We publish descriptions of our research not simply to tell others what we've done but also to persuade them that what we've done is valid and useful. In terms of form, then, the research report is more than a narrative; it is a careful argument. The authors of a research report find themselves in the position of building a case for their research, not simply recounting actions and observations.

From this perspective, the research report represents an extended argument in which researchers seek to convince readers that their research questions are important, their methods were sensibly chosen and carefully carried out, their interpretations of their findings are sound, and their work represents a valid contribution to the developing knowledge of the field. These basic goals are clearly reflected in a common research report format, which consists of four standard parts or sections: introduction, methods, results, discussion. Each section of the report contains an argument, and each section plays a part in supporting the larger argument of the whole.[1] This structure, sometimes referred to as the *IMRAD format*, has been described in a number of science writing guides (Katz 1985; Biddle and Bean 1987; Olsen and Huckin 1991; Day 1994) and is summarized by the Council of Biology Editors (1994) in Figure 3.1.

Sections of a research report: typical headings and functions

Heading for section	Function of section; comments
Introduction	Describes the state of knowledge that gave rise to the question examined by, or the hypothesis posed for, the research. States the question (not necessarily as an explicit question) or hypothesis.
Methods and Materials	Describes the research design, the methods and materials used in the research (subjects, their selection, equipment, laboratory or field procedures), and how the findings were analyzed. Various disciplines have highly specific needs for such descriptions, and journals should specify what they expect to find in a methods section.
Results	Findings in the described research. Tables and figures supporting the text.
Discussion	Brief summary of the decisive findings and tentative conclusions. Examination of other evidence supporting or contradicting the tentative conclusions. Final answer. Consideration of generalizability of the answer. Implications for further research.
References	Sources of documents relevant to elements of the argument and describing methods and materials used.

FIGURE 3.1 **Sections of a research report as described by the Council of Biology Editors. Scientific Style and Format: The CBE Manual for Authors, Editors, and Publishers. 6th Edition. Cambridge University Press. 1994. p. 590.**

---

[1] Notice that we use the term *argument* in the sense of "case building" here, as opposed to the everyday sense of "confrontation" or "debate." Thus, the question "What is your argument?" means "What is your line of reasoning?"

As the descriptions of these sections indicate, the IMRAD format consists of two sections in which the new study is actually described (methods and results), framed by two sections that place the new work in the context of previous knowledge (introduction and discussion). Both of these "framing" sections describe the current state of the field's knowledge. The introduction describes the current state at the start of the report—the state that created the need for the study—and the discussion describes the new state of the field's knowledge at the end of the report—now that the new results have been added to the knowledge pool.

The distinction between framing sections and describing sections is often signaled by verb tense: framing sections usually use present tense to describe the field's current knowledge ("Factors that limit the distribution of the cougar *are* not known entirely but *include* climatic features, availability of prey, and habitat features"), while describing sections typically use past tense in order to describe actions already taken and data already recorded ("We *searched* for cougar tracks between December and April each year"; "Summer and winter home ranges for individual females *overlapped* extensively").[2] One of the effects, and thus uses, of past tense, then, is to localize and limit findings to particular researchers or labs, while present tense identifies a claim or conclusion as part of the field's current understanding.[3]

EXERCISE 3.1   Choose two or more full-length research reports from your field or from those included in this text: Marshall and Warren (1984) or Graham et al. (1992) in Chapter 9; Mallin et al. (1995) in Chapter 10; or Fulbright and Reynolds (1990) in Chapter 11. In each major section of the paper, circle the main verb in the first 10 or 12 sentences. Note whether and how verb tense shifts across sections and whether and how these verbs tend to localize findings or incorporate them into general knowledge claims.

3.2 The logic(s) of scientific inquiry

We have included in this text three journal articles that follow the IMRAD organizational pattern: Marshall and Warren (1984) and Graham, et al. (1992) in Chapter 9, both reporting clinical studies of the stomach bacterium *H. pylori*, and Mallin, et al. (1995) in Chapter 10, reporting experimental tests of the relationship between the toxic dinoflagellate *P. piscicida* and other estuarine predators. The IMRAD format is a natural choice for reporting results from these studies, for in each case the research involved

[2] Quoted sentences from Ross and Jalkotzy (1992; 417, 418, 419).

[3] You are already quite familiar with the effects of present and past tense verbs, for you use these distinctions in everyday conversation. Compare the statement "I mowed the lawn myself" with "I mow the lawn myself." The first sentence clearly refers to a specific occasion on which you mowed the lawn (a localized event), while the second refers to a general practice you've adopted (a recurring or general pattern).

posing a question, choosing and carrying out experimental procedures, interpreting the resulting data, and drawing some conclusions about the significance of the outcome. Thus the IMRAD format mirrors the basic logic of scientific method and is therefore common in many scientific fields.

However, the emphasis on experimental methods and the presentation of new data in the IMRAD format is clearly unsuitable for papers reporting theoretical or historical research—in which the primary goal is to present new interpretations, theories, or models for understanding phenomena previously observed (Harmon, 1992; Miller and Halloran, 1993). Much of the research conducted in fields such as astrophysics, evolutionary biology, and geology, for example, is best described as theoretical or historical, rather than experimental. Papers in these fields may provide some methodological details (e.g., the orientation and time frame of pertinent satellite observations or the location and depth of core samples), but the primary goal of such research is not generally to test hypotheses but to formulate hypotheses—to propose theories or models that account for the field's observations to date.

The structure of papers in the historical and theoretical sciences is therefore more variable than that of experimental papers. For example, in analyzing a report published in *Sedimentary Geology,* student David Brock found an eight-part structure in which three sections describe characteristics of the geological formation under study, and two sections propose descriptions of the formation process (see Figure 3.2). Another student, Douglas Schneider, found a six-part structure in a report published by the *Journal of the Atmospheric Sciences,* in which separate sections are devoted to recent observations, assumptions of the computer model used to simulate the atmospheric conditions under study, and comparisons of the model's predictions with actual observations (Figure 3.3). Notice that these report structures provide explicit opportunities for authors to describe the assumptions and implications of their theories, in a sense the "methods" and "results" of theoretical work. Another good case in point is the 1990 paper on supernova remnants by Fulbright and Reynolds published in the *Astrophysical Journal* (Chapter 11), in which the authors present a mathematical model of supernova formation and structure and test it against observations of supernova remnants.

It is important to notice that the Fulbright and Reynolds paper, as well as those described by Brock and Schneider, still include the all-important framing sections—introduction and discussion—which contextualize the argument being presented just as in the standard IMRAD format. In stating the purpose of the study, the introduction lets readers know what kind of research is to be presented. The American Institute of Physics *Style Manual* explicitly reminds authors that the type and scope of the work, whether theoretical or experimental, should be clear from the introduction (AIP 1990).

In short, all research reports will include a framing introduction and discussion (labeled or unlabeled), but it seems obvious that the form of the body of the paper, the actual description of the work itself, will be determined by the type of work to be described. Because written texts communicate not only facts and observations but also the ideas and logic of their authors and their fields, texts will take different forms in different research communities. You will want to notice how reports in your field tend to be organized and how that organization compares with the organizational logic of the IMRAD format. As you read, pay close attention to the purpose and logic of each section. Notice how authors in your field convince readers of the need for the study and

Reporting Observations in Sedimentary Geology

David Brock

The article "Morphology and sedimentology of two contemporary fan deltas on the southeastern Baja California Peninsula, Mexico," by E. Nava-Sanchez, R. Cruz-Crozco, and D.S. Gorsline only partially follows the generic IMRAD structure. This article, as it appears in the journal *Sedimentary Geology* (98(1995):45–61), is divided into eight general sections titled as follows: introduction, general physiographic setting, geologic setting, oceanographic setting, geomorphology, sedimentology, discussion, and conclusions. The purpose of this research is to describe the method of formation and the source of the formation sediments related to modern fan deltas in southeastern Baja California. From this information the authors concluded that a large section of this coastal area is dominated by similar active fan deltas.

The introduction of this article is consistent with the goals served by introductions in the IMRAD format. In this first section, the authors define the geologic structures that their research focuses on and the factors that influence these formations. The first several paragraphs state the nature and scope of the research as well as give a review of the pertinent information to be discussed throughout the remainder of the paper. This article introduction is definitely aimed at specialists and is composed almost entirely of jargon specific to sedimentary geology. Near the end of the introduction, the objective is clearly stated so as to create a fusion of the introduction and purpose into one segment. This is also characteristic of the general IMRAD format.

A section devoted to methods is absent from this research article—most likely due to the nature of the science of geology, which relies heavily on observations and descriptions and the subsequent correlation of this information with historic geologic episodes. In order to clarify this concept, the nature of geology as a science compared with other sciences should be considered. Geology is a speculative science that does not rely on cookbook experimentation due to the fact that geological processes occur on average over hundreds of thousands and millions of years. A short period of time to a geologist is 100,000 years.

In lieu of a methods section, this article contains three sections devoted to the description of the settings in which the observations occurred, and two sections which describe the surface processes and sediment makeup in the area. The first section of setting description is labeled "General physiographic setting" and describes the location and weather patterns and provides a landscape evaluation of the study area. This type of regional description is valuable in geology because processes are highly dependent on weather patterns, climate, and existing topography. The second section of setting description is "Geological setting." This section focuses on when and how the Baja Peninsula formed and a definition of the topography in geologic terms. The regional geologic fea-

(continued)

FIGURE 3.2 Analysis of a research report in the field of sedimentary geology. Student sample.

tures are mentioned, such as faults, regional spreading, and subduction, which all characterize this area. This format is not in concordance with the general IMRAD formula; however, it substitutes sufficiently for a methods and materials section in a traditional research paper. Formatting the article in this manner is more suited to the field of geology, and particularly suitable in sedimentary geology, because this sort of information is crucial to understanding the remainder of the paper. The last section describing the setting is "Oceanographic setting." In these paragraphs the morphology of the underwater basins, specific formations, and the water circulation in the area are mentioned. This serves the same purpose as the previous sections on the setting, the main difference being that it concentrates more specifically on the facets of the landscape influenced by the ocean.

The remaining two sections before the discussion and conclusion become more specific and technical. In relation to the general IMRAD format, these two sections are most clearly related to a results section. The fifth section, "Geomorphology," is a technical discussion of the surface processes responsible for alluvial fan formation in the area of research. This section becomes technical enough to make major generalizations about the study area as well as present pertinent observations or data to support them. The sixth section, "Sedimentology," focuses on exactly what the paper is about. It describes the sediments which compose the modern fan deltas, the percentage makeup of these sediments, and their thicknesses. The sedimentology discussion is the final descriptive section before the discussion and conclusion. It finalizes the observations and narrows the scope of the paper to focus on the structures specifically mentioned in the title.

The final two segments of the paper are the discussion and conclusion. The discussion in this case classifies the two major fan deltas that were studied, and supports this classification. This effectively fits the results into the context of the field by relating them to an already existing classification scheme. Overall, this section does not state how these observations contribute to the advancement of the field. It simply brings together the previous observations and classifies the fan deltas. Although this is not in accordance with the general IMRAD format, it is characteristic of the descriptive nature of geology. The last section, the conclusion, summarizes the five major findings by the authors. In general, the findings pertain to how the deltas were formed, the source of formation sediments, the modifications of these sediments, and the character of the resulting fan deltas. Again, this format is specific to the descriptive nature of geology. Rather than discuss the results of an experiment, the paper generalizes and classifies observations.

Throughout this research article in sedimentary geology, it is apparent that the main focus is on observation and description of landforms. Although the introduction and discussion sections tend to follow the general IMRAD format, the other sections of the article are geology specific. Despite this specificity, the article accomplishes the same general goals as other research reports. The majority of differences can be attributed to the difference in the scientific process between geologists and classical scientists.

FIGURE 3.2 Analysis of a research report in the field of sedimentary geology. Student sample.

Analysis of an Atmospheric Modeling Report

Douglas Schneider

The article "A Modeling Study of the Dryline" by Ziegler, et al. in the *Journal of the Atmospheric Sciences* (15 January 1995) does not exactly conform to the generic IMRAD format for a research article. This is due to the type of research that was conducted. A mesoscale model of a dryline was created by computer simulation. The model was then compared with an actual case study of an observed dryline, with different variables being held constant for each model run. While the article contains clearly defined introduction and conclusion sections, the methods, results, and discussion sections are less obvious.

The introduction section begins by citing previous studies of drylines that have used modern data analysis techniques. The scientists' hypothesis about dryline formation is stated, and the methods of proving the hypothesis are briefly summarized. The authors describe the state of knowledge that gave rise to the research by referring to previous modeling studies of the dryline, and what processes have been addressed as well as ignored in these studies. Then the authors point out the contributions that their study will make to the current state of knowledge about drylines. The researchers intend to compare the results of the model with an actual dryline that occurred in Oklahoma in 1989. The purpose of the investigation is restated at the end of the introduction

The next two sections of the article, "Observations" and "Mesoscale Model," could be considered parallels to the "methods and materials" section of the IMRAD format. The observations section gives a brief synoptic overview of the region in which the observed dryline formed. A more detailed analysis of this dryline was given in a previous article. The third section states the atmospheric assumptions and the physical properties that were used in creating the mesoscale model. For example, the authors assume the atmosphere to be nonhydrostatic, and the geostrophic wind is assumed to be equal to the surface wind below 700 millibars due to a lack of data. The authors tell where "nudging" of the data was used to smooth extrapolations without affecting the results. Grid configuration and the initialization of atmospheric and vegetation fields are described in detail, and important mathematical formulas that were used are given. The differences between seven model runs are described. For example, one run held Coriolis force at zero, while another run was performed over a short grass domain.

The next two sections reveal the results of the control case, and the second section contains the results of sensitivity tests, with different parameters being changed each time. A description of variable gradients is presented, and the authors tell how these gradients can be used to determine dryline movement. The control model results of the dryline movement were consistent with the movement of the observed dryline. The researchers point out that certain gradients and updraft speeds at the dryline were considerably greater in this study than in previous studies, possibly due to frontogenic effects. The

(continued)

FIGURE 3.3 **Analysis of a research report from the field of meteorology. Student sample.**

authors continue by describing the movement of typical drylines and giving explanations for the movement pattern based on their model, but they admit that they cannot conclude why drylines move westward during the evening. The second results section describes the runs made by changing soil moisture, surface texture, Coriolis force, and wind components, and how these runs differ from the control case. Several tables and figures are presented to support the text. In this particular article, the results and discussion sections of the IMRAD format overlap with one another in two results sections.

The conclusion section of the article, "A Modeling Study of the Dryline," summarizes the purpose and intent of the modeling study. The authors state that "there is broad agreement between the morphologies of the modeled and observed drylines." The results of the sensitivity tests are briefly compared, and the authors use these data to speculate on the effects of soil moisture and vegetation types in dryline formation. Finally, the authors suggest other dryline characteristics that could be investigated in the future with the three-dimensional model.

FIGURE 3.3 Analysis of a research report from the field of meteorology. Student sample.

the significance of the outcome, as well as how they describe the methods, assumptions, and observations that serve as the evidence in their arguments.

In the next sections, we describe the logic of each of the standard IMRAD sections in more detail.

• • • • • • • • • • • • • •

3.3 Introducing the research problem

All research reports begin with an introduction of some sort, no matter what structure is followed in the rest of the paper. This is where you explain your research objectives, argue that the research is important, and place your study in the context of previous research. As Figure 3.1 demonstrates, editors and other readers expect the opening paragraphs of a journal article to describe the state of knowledge that motivated the research in the first place and to introduce the purpose of the study. We have referred to this opening section as a "framing" section above, to underscore its role in establishing a context or framework for interpreting the new research.

• •

EXERCISE 3.2   Take a minute to read the scrambled introduction in Figure 3.4. How would you assemble those sentences into a focused and readable one-paragraph introduction? List the sentence letters in an appropriate sequence in the margin before reading on.

• •

Blood flow through the human arterial system in the presence of a steady magnetic field

A. Such studies of flow through single arteries, however, have somewhat limited practical applicability because the actual human arterial system is composed of a large number of interconnected vessels of different lengths and cross sections which cannot be treated as independent.

B. In the present work, the finite-element method (FEM) is used to analyse the effects of a magnetic field on blood flow through a model of the human arterial system.

C. In recent years some studies have been reported on the analysis of blood flow through single arteries in the presence of an externally applied magnetic field.

D. It is known from magnetohydrodynamics that when a stationary, transverse magnetic field is applied externally to a moving electrically conducting fluid, electric currents are induced in the fluid.

E. These include the work of Belousova (1965), Korchevskii and Marochunik (1965), Vardanyan (1973) and Sud *et al* (1974, 1978).

F. Quantitative results on the effects of field intensity and orientation on flow and arterial pressures are presented and discussed.

G. This could occur for instance in nature or in physics laboratories, during space travel, or in hospitals.

H. The interaction between these induced currents and the applied magnetic field produces a body force (known as the Lorentz Force) which tends to retard the movement of blood.

I. Human subjects could by accident or design be made to experience magnetic fields of moderate to high intensity.

FIGURE 3.4 Scrambled introduction to Sud VK and Sekhon GS. 1989. Blood flow through the human arterial system in the presence of a steady magnetic field. Physics in Medicine and Biology 34(7):795–805.

In Exercise 3.2, many of you probably chose to start the paragraph with sentence C, D, or I, presumably because you felt some background or announcement of the topic was warranted at the beginning, or perhaps because you were following the familiar "funnel" introduction pattern in which you start "broad" and then narrow to the particular issue at hand. If you placed sentence B or F at the end of the paragraph, you chose to end by introducing the particular study to be reported in the paper, also in keeping with the inverted pyramid pattern. The actual introduction is reproduced in Figure 3.5.

Notice in Figure 3.5 that the authors Sud and Sekhon begin as we've just described, using sentence I to announce the general topic or issue (human beings occasionally encounter magnetic fields), briefly elaborated with sentence G (which tells us

Blood flow through the human arterial system in the presence of a steady magnetic field

Human subjects could by accident or design be made to experience magnetic fields of moderate to high intensity. This could occur for instance in nature or in physics laboratories, during space travel, or in hospitals. It is known from magnetohydrodynamics that when a stationary, transverse magnetic field is applied externally to a moving electrically conducting fluid, electric currents are induced in the fluid. The interaction between these induced currents and the applied magnetic field produces a body force (known as the Lorentz Force) which tends to retard the movement of blood. In recent years some studies have been reported on the analysis of blood flow through single arteries in the presence of an externally applied magnetic field. These include the work of Belousova (1965), Korchevskii and Marochunik (1965), Vardanyan (1973) and Sud *et al* (1974, 1978). Such studies of flow through single arteries, however, have somewhat limited practical applicability because the actual human arterial system is composed of a large number of interconnected vessels of different lengths and cross sections which cannot be treated as independent. In the present work, the finite-element method (FEM) is used to analyse the effects of a magnetic field on blood flow through a model of the human arterial system. Quantitative results on the effects of field intensity and orientation on flow and arterial pressures are presented and discussed.

FIGURE 3.5 Original (unscrambled) introduction to Sud VK and Sekhon GS. 1989. Blood flow through the human arterial system in the presence of a steady magnetic field. Physics in Medicine and Biology 34(7):795–805.

where this might happen). Next, sentences D and H summarize some basic principles of the field in order to show readers why this phenomenon is of interest or concern (exposure to a magnetic field might have adverse health effects: it may retard the flow of blood). Sentences C and E then refer very briefly to the research that has been done on the topic so far (five studies, including two of Sud's own, which have looked at the effects of magnetic forces on blood flowing through single arteries).

This quick review of prior knowledge on the topic sets up the critical next move: in sentence A, Sud and Sekhon point out that the previous research is limited (focusing on single arteries does not provide an accurate picture of effects on blood flow through the entire, interconnected arterial system). Hence the need for the new study reported in this paper. In the next move, sentence B announces the purpose and value of the new study (it presents a model for analyzing effects of magnetic fields on the whole arterial system, not just on single arteries). Finally, sentence F previews the types of information that will be reported in the paper to follow. Notice that both independent variables (field intensity and field orientation) and dependent variables (blood flow and arterial pressures) are announced in this final sentence. Brief as it is, it gives us an overview of critical features of the research design. The introduction section in most journal reports is quite compact, conveying a great deal of information in a relatively small space.

a common approach

In asking you to try the "scrambled intro" exercise, we have borrowed an instructional technique from John Swales, a linguist who has studied the logic and form of scientific papers (Swales 1984). In a study of introductions from 48 scientific papers published in three research fields, Swales discovered a remarkable degree of consistency across fields and journals. He found, as we did in the Sud and Sekhon piece, that authors use introductions to demonstrate that their research responds to a gap in the field's knowledge. They typically begin by announcing the topic at hand and then proceed to give an overview of recent research on the topic, in order to point out the gap or question that their study addresses. Swales thus identified four interconnected moves commonly found in journal article introductions, which we have summarized in Figure 3.6.

Common moves in research article introductions

Move 1   Announce the topic.
Move 2   Summarize previous knowledge and research.
Move 3   Prepare for present research
     by indicating a gap in previous research and/or
     by raising a question about previous research.
Move 4   Introduce the present research
     by stating the purpose and/or
     by outlining the research.

FIGURE 3.6 Based on John Swales, "Research into the structure of introductions to journal articles and its application to the teaching of academic writing." In *Common Ground: Shared Interests in ESP* **[English for Specific Purposes]** *and Communication Studies.* **Williams R, Swales J, and Kirkman J, editors. 1984. Pergamon.**

These four rhetorical moves can easily be seen even in Sud and Sekhon's brief introduction. (Take a minute to find them in Figure 3. 5.) Notice that moves 3 and 4 in particular are clearly signaled by Sud and Sekhon. We can't miss move 3 in sentence A because it is signaled by the contrastive adverb *however*: "Such studies of flow through single arteries, *however,* have somewhat limited practical applicability because the actual human arterial system is composed of a large number of interconnected vessels of different lengths and cross sections which cannot be treated as independent." As we have noted, Sud and Sekhon must point out this limitation of previous studies in order to show why their new work is needed—to "argue" its significance to the field. They move logically from move 3 to introduce their work, move 4, in sentence B. Move 4 is also clearly signaled, in this case with an introductory phrase: "*In the present work,* the finite-element method (FEM) is used to analyse the effects of a magnetic field on blood flow through a model of the human arterial system." There's no mistaking what the authors consider their primary contribution to be.

● ●

EXERCISE 3.3   Read the introduction to the study by Graham et al. (1992) reprinted in Chapter 9. Mark off the four basic moves, circling any clear signals included in the text. Paraphrase the line of reasoning presented in this intro, as we did for the Sud and Sekhon introduction.

● ●

As you continue to read formal research reports in your field and others, you will come across many variations on these four moves. Sometimes the moves will be made in a different order, one or more moves may be only implied, moves may be made more than once, or moves may overlap. For example, authors often cite previous research *while* announcing the topic. Citations of prior studies may appear in any number of places in an introduction; in such cases it would be difficult to distinguish move 2 from other moves. In fact, Swales (1990) later revised his model to underscore this variability. We believe the original model in Figure 3.6 is quite useful in identifying the basic gestures authors tend to make in introductions, as long as these gestures are understood as a set of flexible rhetorical moves that authors combine in diverse ways to achieve their rhetorical goals.

The Fulbright and Reynolds (1990) report in Chapter 11 represents a good case in point. Fulbright and Reynolds begin with a six-paragraph introduction. The first paragraph clearly announces the topic, move 1 ("Supernova remnants can serve as prime laboratories for the study of the acceleration of energetic electrons and the turbulent amplification of magnetic field"), indicates a gap in prior research, move 3 ("However, little agreement has been obtained on basic questions such as . . ."), and announces the purpose of the study, move 4, ("We shall address some of these questions by . . ."). Move 2, the review of previous knowledge, is made in the second, third, and fourth paragraphs. After this extended review, the authors come back to move 4, this time providing an extended description of the purpose and methods of the research in the fifth and sixth paragraphs ("We shall address the same problem using a somewhat different technique").

Despite variation in length, organization, headings, documentation format, and other text features, effective introductions in all fields include similar rhetorical moves because they share the same rhetorical goal: the authors want to convince readers that the topic is important and that their work on the topic will advance the field's knowledge. Thus the introduction serves several functions: It orients the reader to the research topic, reviews pertinent literature (and in doing so helps establish the authors' credibility), and sets up the argument for significance that is a goal of the discussion section.[4] The effectiveness of this argument rests largely on the authors' description of what the field knows now. Conventional ways to summarize the state of knowledge in the field and to acknowledge the work of others are discussed at more length in Chapter 4.

[4] In fact, the structure of the introduction, in which the research gap or problem is made prominent, can be understood to create the psychological desire for a solution (see Olsen and Huckin 1991). As rhetorician Kenneth Burke puts it, "form is the creation of an appetite in the mind of the auditor, and the adequate satisfying of that appetite" (1968, 31).

• •

EXERCISE 3.4   Read the multiparagraph introduction to the study by Mallin et al. (1995) reprinted in Chapter 10. Mark off the four basic moves in this section, circling any clear signals included in the text. Write a paragraph in which you paraphrase the authors' line of reasoning.

EXERCISE 3.5   Choose a published research report from a journal in your field. Type up a scrambled version of the introduction to this report, as in Figure 3.4. (If you choose a multiparagraph introduction, break it into "packets" or clusters of sentences instead of individual sentences.) In class, have a partner read and unscramble the intro, with the goal of reassembling the original line of reasoning.

• •

• • • • • • • • • • • • • •
3.4 Describing methods

The second major component of the research report is a description of the methods and materials you used in your research. The "materials and methods" section traditionally follows the introduction, though some journals have begun to move these details out of the main body of the article to the end of the paper and/or to set this section in smaller type. Berkenkotter and Huckin (1995) speculate that this trend toward deemphasizing the methods section indicates that journal readers rely heavily on peer reviewers' assessment of methodological details, saving their own reading time for inspection of a study's results. It is also generally recognized that the methods section cannot describe every action of the experimenters. Readers who need further methodological detail (for example, in order to replicate the study) are likely to contact the experimenters directly. Wherever it appears in the journals in your field, the methods section remains an important component of the overall argument of the research report, for it explains the parameters within which the study's results were generated.

As noted earlier, the concrete information in the methods section is usually presented in simple past tense, either active voice ("We *collected* water samples every three days") or passive ("Samples *were collected* . . ."). Although the "scientific passive" has a long and venerable tradition, it is often easier and more direct to write in active voice, and this mode is preferred by many journal editors in the interests of brevity and clarity. The editors of the *Journal of Heredity* directly inform contributors that "first-person active voice is preferable to the impersonal passive voice" (*J. Heredity,* Information for Contributors).

On the other hand, it is observed in the *AIP Style Manual* that "the passive is often the most natural way to give prominence to essential facts," as in sentences like "Air was admitted to the chamber," where it is not important to know who turned the valve (AIP, 1990, 14–15). The AIP does recommend shifting to active voice where necessary to avoid confusion or awkwardness. You'll notice that active voice naturally leads to the use of first person ("*We* collected . . ."), which is increasingly common in scientific prose, a trend also endorsed by the AIP (14) and other editorial panels. In addition of the use of active and passive voice for clarity and emphasis, scientists sometimes use these stylistic features strategically to highlight or diminish different aspects in their research—to make "stylistic arguments." For example, note how Mallin et al. (1995) use passive voice in their introduction when describing their own research, but active voice when describing the new dinoflagellate they are investigating (Chapter 10). The effect is to put the new dinoflagellate in the forefront, to call attention to its "active" existence and toxicity, while downplaying the role of the researchers in discovering it.

• •

EXERCISE 3.6   Look for instances of active and passive voice in one or more of the research reports in Chapters 9, 10, and 11, or in sample articles from your field. Do authors tend to use these modes consistently? Do you perceive any patterns or strategies in the use of active and passive voice? Compare the types of information that tend to be presented in each mode.

• •

Obviously, the content of the methods section will vary according to the type of research to be described. This variation is illustrated in Figure 3.7 with a simple comparison of subheadings used by different authors to organize this section.

Although the logic of the methods section is dictated by the logic of the research and therefore varies widely across fields, notice in Figure 3.6 that there are some basic organizational similarities. Methods descriptions begin by identifying the subjects of study, whether they be bacterial strains, fish, forests, or human beings. It is conventional in many environmental science fields, for example, to include a separate "Study area" section in which the ecosystem under study is identified and located and its distinctive features described. Papers reporting research involving human or animal subjects typically include a subsection labeled "Subjects" or "Population characteristics" at the start of the methods section. In Example E, from a study which tests a new forest plot mapping technique, the technique itself is the object of study, so the authors inserted a separate section describing the mapping equipment before going on to describe how it was tested.

Once the subjects of study have been identified, the materials and procedures used to study them can be described. This description should be detailed and complete enough to enable knowledgeable colleagues to repeat the experiment, observation, or calculations successfully. Standard procedures that will be familiar to your in-field readers can be identified quickly without citations, further explanation, or justification. Procedures established in previous studies may appear with citations. And new procedures or substantial modifications will be explained and justified. Matthew Barker noticed several levels of procedural explanation in a report published in the *Journal of Protein Chemistry* (see Figure 3.8). As Barker's analysis illustrates, the degree of justification needed for a procedure depends on the status of these procedures in the research community.

Example A, from Marshall and Warren (1984). *Lancet* (reprinted in Chapter 9).
Patients and Methods
  Patients
  Questionnaire
  Endoscopy
  Histopathology
  Microbiology
  Analysis of Results

Example B, from Lorimer, et al. (1994). *Journal of Ecology* 82:227–237.
Study areas
Methods
  Experimental design
  Plot measurements and analysis

Example C, from Lim, et al. (1995). *Journal of Bacteriology*.
Materials and Methods
  Bacterial strains, plasmids, and culture conditions
  DNA manipulation and sequencing
  Data bank analyses
  Plasmid constructions
  Construction of *M. tuberculosis* genomic libraries
  Alkaline phosphatase assay
  Antibody preparations, SDS-PAGE, and immunoblots

Example D, from Fabrizio (1987). *Transactions of the American Fisheries Society*.
Methods
  Sample collection and treatment
  Data analysis

Example E, from Quigley and Slater (1994). *Southern Journal of Applied Forestry*.
Equipment
Methods
  Field Procedure
  Laboratory Procedure

FIGURE 3.7 Subheadings from the methods sections of selected journal articles.

● ●

EXERCISE 3.7   In the article by Mallin et al. (1995), in Chapter 10, look for examples of each of the three levels of procedural explanation described above: routine procedures, procedures established in previous studies, new procedures or substantial modifications. How much explanation or justification is provided in each case? Why?

● ●

Analysis of a Research Report in the Field of Biochemistry

Matthew Barker

The article "Chemical Modification of Cationic Residues in Toxin α from King Cobra (*Ophiophagus hannah*) Venom" by Shinne-Ren Lin, Shu-Hwa Chi, Long-Sen Chang, Kou-Wha Kuo, and Chun-Chang Chang reflects the general IMRAD format quite closely with only a slight modification to the form. The article appears in *Journal of Protein Chemistry, Vol. 15, No. 1, 1996,* pages 95–101. The research report is divided into three main sections. The introduction is section one, the materials and methods is section two, and the results and discussion make up section three. In all these sections, the use of citation based on the knowledge of the audience can be seen.

In the first two sentences of the introduction, and the topic of research, snake toxins, is announced. Although these sentences use technical jargon that might be confusing to someone outside the field, the information they furnish is actually quite general and requires no citations. The authors then use eight extensively detailed sentences supported by ten citations to summarize the current knowledge and research in the field. In the first sentence of the second paragraph, the topic is more tightly focused on the toxins of the king cobra. The second sentence of the paragraph points out the low number of studies on the α-neurotoxins of the king cobra. The next three sentences list the research previously done in this subfield, mostly by the authors, and are supported by four citations. The fact that all of the current research cited was performed by members of the research team greatly enhances the credibility of this study. The final sentence of the introduction states the objective of the research.

The materials and methods section is divided into six subsections. The section opens with a paragraph describing the materials used. The rest of the section is broken down into five enumerated subsections as follows: "2.1 Modification of Arginine Residues with HPG, 2.2 Modification of Amino Groups with TNBS, 2.3 Localization of the Incorporated Groups, 2.4 Assay for Lethal Toxicity, and 2.5 nAChR-Binding Assay."

The materials paragraph describes the materials used in the experiment and gives the names and locations of the chemical and biological companies that supplied the materials. The paragraph also refers to a procedure used to isolate and purify cobra venom and supports it with two citations.

The methods subsections describe the two different modification procedures used, the procedure by which the sites of those modifications occurred, and the two different assay procedures employed to assess the effects of the modifications. Of the ten citations within the materials and methods section, four of the research reports cited were authored by members of the research team, again indicating the expertise possessed by this group of scientists.

The methods sections themselves briefly describe the procedures and parameters under which the research was carried out. Citations are used where possible rather than

(continued)

FIGURE 3.8 Analysis of a research report from the field of biochemistry. Student Sample.

writing summaries of the procedures. A procedure from 1938 is even cited. Familiarity with a procedure from so long ago demonstrates extensive knowledge of the field. It is possible that the procedure is actually fairly common knowledge in the field, but citing it suggests awareness of research and procedures discovered between then and the present.

As stated earlier, the results and discussion sections are consolidated into one section rather than separated. Even though that is the case, the section is divided into two subsections. The first, titled "3.1 Characterization of the Modified Derivatives," is actually a results section; and the second subsection, titled "3.2 Biological Activity of the Modified Toxins," discusses the results.

Subsection 3.1 presents the results and supports the data with four figures and three tables. The figures and tables are clearly labeled and complement the corresponding textual support for the data. The data from a polyacrylamide gel that was run was omitted and is so noted in the report. The reason for such an omission can only be guessed, but perhaps the reputation of the researchers has been so well established that they could forego visual support in this instance. It is also possible that the data can stand alone and did not necessitate displaying the gel, or the omission of such a figure is commonplace in the field. Aside from the one anomalous omission of a figure, the results subsection otherwise conforms to the IMRAD format.

The discussion section is contained in subsection 3.1. Here the significance of the results and the conclusions of the research are displayed. Four other studies are cited as having carried out similar research, and this report is offered as an extension of the information that those previous studies first elucidated. Conclusions are further supported by citing another study that supports another aspect of the current study.

Unlike the citations found in the other sections of the report, none of the authors contributed to any of the citations found in the discussion section. Because all of the current research on the α-neurotoxins of the king cobra have been conducted by members of the team, it seems that those studies could easily have been chosen to lend support to the conclusions made in this report. Such support might be construed as being biased. By using outside support, however, the possibility of bias is avoided, and the conclusions gain credence. The researchers also successfully place their research within the context of other preexisting, accepted research in the field.

[handwritten margin notes: "speculating about something he found"; "Good analytical comments"; "bring in stuff we've discussed in class"]

FIGURE 3.8 Analysis of a research report from the field of biochemistry. Student Sample.

Subheadings are often used to subdivide the methods section to highlight the various types of analyses that were performed, as in examples A and C in Figure 3.7, or to distinguish groups of related procedures, as in example E. It is often helpful and sometimes essential to use graphics in describing methods, such as maps of study sites or diagrams of experimental apparatus. In Section 3.5 we will present some guidelines for incorporating graphics into your text.

• •

EXERCISE 3.8   Examine the organization of the methods section in the research report by Graham et al., 1992 (Chapter 9), or by Mallin et al., 1995 (Chapter 10). Both of these papers follow conventional formats prescribed by the journals' editors; neither journal encourages the use of subheadings within sections. Suppose the editors had a change of heart and decided to allow authors to subdivide their methods section. What subheadings would you recommend, and where would they appear?

• •

Notice that in describing the subjects, materials, locations, and procedures of your study you are also explaining why you chose them. Just as the introduction section argues that the study was needed, the methods section argues that the study was sensibly designed and carefully conducted. This is the occasion to defend the decisions you made in designing and carrying out the research. Often these explanations are simply offered in passing, as in, "We increased the volume to 30ml for the copepod *to minimize containment effects.*" At other times, an explicit justification is offered to explain a methodological choice: "The dinoflagellate's TFVCs require an unidentified substance in fresh fish excreta; *hence, it was necessary to maintain cultures using live fish.*" (Both examples are from Mallin, et al., 1995, emphasis added.)[5]

The methods section thus establishes the conditions under which the results of the study were generated, and therefore the context in which they must be interpreted. In this sense, the methods "frame" the results (and represent a frame within a frame). Notice that describing the methodological features of the study is comparable to the articulation of assumptions required in theoretical papers. In describing our methods, we articulate the assumptions we made about the phenomenon under study: for example, that it will or will not vary over time or as a function of temperature or nutrient content or medical treatment or altitude or other factor; that the effects of these variables are observable with these instruments or observational techniques; that these techniques are reasonably free of bias; that these subjects are representative of the larger population to which we wish to generalize; and so forth. The validity of the study rests on the persuasiveness of our description of the methodological decisions we made.

• •

EXERCISE 3.9   Turn the methods section from one of your recent class lab reports into a methods section suitable for journal audience in your field.[6] Where appropriate, state the rationale for your methodological choices. Target a specific journal, and use informative subheadings if editorial guidelines permit.

• •

[5] Methods are sometimes described without such justification in student lab reports (because the primary audience, the instructor, has chosen the methods and does not need to be convinced of their validity), but in most other scientific contexts you will be describing methods *you* have chosen to readers who were not privy to your methodological decision-making. How you explain and justify these choices is crucial to your *ethos* as a professional scientist.

[6] We are indebted to Olsen and Huckin (1991) for this exercise.

3.5 Reporting results

Graphics represent the logic of your research design. (handwritten margin note)

The presentation of results plays a critical role in the developing argument of the research report, for it is here that new evidence is presented to address the gap or question outlined in the introduction. At first glance, the content of this section of the IMRAD format seems obvious: this is where we present the data. But presenting data is not a simple matter. We cannot display all the information we accumulated during the study, which may consist of pages and pages of lab or field notebooks or computer printouts or disk space. Journals do not have room for us to report each day's striped bass capture rate in a three-month study period, for example, or pH levels from the 200 soil samples we collected.

Summarize (handwritten margin note)

Instead, we summarize the data. We summarize first by *reducing* the data to a manageable size for presentation; in experimental studies this is often accomplished by converting raw data to means (averages) and by using figures and tables. We might report mean capture rates for each week or month, depending on how much variation we see over those time periods; we might group and average the results of our soil tests according to the location of the samples or the distance from a waste discharge point, depending on the design of our study. Once we've reduced the data in these ways, we can make comparisons; e.g., we can look for changes in capture rate over time or differences in pH among sampling sites. It is these observed changes or differences (or the lack of change or difference) that constitute the answers to our research questions.

generalize (handwritten margin note)

Reducing the data thus enables us to take a second step in presenting results, that of *generalizing* from the data. We want to point out the trends we've noticed in the data so our readers can see why we drew the conclusions we did. If we are to convince readers that our conclusions are valid, we must first ensure that the patterns we saw in the data are readily apparent. Olsen and Huckin (1991) point out that both levels of information must therefore be presented: "(1) the major generalization(s) you are making about your data and (2), in compact form, the data supporting the generalization(s)" (363).

Compare the following sentences, both of which refer to the data contained in a table:

Example A
Bioassay trials showed that all 10 rotifers *(B. plicatilis)* exposed to non-toxic green algae survived; all 10 exposed to dinoflagellates survived; all 10 exposed to both survived; 3 of the unfed group survived (Table I).

Example B
Bioassay trials with the rotifer, *B. plicatilis,* showed no mortality among any of the three algal food treatments (Table I). In the unfed treatment, 7 of 10 replicate animals died during the 9 day experimental period, and survival of unfed rotifers was significantly lower than in the other treatments (Fisher's exact test, $P = 0.0015$).

Example B is from the report by Mallin, et al. (1995) in Chapter 10. Example A is invented. If you look at Table I in the report in Chapter 10, you'll see that example A simply repeats data that can easily be read in the table itself. Example B, on the other hand, generalizes from the data, emphasizing the similarity among the three treatment groups and the difference between those groups and the control.

The relationship between data and generalizations is easily demonstrated by looking at how tables and figures are referred to in well-written reports, as in example B, above. Tables, graphs, and other visual aids are essential tools for reporting scientific results, but it is important to keep in mind that graphics only present the data; the generalizations needed to interpret those data must be provided in the text. We recommend two essential steps in integrating visual and verbal information:

- Refer readers to the graphic explicitly.
- Tell them what patterns to notice.

In example B, the researchers refer us to Table I by referencing it in parentheses at the end of the first statement describing data in that table. Other references to tables and figures are even more explicit, as in the first sentence in the following example from the Fulbright and Reynolds (1990) paper in Chapter 11:

Example C

Table 1 lists the values of A at 90, 32, and 10 beams per remnant diameter, and these values are plotted in Figures 5a–5c.

At the resolution of 90 beams per diameter, the three models have distinctly different A versus ϕ curves. In the isotropic particle case, A varies from 1 to 6.3. In the quasi-parallel case, A varies from 1 to basically infinity, while in the quasi-perpendicular case, A varies from 1 to 15.

This initial statement points us to the table, identifying the kind of information to be found there (turn to Table 1 in the Fulbright and Reynolds paper in Chapter 11). The authors then go on to highlight the trends to be noticed in this complex table. They summarize the results by emphasizing the differences among the three models, illustrating the scope of the difference by highlighting the range of values in each case. It is often useful to illustrate generalizations with representative or distinctive cases, or by highlighting the range of values or the size of the differences among groups.

. . . .

EXERCISE 3.10   Look for explicit references to figures and tables in sample articles in this text and from journals in your field.

1. Notice that the parenthetical reference to Table I in example B is different from the direct reference to Table 1 in example C. How common are these two "reporting formulae" in the articles you examined? Do these forms vary across fields or across journals? Pay special attention to the ways in which authors in your field refer to figures and tables.

2. Describe how the raw data have been transformed in these figures or tables; that is, how have the raw data been reduced into a summary visual form?

3. Develop a set of categories to describe the types of generalizations you see these authors making about the data in figures and tables, e.g., "comparing groups" as in example B or "describing the range of values" as in example C. Do the authors use representative cases or distinctive cases from the data to illustrate these generalizations?

EXERCISE 3.11   In Figure 3.9 we have reprinted a table and a figure representing the same set of data. What do you notice about the two forms of visual presentation? List the types of information highlighted in each form, as well as the types of information that are subordinated or not available. What is gained or lost in the transformation from one representation to the other?

• •

As you will have noticed in Exercise 3.11, different types of visuals serve different purposes. Tables are used to summarize data when the exact values of the data are important or when there are no clear patterns that would lend themselves to graphical presentation (Monroe, Meredith and Fisher 1977). Tables are the least "visual" of visuals insofar as they don't actually transform data into a pictorial form but rather present numerical data in an organized matrix; however, tables as grids are extremely valuable for reporting and categorizing individual datapoints or group means, which may account for their frequent use in science.

Graphs help highlight trends and patterns. For example, line graphs illustrate chronological trends well, for lines as a visual form naturally illustrate movement and direction over time. Line graphs are included in the papers by Graham et al. (1992) in Chapter 9, Burkholder et al. (1992) in Chapter 10, and others in this text; you'll notice in each case that the graph's x-axis represents time, measured in units such as weeks or days. Compare these x-axes with the x-axis in the bar graph (Figure 5) Burkholder and Rublee (1994) included in their proposal in Chapter 10. In this bar graph, eight sampling sites—four control sites and four discharge sites—are arrayed along the x-axis. Each bar represents one observation (the number of zoospores counted at the site), rather than a sequence of observations over time. The bar graph enables readers to see at a glance the relative magnitude of the counts at each site. Similarities among control sites and among the discharge sites are easily apparent, as is the striking difference between the two conditions.

Graphs are labeled as "figures" in most scientific texts. Other types of figures include diagrams, cross-sections, maps, photographs, and flowcharts, each of which serves a distinctive purpose. For example, photographs and similar media are useful when observational detail of the object is important; Warren and Marshall use electron micrographs in their 1983 *Lancet* letters to show the shape and location of *Campylobacter* bacteria (Chapter 9). Diagrams are useful for highlighting structural features and relationships (Olsen and Huckin 1991), as in Burkholder and Rublee's Figure 2 (Chapter 10), in which the authors illustrate the many life stages of the toxic dinoflagellate, the relationships among these stages in the life cycle, and interactions with other factors in the environment (finfish and phosphates). In addition to summarizing results, visuals such as maps, diagrams, and photographs are often used to illustrate methods or to locate study sites.

Table 4. Gastritis Scores of the 175 C. pyloridis Culture-Positive Patients Before and After Therapeutic Regimens

		Treatment									
		CBS		Amoxicillin		CBS + amoxicillin		Cimetidine		Sucralfate	
	Culture	m^a	n^b	m	n	m	n	m	n	m	n
Before treatment	+	5.9	67	5.5	22	6.2	20	5.5	53	5.8	13
After treatment	+	3.8	37	2.9	7	6.5	2	5.4	52	5.3	13
	−	1.2^d	30	0.9	15	1.1^c	18				
1 mo after treatment	+	5.6	55	4.2	17	5.0	12	5.8	51	5.3	13
	−	0.7^d	12	0.7^d	5	0.5^d	8				
3 mo after treatment	+	5.8	54	5.1	16	5.3	12	5.5	52	5.7	13
	−	0.5^d	10	0.2^d	5	0.9^d	8				
6 mo after treatment	+	5.7	54	5.2	16	6.0	12	5.5	48	5.6	13
	−	0^d	10	0.6^d	5	0.6^d	8				
12 mo after treatment	+	6.6	52	5.5	16	5.5	13	5.3	51	5.9	13
	−	0^d	10	0^d	5	0^d	7				

Gastritis was assessed by scoring for four characteristic pathological parameters of chronic active gastritis (see Methods). +, positive; −, negative. [a] Mean score. [b] Number of patients. [c] $p < 0.05$ vs. culture-positive patients by Wilcoxon rank-sum test. [d] $p < 0.01$ vs. culture-positive patients by Wilcoxon rank-sum test.

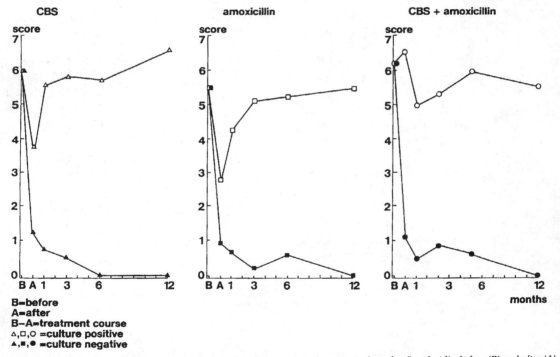

Figure 2. Mean gastritis scores in patients with chronic active gastritis and positive culture for *C. pyloridis*, before (B) and after (A) treatment with CBS, amoxicillin, or a combination of CBS and amoxicillin. Note the ultimate complete disappearance of inflammatory changes of gastric mucosa (gastritis score = 0) after persistent eradication of *C. pyloridis* during follow-up of 1 yr. (See Table 4 for actual values.)

FIGURE 3.9 Two different representations of a data set. Table and figures from Rauws et al. (1988).

Some research fields are <u>characterized</u> by the use of particular types of nonverbal representations. For example, <u>papers in organic chemistry rely heavily on diagrams of molecular models,</u> whereas modeling studies in physics often include mathematical equations and formulas. When such visual or numerical information is part of a theoretical argument, it is typically inserted in the text itself rather than in a separate figure, as can be seen in the Fulbright and Reynolds (1990) paper in Chapter 11. But theoretical studies may include separate figures as well; in the same paper, Fulbright and Reynolds included a typical map generated by their model in a Cartesian coordinate system and a version of the same image converted to a polar coordinate system, in order to show how the parameter A was measured from the models.

• •

EXERCISE 3.12   Go back to the sample articles you examined in Exercise 3.10. List the types of graphics used in these articles, the kinds of data contained in them, and the patterns revealed. Why were these visual forms selected, and not others? Pay special attention to the purpose of the visuals and the information contained in table titles and figure captions.

• •

Guidelines for designing figures and tables can be found in style manuals (e.g., AIP Style Manual 1990; CBE 1994), writing guides (Monroe, Meredith, and Fisher 1977; Katz 1985; Pechenik 1987; Day 1994), and in the instructions to authors published by individual journals, particularly in fields which rely heavily on specialized types of visuals. For example, the *Astrophysical Journal,* in which the Fulbright and Reynolds paper appeared, provides guidelines for preparing tables and five different kinds of illustrations: halftones, color graphics, line drawings, computer graphics, and planograph tables (1990 July 1, 357(1):ii–iii). These varied sources all agree on the following general guidelines for preparing visuals:

- <u>Each table or figure must be independent,</u> that is, self-explanatory. Readers should be able to understand what kind of information is presented in the table without having to find the description in the text.
- For this reason, table titles (which appear *above* the table) should be as informative as possible: not "Table 1. Vegetation in trial plots" but "Table 1. Percentage coverage by low vegetation (< 1.5 m high) in midsummer by treatment, year, and site" (Lorimer, Chapman, and Lambert 1994). Titles should identify the variables being compared in the table. All labels in the table (column headings, row headings, notes, etc.) should be clear and brief.
- <u>Similarly, figure captions</u> (which appear *below* the figure) should be as informative as possible: not "Figure 1. Study area" but "Figure 1. Location of Waccabassa Bay study area, sampling sites, and distribution of storm deposit" (Goodbred and Hine 1995). Symbols should be clearly identified in a key; axes on graphs should be clearly labeled.
- Despite their independent status, figures and tables must also be *inter*dependent with the text. The visual and verbal presentations of results must work

together. Visuals should be clearly referred to in the text, as discussed above, and they should appear in the text as soon after the textual reference as possible (visuals should never precede their text mention, as this can confuse readers).

Keep in mind that even though graphics are intended to be self-explanatory, readers will not know what *you* find interesting or noteworthy in the table or figure unless you tell them. Whether your data are presented graphically or discursively, that is, visually or verbally, be sure to highlight the trends you see and explain where they came from.

EXERCISE 3.13   Read the introduction and methods sections (only) of the study by Graham et al. (1992) in Chapter 9.

1. Before reading the results section, study Table 1 in this paper. What trends do you see in these data? Draft a few sentences describing the patterns in this table as though you were writing up these results.

2. Then study Figure 1 in the same paper, and again draft some sentences about the patterns in the data. How would you describe the results presented in this figure?

3. Compare your descriptions with the authors' descriptions of this table and figure.

It is conventional in many fields to subdivide the results section to highlight the different types of analyses reported. If subheadings are used in both the methods and results sections, they typically reflect the same principle of division, as in the Marshall and Warren (1984) paper in Chapter 9. Both the methods and the results sections of the Marshall and Warren paper include subsections describing the questionnaires administered in the study and sections describing each of the three types of analysis conducted: endoscopy, histopathology, and microbiology.

In sum, though we present our formal conclusions in later sections of the research report, a great deal of interpretation goes on in results sections as well (Thompson 1993). Reducing the data, generalizing from the data, and highlighting specific cases are all highly interpretive processes. It should be clear by now that we don't let the data speak for themselves in research reports; in summarizing our results, we interpret them for the reader (Fahnestock 1986; Thompson 1993). We must present our data clearly, in the text and/or in tables and figures, to demonstrate to readers that our interpretations are warranted.

EXERCISE 3.14   Choose a sample research question for which you can gather some actual data quickly from your classmates, friends, or fellow dorm residents. For example, you might be interested in who spends more time studying: women or men, seniors or first-year students, on-campus or off-campus residents. Or you might gather information about time spent writing term papers for different classes, or number of visits to the library in a semester. To enable comparisons, be sure to include more than one group in your sample (e.g., freshmen versus seniors) or more than one variable (e.g., time spent writing versus grade on paper).

Present the data in a table or graph with an informative title or caption. Write a one-paragraph description of the purpose and research question (or hypothesis) of your "ministudy." After you've reviewed the data yourself, exchange research questions and data sets with another student. Have that student describe the major trends he or she sees in the data and tell you the answer to your research question. If you've stated a formal hypothesis, have your reader tell you whether or not the hypothesis was supported.

3.6 Discussing trends and implications

As noted earlier, the discussion section of the IMRAD format completes the frame opened in the introduction by returning to the significance argument. The purposes of introduction and discussion sections are inversely related. Whereas the introduction introduces the research question and reviews the state of knowledge in the field that motivated the question, the discussion explains how the question has been answered (at least in part) by the new research and shows how the field's knowledge is changed with the addition of this new knowledge. Thus, the discussion describes a new state of knowledge in the field, one that in turn motivates further research questions, which are typically highlighted in this final section.

Structurally, the introduction and discussion are mirror images as well. Berkenkotter and Huckin (1995) describe the introduction as proceeding "from outside in" in that it begins by talking about the general topic and reviewing the existing research in the field, and then it narrows to focus on the particular work at hand. Conversely, the discussion proceeds "from inside out," by first summarizing major trends in the new results and then situating those results in the context of the field (41).

We "situate" results by comparing them with the results of previous studies and with the field's general understanding of the phenomenon under study. This is where you offer your interpretation of your findings. How do they extend, refine, or challenge previous findings or assumptions (Olsen and Huckin 1991)? The discussion section is the place to comment on

- The magnitude and direction of the effects observed (compared with what others found or compared with what might be expected)
- The advantages and limitations of the methods used in the research (and how these features may have influenced the observed effects)
- The implications of the findings for current practice or theory
- The research questions that remain

Notice that these discussion issues are interdependent. In commenting on the magnitude and implication of their findings and the research questions that remain, it is important that scientists recognize and acknowledge the limitations of their study. In doing so, scientists must anticipate the questions that they know other researchers might have about their methods, findings, and conclusions, and answer these questions in the text before they are asked. It must be clear that the researchers have considered "rival hypotheses," that is, possible alternative explanations for their findings. Philosopher Stephen Toulmin, an expert on scientific argumentation, calls these answers to potential questions, objections, and counterarguments "rebuttals" (Toulmin et al. 1984). According to Toulmin, rebuttals are the recognition of the circumstances or exceptions that might undermine the argument of the research.

Rebuttals often consist of further qualifications or acknowledgments of limitations, exceptions, and the need for further study and research. Rebuttals therefore are a way for a writer to verbally close loopholes, seal leaks, and tie up loose ends in the logic of the report. The last sentence of the research report by Marshall and Warren (1984) is a perfect example: "Although cause-and-effect cannot be proved in a study of this kind, we believe that pyloric campylobacter is aetiologically related to chronic antral gastritis and, probably, to peptic ulceration too" (1314). Also note the paragraph of qualifications that appears at the end of the report by Fulbright and Reynolds (1990).

One of the disadvantages of the research letter genre is that it does not allow for the extensive reporting of results that support claims, lengthy explanations of unexpected finding, or extensive rebuttals (as illustrated by the responses elicited by Fleischmann and Pons's letter (1989) in the *Journal of Electroanalytical Chemistry* announcing the creation of cold fusion). All arguments, of course, are open to rebuttal; as we discussed in Chapter 1, scientific statements must be falsifiable to be accepted as valid. But in anticipating objections and answering them, scientists show they understand the limitations of their research, the nature of research problems in their field, and the expectations of their colleagues. Thus, the professional use of rebuttals is another way scientists enhance the credibility of their research as well as their *ethos* as researchers.

• •

EXERCISE 3.15   In the research report by Graham et al. (1992) contained in Chapter 10, note the five rebuttals starting in the middle of the last paragraph and running up to the final recommendation supporting Marshall's ulcer therapy. What counterarguments are these rebuttals to? Why do you think Graham felt he had to include these rebuttals in his report?

Rebuttals are in last 2 paragraphs.

• •

As with the other sections of the report, the emphasis in discussion sections varies somewhat across fields. It is conventional in journals in applied fields such as the *Journal of Wildlife Management* to include a separate section on management implications after the main discussion. Similarly, discussion sections in agricultural journals may include advice for producers on feeding practices or feed composition. This is where authors make recommendations for practice or policy based on their new results. The audiences for these journals typically include resource managers, agricultural producers, and/or researchers and other professionals employed by regulatory agencies and other policy-making bodies.

In basic research areas the focus is not on practical applications but on the theoretical or methodological implications of the new results. This is certainly the case in reports of theoretical studies like the Fulbright and Reynolds (1990) piece included in Chapter 11. Notice that Fulbright and Reynolds use their extensive discussion section to compare the predictions of their model with actual radio images of supernova remnants. The implications of this test of their model are discussed in a separate "Conclusions" section, in which you will recognize several of the standard discussion moves listed earlier.

Thus discussion and conclusions sections, whether combined or labeled individually, represent the conclusion to the broader argument of the research report. We do not leave it up to readers to draw their own conclusions about our data. We tell them *our* conclusions, the ones we believe the data support, and we provide enough information about the data and the conditions under which they were collected to enable readers to determine whether our conclusions are justified. The American Institute of Physics (1990) describes conclusions as "convictions based on evidence" (4). The discussion section is the place to state our convictions and demonstrate how they follow from the evidence presented in other sections of the report.

• •

EXERCISE 3.16   1. Analyze the discussion and conclusion sections of ~~three or more~~ one full-length research reports from journals in your field (include any separate sections describing recommendations or management implications). List the types of discussion issues raised in these sections. Use the categories listed in this section as a guide (size and direction of effects, advantages and limitations of methods, implications for practice or theory, remaining questions), but be prepared to expand the list as necessary to best describe the kind of issues raised in your field.

2. In groups of three or four students from different fields, compare lists, and prepare to discuss similarities and differences across these research areas. How do these similarities and differences reflect the nature of research in these fields?

EXERCISE 3.17   Plan a discussion section to contextualize the results of the informal study you conducted in Exercise 3.14. What kinds of previous research would you want to be familiar with if you were to conduct this study on a larger scale and prepare a formal report for a research audience? What implications can you draw from your results? What methodological limitations would you discuss? If your study were more formal, would you recommend changes in practice or policy on the basis of your results? What further research directions would you suggest? List, as specifically as you can, (1) the issues you would raise in such a discussion and (2) any further information you would need to complete this task.

• •

We have focused thus far on the logic and structure of the arguments presented in research reports, but it is also important to notice distinctive features of the language in which these arguments are expressed. The persuasive and interpretive nature of scientific texts can be detected even in the language of individual sentences if we look carefully at how scientists phrase their claims and conclusions. Take a minute to fill in the blanks in the sentences listed in Figure 3.10.

DIRECTIONS

The following selections are from a variety of research reports published in scientific journals. Read the selections carefully. When you come to a blank, fill in an appropriate word or phrase to complete the sentence. You may insert more than one word in the blank if necessary, and you may use the same word in more than one sentence if you wish.

1. Eleven of the trials have shown the treatments to be ineffective, yielding an overall response rate of 4/278 (1.4%). These data _____ that the minimal response rate of interest should be .15.

2. The above observations _____ that (1) fertilized soils tend to attain apparent equilibrium with orthophosphate solid phases and (2) soils with moderate to high P-fixing capacity tend to have limited movement of P when fertilized with inorganic P sources.

3. Statistical analysis _____ that corn yields were not influenced by N fertilizer rate in 1980, but were in 1981 (Table 1). The lack of influence of N fertilizer in 1980 was attributed to high levels of native N in the soil and climatic conditions unconducive to high corn yields (Fig. 2).

4. More recent studies of modern thickly sedimented convergent margins _____ that the Washington margin is anomalous. For example, the Makran (Platt, et al., 1985) and Barbados (Westbrook, 1982) convergent margins are thickly sedimented and have convergent rates similar to the Washington margin (about 5 cm/yr). However, only the Washington margin is dominated by landward-verging structures.

5. Results of this study _____ that significant genetic divergence has occurred among geographically separated groups of raccoons. The average differentiation among the 14 localities examined (37.4%) is similar to the value obtained among populations of pocket gophers (41.0%; Patton and Yang, 1977).

FIGURE 3.10 Sample claims from Penrose and Fennell (1992) study of scientific discourse. Sources: Sylvester (1988); Schwab and Kulyingyong (1989); Clay et al. (1989); Byrne and Hibbard (1987); Hamilton and Kennedy (1987).

In a recent study of scientific discourse, Penrose and Fennell (1993) asked experienced scientists from a variety of fields to complete the sentences in Figure 3.10. The verbs *suggest, indicate, show,* and *demonstrate* were the most common responses from these experts, regardless of field. The finding that responses from these individuals were fairly uniform indicates that these terms are conventional across disciplines and are not unique to writing in, say, geology or botany. Notice that we've used one of these conventional verbs, *indicates,* in the preceding sentence to signal our interpretation of Penrose and Fennell's results. The sentence tells you that from the responses of these individual scientists we are generalizing about scientific discourse at large.

When we say that verbs like *suggest* and *indicate* are "conventional" in scientific discourse, we mean that they are commonly used constructions which carry particular, agreed-upon meanings in the community. These conventions enable scientists to make claims within the established parameters of knowledge-making in science. That is, these expressions enable a researcher to put forth a conclusion—to contribute to the advancement of knowledge in the field—while at the same time acknowledging that it *is* a conclusion—an interpretation of the facts and not a fact itself. Use of these common expressions signals to other scientist readers that this author is aware of the interpretive nature of scientific knowledge and is cognizant of the reasoning behind his or her claims.

• •

EXERCISE 3.18　1.　Penrose and Fennell (1993) also asked students from different majors and levels of experience to complete the sentences in Figure 3.10. They found that first-year students and non-science majors were more likely to supply the word "prove" in these sentences than were science majors or expert scientists. Why is this usage less conventional among scientists? What is the difference between "suggest" and "prove" in these types of scientific claims?

2.　Another unconventional choice some students made in this study was to use words like "conclude" or "hypothesize" in the sample sentences. What is unconventional about this set of choices? How does "conclude" differ from "suggest" or "indicate"?

• •

Scientists also use other kinds of qualifiers to convey the interpretive nature of their claims. Adverbs and adverbial phrases are often used to note limitations or special conditions; examples include: *possibly, probably, very likely, necessarily, certainly, without doubt, presumably, in all probability, hypothetically, maybe, so far as the evidence suggests, as far as we can determine* (Toulmin et al. 1984). Such qualifiers indicate the strength or extent of the claim being made, as in the following examples from the 1995 article by Mallin et al. (emphasis ours):

> *Pfiesteria*-like species <u>apparently</u> are widespread; recent investigations have demonstrated their presence at "sudden-death" estuarine fish kill sites from the mid-Atlantic . . .

> Since the fecundity of *B. plicatilis* in the two *P. piscicida* treatments was <u>slightly</u> elevated overall, relative to fecundity when given green algae alone, this dinoflagellate is <u>probably</u> a nutritious food source for the rotifer.

A third type of qualifier is the modal auxiliary verb. Verbs like *may, might, would, could, should, must, can,* and *shall* are used to indicate qualifying conditions, as in the following examples:

Given its broad temperature and salinity tolerance, and its stimulation by phosphate enrichment, this toxic phytoplankter <u>may</u> be a widespread but undetected source of fish mortality in nutrient-enriched estuaries. (Burkholder et al. 1992)

The stomach <u>must</u> not be viewed as a sterile organ with no permanent flora. (Warren and Marshall 1983)

Qualifying verbs, adverbs, and modal auxiliary verbs can and do occur anywhere in a scientific text where researchers need to qualify or limit their claims. In a sense, then, qualifiers are related to rebuttals insofar as scientists use them to acknowledge the limitations of their work and to anticipate and head off questions and counterarguments that readers might pose. These linguistic devices thus reflect the interpretive and interactive nature of knowledge-building in science. As you join the community of scientists, you are acquiring knowledge of both the logic of scientific arguments and the language in which that logic is expressed.

EXERCISE 3.19   In the last three paragraphs of Marshall and Warren's (1984) report contained in Chapter 10, note the increased use of qualifiers in response to their own question as to how the bacteria that they suggest is the cause of gastritis survives in the hostile environment of the stomach: "the bacteria were *usually* in close contact with the mucosa, often in grooves"; "the mucus *appeared* to form a stable layer over the spiral bacteria"; "The absence of these bacteria from past reports of gastric microbiology *may be* because gastric juice was cultured"; "*Perhaps* the mucus coating is deficient or unstable near ulcer borders" (1314; emphasis added). Identify the "kind" of qualifiers being used here (verbs, adverbs, modal auxiliary verbs). Look for other examples of qualifiers in the last paragraph. What purpose do they serve? Why do you think there are fewer qualifiers in the introduction?

EXERCISE 3.20   The original title of the Burkholder team's letter to *Nature* (1992) was "New 'phantom' dinoflagellate is implicated as the causative agent of major estuarine fish kills." This title was shortened during the publication process (see Chapter 10), much to Dr. Burkholder's dismay. What is the effect of the title change, and why do you think Dr. Burkholder was dismayed?

3.7 The research report abstract

The abstract is a summary of a document's major points, appearing at the start of each article in most journals. Along with the title of the report, the abstract helps readers decide whether the paper is pertinent to their research interests. The abstract is sometimes used in indexing and may even appear on its own, detached from the full paper, in abstract journals and on-line information services (AIP 1990). It therefore must represent the full argument contained in the report: the topic and purpose of the study, the methods used, the results obtained, and the conclusions drawn from those results. However, this argument must be outlined in extremely condensed form, usually only about 5 percent of the length of the full paper (AIP 1990).

Given its role in representing a larger document, the abstract can be described as a "contingent" genre—its content is contingent on the content and organizational logic of the document being abstracted. Thus, we will talk about research report abstracts in this chapter and will provide guidelines for other sorts of abstracts in later chapters, specifically the conference presentation abstract in Chapter 5 and the research proposal abstract in Chapter 6. Some journals provide highly specific guidelines for the content of abstracts. *Annals of Internal Medicine,* for example, requires the structured eight-part list illustrated in the paper by Graham, et al. (1992) in Chapter 9. This structure—with separate sections for the study's objective, design, setting, participants, intervention, measurements, results, and conclusions—is used in other journals reporting clinical trials as well.

Most journals, however, specify only a maximum length for the abstract (typically 100 to 200 words) and leave its structure up to the author. Because the abstract mimics the form and logic of the full report, the structure should be relatively straightforward. If the full report follows the IMRAD format, the abstract is likely to include four basic moves representing the major parts of this pattern, and verb tense will shift accordingly (Olsen and Huckin 1991): the topic will be introduced in present tense, usually in a sentence or two; the background and/or need for the study will be outlined in another few sentences (but without references to individual prior studies (AIP 1990; CBE 1994); methods and results will be briefly described in past tense; and the major conclusions and implications of the study will be stated briefly in present tense.

If the report to be abstracted follows a logic other than the IMRAD pattern, the structure of the abstract should be adjusted accordingly (for an example, see Fulbright and Reynolds (1990) in Chapter 11), but most research report abstracts can be easily generated using the four-part structure. This type of abstract has been labeled "informative," to distinguish it from the "descriptive" abstract sometimes used for research reviews, conference reports, and other genres (Olsen and Huckin 1991). Descriptive abstracts (also called "indicative" abstracts [Day 1994]) simply describe the kinds of information that will be contained in the paper, rather than providing a summary of that information. For example, compare the following hypothetical descriptive abstract with the informative abstract ("Summary") that accompanies Marshall and Warren's 1984 *Lancet* paper, contained in Chapter 9:

> The purpose of this report is to examine the relationship between bacteria and gastritis. Results of a clinical study of the presence of *Campylobacter* bacilli in gastroscopy patients are reported.

Because the descriptive abstract provides little information about the study itself or its results, it is less useful to readers (Olsen and Huckin 1983) and thus less common in the research literature. The informative abstract is generally preferred in research journals.

EXERCISE 3.21   Read the abstract for the paper by Mallin et al. (1995) in Chapter 10. Is this an informative or a descriptive abstract? Identify major moves and corresponding verb-tense shifts.

EXERCISE 3.22   Write a 100-word informative abstract for the research report by Brody and Pelton (1988), presented in Figure 3.11.

3.8 Brief report genres: Research letters and notes

In addition to the full-length research report, many journals publish brief reports in the form of *letters,* or research notes to enable researchers to communicate important findings to the research community as quickly as possible. The Brody and Pelton (1988) piece in Figure 3.11 is classified as a research "note." The journal *Physical Review Letters* is devoted exclusively to "short, original communications" (see Figure 2.1), and many other journals solicit similar types of articles: the *Journal of Biotechnology* publishes "short communications," which "must be brief definitive reports and not preliminary findings"; *Analytical Chemistry* publishes "Correspondence" ("brief disclosures of new analytical concepts of unusual significance") and "Technical Notes" ("brief descriptions of novel apparatus or techniques"). Theoretically, letters and notes can be reviewed and published more quickly than full-length reports, though this advantage may be negated by the large number of submissions editors receive in this increasingly popular category. Most of these forms have evolved from earlier, less formal forms: *Physical Review Letters* essentially began as a "Letters to the Editor" section in the journal *Physical Review* (Blakeslee 1994).

Though this category includes forms serving a variety of purposes—from presenting preliminary results, to introducing significant analytical concepts, to describing new techniques—research letters are typically used to describe innovative research or to announce novel findings. Watson and Crick's 1953 letter to *Nature* on the double helix structure of DNA is a case in point. Both *Nature* and *Physical Review Letters,* two of the most prestigious letter journals in publication, use the letter forum for this purpose.

We noted in Chapter 2 that the Burkholder team first announced their discovery of the "phantom" dinoflagellate in a letter to *Nature* (Burkholder et al. 1992). Similarly, Marshall and Warren initially reported the presence of bacteria in the stomach in joint letters in the *Lancet* (Warren and Marshall 1983). These letters are included in Chapters 9 and 10. Notice that neither the dinoflagellate, *P. piscicida,* nor the "ulcer-causing" bacterium, *H. pylori,* had yet been named at the time the letters were published, indicative of the early stage at which these reports appeared. Both might be considered cases of "revolutionary" or extraordinary science (Kuhn 1970; Toulmin et al., 1984). That is, both research teams were reporting findings that, if further supported, could signifi-

Seasonal changes in digestion in black bears

ALLAN J. BRODY[1] AND MICHAEL R. PELTON

Department of Forestry, Wildlife, and Fisheries, The University of Tennessee, P.O. Box 1071, Knoxville, TN 37901, U.S.A

Received June 23, 1987

Introduction

Winter dormancy in black bears (*Ursus americanus*) is likely an adaptation to predictable seasonal food shortages. Additionally, embryo development and parturition in bears are physiologically linked to hibernation (Herrero 1978). Fat is the major source of calories during hibernation and bears may lose as much as 25% of their body weight during the winter while maintaining lean body mass (Nelson et al. 1973). Thus, the ability to increase fat reserves before denning is necessary for winter survival and successful reproduction. Fat storage occurs during the fall (the "hyperphagia stage" of Nelson et al. (1983)), and many field studies have documented pronounced seasonal shifts in the diets of free-ranging bears (e.g., Tisch 1961; Hatler 1972; Beeman and Pelton 1980; Eagle and Pelton 1983; Grenfell and Brody 1983). Dietary shifts track plant phenology, and generally involve a transition from green forage and soft mast during spring and summer to hard mast in the fall.

Despite a strong dietary dependence on vegetable matter, bears exhibit only minor dental adaptation to herbivority and have retained the short, unspecialized gut of their carnivorous ancestors. Bunnell and Hamilton (1983) suggest that in grizzly bears (*Ursus arctos*) the evolution of a few morphological adaptations to herbivority combined with the conservation of physiological adaptations to carnivority make possible the rapid weight gains that occur before denning. They assume an evolutionary trade-off between the ability to digest food rapidly (a trait of carnivores) and the ability to digest low quality food efficiently (a trait of herbivores), and conclude that, in grizzlies, rapid processing at the expense of efficient digestion of fiber allows bears to take advantage of the large amounts of food available during the foraging period. Their experiments demonstrated that the digestive efficiencies of grizzlies are not markedy different from those of obligate carnivores. This note describes an effort to determine the digestive abilities of black bears, and to determine any seasonal differences in those abilities.

Methods

Six adult bears, ranging from 91 to 178 kg (mean weight = 63.4 kg, SD = 28.6 kg) at the Ober-Gatlinburg Black Bear Habitat in Gatlinburg, Tennessee, were used in digestion trials in August and November 1983. The bears were housed in three enclosures, each holding one male and one female. The normal ration for the bears consisted of dry dog food (Tennessee Farmers Cooperative, Lavergne, TN 37086, U.S.A), fed *ad libitum*, supplemented by produce discarded by local grocers. For the trials, all produce was withheld and the dog food ration was reduced to approximately 95% of *ad libitum* consumption for 7 days; portions of three feces in each enclosure were collected on the 7th day.

Fecal and ration samples were frozen until laboratory analysis was performed, at which time the samples from each enclosure in each trial were thawed and dried at 101°C. three subsamples were drawn from the fecal samples from each enclosure. Crude protein content was estimated by the macro-Kjeldahl technique (Association of Official Analytical Chemists 1970) (three replicates per subsample). Gross energy was estimated in an adiabatic oxygen bomb calorimeter (Parr Instrument Company, Moline, IL 61265, U.S.A.) (two replicates per subsample). Acid-insoluble ash content was estimated by the method of Van Soest (1966) (three replicates per subsample).

[1] Present address: Department of Ecology and Behavioral Biology, University of Minnesota, 318 Church Street SE, Minneapolis, MN 55455, U.S.A.

Printed in Canada / Imprimé au Canada

(continued)

FIGURE 3.11 Text (without abstract) from Brody and Pelton (1988), *Canadian Journal of Zoology.*

cantly alter basic knowledge paradigms in their respective fields: the Burkholder team described a new family, genus, and species of algae that evidenced predatory behavior and was found to be responsible for numerous previously unexplained fish kills; Marshall and Warren described a bacterium which not only was found to live in the supposedly sterile environment of the stomach but also was potentially linked to illnesses the field firmly believed to be caused by psychological rather than physical factors. In both cases, these early announcements were (and continue to be) followed by a wealth of further research in which these preliminary findings are tested and extended.

TABLE 1. Composition of experimental diets and fecal samples from bears, on a dry-weight basis

	Crude protein (%)	Gross energy (kJ/g)	Acid-insoluble ash (%)
This study			
Dog food ration			
August trial	36.7	19.292	1.12
November trial	30.9	19.212	1.22
Fecal samples*			
August mean ($n = 9$)	20.4 (0.60)	17.403 (1.333)	2.58 (0.002)
November mean ($n = 9$)	19.3 (0.76)	15.038 (0.063)	2.72 (0.001)
Grizzly rations used by Bunnell and Hamilton (1983)			
Basal ration	36.3	5577	0.11
Basal and beet pulp	21.2	4729	1.42

*SE given in parentheses.

TABLE 2. Apparent digestibilities in bears

		Apparent digestibility coefficients†	
	Approximate consumption*	Crude protein	Gross energy
August trial			
Enclosure 1	4.4 (0.33)	0.765	0.645
Enclosure 2	3.3 (0.24)	0.745	0.626
Enclosure 3	2.5 (0.41)	0.764	0.645
August mean	3.4 (0.70)	0.758 (0.0113)	0.638 (0.0110)
November trial			
Enclosure 1	7.5 (0.45)	0.729	0.651
Enclosure 2	5.0 (0.34)	0.708	0.640
Enclosure 3	7.0 (0.21)	0.723	0.656
November mean	6.6 (1.12)	0.720 (0.0108)	0.649 (0.0082)
Grizzlies on basal and beet ration (Bunnell and Hamilton 1983)‡		0.751	0.620

*Amount provided − orts, in kg/day for each enclosure. SD given in parentheses.
†SE given in parentheses.
‡Mean values for two bears.

Apparent digestibilities of crude protein and gross energy were estimated using the indicator method (McCarthy et al. 1974; Bunnell and Hamilton 1983), with acid-insoluble ash serving as the indicator:

$$\text{apparent digestion coefficient} = 1 - \frac{(\text{AIA in feed}) \times (Y \text{ in feces})}{(\text{AIA in feces}) \times (Y \text{ in feed})}$$

where AIA is the dry weight proportion of acid-insoluble ash and Y is the dry weight energy content or proportion of crude nitrogen.

Results and discussion

The dog food ration used in this study was substantially higher in crude protein than the natural food plants normally eaten by black bears, which typically contain $2-19\%$ crude protein (Mealey 1975; Eagle and Pelton 1983). Composition of the ration used in this study (Table 1) was most similar to the "basal and beet pulp" ration used by Bunnell and Hamilton (1983) and the apparent digestibilities (Table 2) were correspondingly similar.

Digestibility of crude protein decreased (paired t-test, $t = 26.41$, $P = 0.0014$) while the digestibility of gross energy increased (paired t-test, $t = 4.58$, $P = 0.0445$) from August to November. Increased food consumption in November implies an increased transit rate which in turn could have caused the decrease in apparent protein digestion (Castle and Castle 1956; Rerat 1978). If transit rate were the only factor affecting seasonal changes in digestibility, however, a simultaneous decrease in apparent gross energy digestion would be expected. Instead, we found an increase in apparent energy digestion, indicating that bears were selectively digesting and (or) absorbing carbohydrates and fats at the expense of protein.

Inhibition of amino acid absorption by carbohydrates has been well documented in several monogastric animals (Alvarado 1971), but there is little evidence that the degree of inhibition is controlled by factors other than concentrations of specific substrates in the gut. This mechanism could operate in the wild, where summer foods are typically much higher in protein than fall foods (Eagle and Pelton 1983), but cannot explain our experimental results because rations in both trials were similar. We suggest that a systemically, possibly hormonally, mediated increase in carbohydrate and fat assimilation and decrease in protein assimilation occurs during the predenning hyperphagic period. If lean body growth ceases in the fall, as data from Nelson et al. (1983) imply, protein

(continued)

FIGURE 3.11 Text (without abstract) from Brody and Pelton (1988), *Canadian Journal of Zoology.*

requirements would be reduced and preferential assimilation of dietary substrates most efficiently converted to fat would appear to be adaptive.

This type of hormonal control of assimilation has yet to be documented, but would be consistent with the array of physiological adaptations, particularly those of nitrogen metabolism (Nelson et al. 1973, 1975, 1983), already described in bears.

Acknowledgements

We wish to thank J. R. Carmichael and D. S. Carmichael for technical assistance. We benefitted from conversations with T. C. Eagle, R. L. Moser, R. A. Nelson, and W. D. Schmid. Comments from S. Innes and an anonymous reviewer improved an earlier draft of this paper.

ALVARADO, F. 1971. Interrelation of transport systems for sugars and amino acids in the small intestine. *In* Intestinal transport of electrolytes, amino acids, and sugars. *Edited by* W. M. Armstrong and A. S. Nunn. Charles C. Thomas, Springfield, IL. pp. 281–315.

ASSOCIATION OF OFFICIAL ANALYTICAL CHEMISTS. 1970. Official methods of analysis. 11th ed. Washington, DC.

BEEMAN, L. E., and PELTON, M. R. 1980. Seasonal foods and feeding ecology of black bears in the Smoky Mountains. Int. Conf. Bear Res. Manage. 4: 141–147.

BUNNELL, F. L., and HAMILTON, T. 1983. Forage digestibility and fitness in grizzly bears. Int. Conf. Bear Res. Manage. 5: 179–185.

CASTLE, E. J., and CASTLE, M. E. 1956. The rate of passage of food through the alimentary tract of pigs. J. Agric. Sci. 47: 196–204.

EAGLE, T. C., and PELTON, M. R. 1983. Seasonal nutrition of black bears in the Great Smoky Mountains National Park. Int. Conf. Bear Res. Manage. 5: 94–101.

GRENFELL, W. E., and BRODY, A. J. 1983. Seasonal foods of black bears in Tahoe National Forest, California. Calif. Fish Game, 69: 132–150.

HATLER, D. F. 1972. Food habits of black bears in interior Alaska. Can. Field-Nat. 86: 17–31.

HERRERO, S. 1978. A comparison of some features of the evolution, ecology, and behavior of black and grizzly/brown bears. Carnivore (Seattle), 1: 1–17.

MCCARTHY, J. F., AHERHE, F. X., and OKAI, D. B. 1974. Use of HCl insoluble ash as an index material for determining apparent digestibility with pigs. Can. J. Anim. Sci. 54: 107–109.

MEALEY, S. P. 1975. The natural food habits of free ranging grizzly bears in Yellowstone National Park, 1973–1974. M.S. thesis, Montana State University, Bozeman.

NELSON, R. A., WAHNER, H. W., ELLEFSON, J. D., and ZOLLMAN, P. E. 1973. Metabolism of bears before, during, and after winter sleep. Am. J. Physiol. 224: 491–496.

NELSON, R. A., JONES, J. D., WAHNER, H. W., McGILL, D. B., and CODE, C. F. 1975. Nitrogen metabolism in bears: urea metabolism in summer starvation and in winter sleep and role of urinary bladder in water and nitrogen conservation. Mayo Clin. Proc. 50: 141–146.

NELSON, R. A., FOLK, G. E., JR., PFEIFFER, E. W., CRAIGHEAD, J. J., JONKEL, C. J., and STEIGER, D. L. 1983. Behavior, biochemistry, and hibernation in black, grizzly, and polar bears. Int. Conf. Bear Res. Manage. 5: 284–290.

RERAT, A. 1978. Digestion and absorption of carbohydrates and nitrogenous matters in the hindgut of the omnivorous nonruminant animal. J. Anim. Sci. 46: 1808–1837.

TISCH, E. L. 1961. Seasonal food habits of the black bear in the Whitefish Range of northwestern Montana. M.S. thesis, Montana State University, Bozeman.

VAN SOEST, P. J. 1966. Non-nutritive residues: a system of analysis for the replacement of crude fiber. J. Assoc. Off. Agric. Chem. 49: 546–551.

FIGURE 3.11 Text (without abstract) from Brody and Pelton (1988), *Canadian Journal of Zoology.*

Nature limits letters to 1000 words and requires that they contain no subheadings and fewer than 30 references (contrasted with the 50 references allowed in full-length reports). The *Lancet* also publishes "preliminary reports and hypotheses" up to 1500 words, but letters in the *Lancet* are even more abbreviated, with a 500-word limit (Lancet 1990, Information for Authors). Notice that the Warren and Marshall letters are formatted like letters to the editor, with the traditional salutation, *Sir,* at the start and the author's name at the end. The Burkholder team's letter to *Nature* looks quite different. No subheadings are used, but the piece includes all the standard components of the IMRAD format, although of course these moves are less elaborated, and they are adapted to reflect the more general purpose of the letter: Burkholder and her colleagues are not describing the methods and results of a single new study; they are reviewing findings to date from a developing line of research.

• •

EXERCISE 3.23   Look for the basic components of the IMRAD format in the Burkholder team's letter to *Nature* (Chapter 10). In what ways does this short report follow the IMRAD pattern? In what ways does it deviate from this pattern? Where are the basic research report moves made (introducing the problem, describing methods, describing results, drawing conclusions)? Notice that table and figure captions are quite long in this report. What kind of information is contained in table and figure captions, and why is it there?

In a recent study of *Physical Review Letters,* Blakeslee (1994) points out that the brevity of these reports may make it difficult for reviewers to evaluate the quality of the science being reported, and indeed the letters forum allows little room for the types of evidence we expect to find in full-length research arguments. The letters genre is well suited, however, for presenting quick summaries of new developments and recent observations, as the letters by Warren and Marshall and the Burkholder team illustrate. When compared with the more formal and fully developed research report, letters have something of a journalistic flavor, consistent with their function of presenting "cutting-edge" research (Blakeslee 1994).

EXERCISE 3.24 The argumentative purpose and persuasive logic of scientific reporting often becomes more obvious in research letters and notes, since the length of these genres does not allow for extensive presentation of data. Read the 1994 research note published by Reynolds, Lyutikov et al. in *Monthly Notices of the Royal Astronomical Society* (Chapter 11). Look for explicit argumentative moves in this piece. Where and how do the authors reveal the logic of their argument? Look for places in the text where explicit claims are made. Describe how these claims are supported and/or qualified. Are any rebuttals present in the argument?

3.9 How scientists write reports

The order in which report sections are written is rarely the same as the order in which they appear in the finished text. Your abstract is the first thing readers will see, but you won't be able to write it until you've at least sketched out the main body of your text. Perhaps you'll sketch out your methods section first, while specific steps and decisions are still fresh; or you may begin by constructing the tables and figures you will use to present your results, filling them in as the data are generated and analyzed (see Monroe, et al. 1977). As noted earlier, because your graphics represent the logic of your research design, the form of your tables and figures should be determined long before you begin to write.

As is the case with many types of writing, the introduction to the research report is often revised throughout the drafting process; some authors even write it last. The focus and argument of the introduction depend in part on the outcome of the study. For example, we may discover that our findings raise questions we did not expect to be important; thus our study makes a somewhat different contribution to the field than we anticipated. The introduction needs to prepare readers for this contribution so that it is recognized and appreciated as important. Even though the purpose and design of the study have not changed, we will need to adjust our review of background issues and research to establish a context for later commenting on these new questions. In short, we may not be able to finalize the introduction until we've thought about the implications we want to raise in the discussion section.

Writing a report is thus a recursive process in which we continuously revisit and rethink our earlier arguments in light of our results and further thinking. In a study of the composing processes of nine eminent biochemists, Rymer (1988) found they employed a wide range of writing strategies and styles. Some subjects reported that they typically spend a great deal of time planning and outlining before drafting sections of the text, while others begin by writing a full "impressionistic" draft of the whole paper, refining it in later stages of the process. Most scientists use a combination of these approaches, adjusting their processes according to the type of research they're reporting, their familiarity with the prior research and the target journal audience, the amount of time they have available, and so forth.

The number of authors involved in a project also influences the structure of the writing process. In many scientific fields collaborative projects are the norm. The National Academy of Sciences reports, for example, that "the average number of authors for articles in the *New England Journal of Medicine* . . . has risen from slightly more than one in 1925 to more than six today" (NAS, 1995, 13). Research teams manage the writing process in different ways. In some teams, one author drafts the entire paper and sends it around to others for their comment and revision. In other working groups, different members of the team may draft different sections according to their role in the research; e.g., those who carried out the procedures and collected the data will draft the methods and results sections pertaining to their parts of the project, while the project leader who initially conceptualized the study will write the initial draft of the introduction and discussion (Rymer 1988).

If you completed the interview assignment in Chapter 2, you no doubt discovered that scientists spend a great deal of time writing, revising, and polishing papers before submitting them to journal review. One of Rymer's biochemists described a "predrafting" stage lasting approximately three years, during which the study was conceived and carried out and the postdoctoral fellows on the project drafted figures and tables as well as the methods and results sections. Once the lab work and analysis were completed, the project leader drafted and revised the introduction and discussion sections over a period of two weeks and then spent another two months pulling the manuscript together, which included revising the methods and results sections provided by the postdocs and soliciting feedback on the draft from colleagues (Rymer 1988).

Clearly the form of the final written product tells us little about the process through which that text came to be. Unlike the conventions of the written form, writing processes are governed not by field-specific convention but by the habits, skills, and preferences of the individuals who make up the research team.

3.10 How scientists read reports

It should be clear from this chapter that research reports are governed by the conventions of their respective research fields and by the conventional practices of journals within fields. It is important to remember that "conventional" practices are those that are customary in a community and thus familiar to all members. Readers in a community use their knowledge of these conventions to facilitate the reading and interpretation of the written texts those conventions govern. Thus, readers familiar with the IMRAD format will know where to quickly find the results in a report, where to look for the authors' interpretations of the results, where to find a description of methods, and so forth. They will not need to read the paper from start to finish in order to discover what kind of research was conducted and what was found.

In fact, few scientists have the time or the inclination to read journal articles carefully in the order in which they appear in print (Bazerman 1985; Berkenkotter and Huckin 1995). Readers rely heavily on the titles and abstracts of published reports, not only to help them decide whether to read the paper or not but also to help them quickly learn the key features of the study being reported. They then can turn immediately to pertinent sections of the report for the information they are most interested in. From consultations with researchers in physics and biology, Berkenkotter and Huckin (1995) determined that readers typically began by scanning the title and abstract of an article and then looking for the data, focusing on the tables and graphs in which the data are summarized. Only after examining the data themselves did these scientist readers read the results section provided by the authors. The rest of the reading process was quite variable; readers selectively read or skimmed the introduction, methods, and discussion sections, depending on how much they already knew about the topic, the methodological approach used, and the results of prior studies, and depending on their familiarity with other work of the authors. In an earlier study of physicists, Bazerman (1985) found similar variation in reading strategies. On some occasions readers would read an article's introduction and conclusion to get a sense of the focus of the study for future reference, skipping over details of methods and results; but in other contexts, as when reading a paper on a very familiar topic, the same readers might invert this pattern, ignoring the contextual information in introduction and conclusion and concentrating instead on methodological details, calculations, and results.

Given these varied reading strategies, we can see why section headings and figure captions are important in research reports, and why the content of titles and abstracts is critical. The final form of the report may reflect neither the order in which it was written nor the order in which it will be read, but it must match the expectations of readers in the research community, who will expect to find particular kinds of information and particular parts of the overall argument in conventional locations. The paper must reflect the logic of presentation with which readers in the field have become familiar.

Though readers who are thoroughly familiar with the line of research being reported may skim or ignore the introduction to a research report, the introduction serves a critical function for new readers in a field in that it establishes the motivation for the study being reported and situates the new work in the context of the field's existing knowledge. This context-setting function is particularly important in interdisciplinary fields because readers' backgrounds may differ widely. In a recent study of scientists reading articles on chaos theory, Paul and Charney (1995) found that chaos researchers from the fields of physics, mathematics, ecology, engineering, and meteorology paid more attention to the context-setting information provided in the introduction than to the description of the new study itself. That is, readers' first concern was "whether they could relate the reading to their prior knowledge and to their own work" (427). Paul and Charney's observations indicated that their readers were generally familiar with the four basic introduction moves we described in Section 3.3 (though they may not realize they know this formula), and they use this structure to help them incorporate the new work into their current knowledge of the topic.

Readers are active participants in the generation of knowledge. Scientists don't just gather facts when they read; they interpret those facts and carefully consider the interpretations proposed by the authors of the reports. They pay attention to how the authors designed and conducted the study, the conditions under which the work took place, and the operating assumptions underlying both the design of the project and the interpretation of its results. In other words, research readers are cautious consumers. Readers continually weigh the merits of the research and of the written argument as they read. The outcome of this evaluation process determines whether or not they accept the research as sound and important.

3.11 How reviewers evaluate reports

In the manuscript review process, this process of weighing the merits of research as we read becomes more formal and explicit. Editors and reviewers must evaluate the merits of the research not just for their own edification but for other members of the field as well. They are making the decision of whether to read or not to read on behalf of the larger community; the outcome of their deliberations determines whether the wider community gets to see the paper at all. As we discussed in Chapter 1, the peer review process is the mechanism via which the scientific community monitors the integrity of its work—in effect, "filters" the work of individuals. Though the system has its detractors (it has been criticized as a cumbersome process, subject to abuses such as favoritism and censorship), the peer review system remains a formidable structure enabling the "volatile microcosm of individual scientists" to contribute to the "stable macrocosm of the scientific enterprise" (NAS 1989). The logistics of the peer review (or "peer referee") system are quite straightforward. A typical timetable is described in *Nature*'s Guide to Authors, which we included in Chapter 2 (Figure 2.3).

There is a fair amount of consensus among editors and reviewers about what constitutes a publishable paper. Recall from Chapter 2 that journals clearly identify their target audience and the research areas they are interested in (review Figure 2.1). It follows, then, that we cannot write a research report until we've decided which journal audience we would like to reach.

Journal editors report that "suitability for the journal" is the primary concern when deciding whether to accept or reject a paper submitted to the journal (Huler

1990). Lack of suitability is the most commonly cited reason for rejection of a manuscript (Davis 1985, cited in Olsen and Huckin 1991). The *Journal of Heredity,* for example, announces to contributors that "acceptance will depend on scientific merit and suitability for the journal" (Information for Contributors). With the enormous number of papers competing for space in scientific journals (*Annals of Internal Medicine* publishes about 19 percent of the more than 2000 manuscripts it receives each year [Annals 117(1) 1992]), there is no room for papers that do not specifically address the needs and interests of the journal's target audience.

The following list of review criteria from the journal *Phytopathology* (78(1)1988:6) provides a good summary of the features that journal readers are looking for and thus a good checklist for writers of research reports:

- Importance of the research
- Originality of the work
- Appropriateness of the approach and experimental design
- Adequacy of experimental techniques
- Soundness of conclusions and interpretations
- Relevance of discussion
- Clarity of presentation and organization of the article

ACTIVITIES AND ASSIGNMENTS

1. Using the IMRAD format as the organizational framework for your analysis, describe the organizational structure of a research report in your field, as illustrated in Figures 3.2 and 3.3. Does this report illustrate the generic features described in sections 3.3 through 3.6, or does it deviate from these guidelines in some way(s)? If it does deviate, what has the author done differently, and why do you think he or she chose a different form? How does the form of the report reflect the goals, objects, or methods of research in this particular field? Keep in mind that your goal is to *analyze,* not just describe, the form of this particular research report. You'll need to think about why it takes the form it does.

Note: In this sort of analysis, it is conventional to refer to the authors in the third person (e.g., "Sud and Sekhon divide their presentation of results into three sections . . ."; "Their findings indicate that . . ."). To orient your reader, include the following identifying information in your opening paragraph: the authors' names, the title of the article, the name of the journal, and the purpose or main conclusions of the research. Be sure to turn in a copy of the article itself with your analysis.

2. In groups of four or five, read the analyses you wrote in Activity 1, looking for similarities and differences among the fields your articles represent. Prepare an oral report for the class in which you identify the major structural variables in this set of papers. You'll want to describe *how* the papers differ and *why* they differ in these ways. What do these formal features reveal about the goals and methods of these different research areas?

3. Either as part of Activity 1, or as a separate activity, analyze the use of explanation, justification, and citations in a research report in your field, as illustrated in Figure 3.8. When, and how much, do the author(s) explain and/or justify statements in each section, and why? When, and how, are citations used (to cite relevant studies, to support an argument, to indicate awareness of a controversial or undecided issue, e.g.)? What is the relationship of these strategies to the logic and purpose of each section of the IMRAD format? What does your analysis tell you about the knowledge and expectations of the author's audience? How does the author(s)' awareness of the knowledge and expectations of the audience help create the author(s)' *ethos?*

4. Turn to Figure 1 in the Graham et al. (1992) study in Chapter 9. From the data in this line graph, try to create a table, a bar graph, and a diagram. What data did you select for each visual, and what data did you exclude? Why? What do your visuals show? Which visual forms were most useful in representing the findings of the Graham team and which weren't as useful? Why?

5. Write a research report or letter on a project in your area of specialization. Your instructor may ask you to work alone or with a partner or group. The results you report may be real (i.e., from a study you are conducting in another course or at work) or simulated (invented for the purpose of this assignment). In either case, you will need to begin by reviewing prior research on the topic in order to identify what the field already knows, how the question has been approached in the past (what methods have been used), and what questions remain for you to address. See Chapter 4 for advice on locating previous research.

Write this report as though you were preparing an article for submission to a research journal in your field. Choose an appropriate journal to target, and develop your paper with that journal's audience and editorial guidelines in mind. Select a sample report from your target journal to use as a model.

• •

Reviewing Prior Research

• • • • • • • • • • • •

4.1 The role of prior research in scientific argument

Scientific work is often conducted under solitary and isolated circumstances: late nights in the lab, long hours in the field collecting samples, visits to remote research sites. However, this work is neither planned nor interpreted in isolation. New studies are conducted because they promise to shed light on issues considered important by the research community; research questions are valid in that they reflect gaps or inconsistencies in the field's understanding; methodological decisions are made with an awareness of what has been done by others and how it has worked; and the outcomes of research are interpreted in light of the theories, questions, methods, and findings that have come before. Research becomes meaningful only when viewed in the context of the field's developing knowledge.

This social or "situated" nature of scientific research is clearly reflected in scientific texts, not only in the adherence to conventionalized forms and terminology but also more directly in the practice of citing previously published reports. Most scientific arguments rely heavily on references to previous research to support the claims and methods reported or proposed. The practice of citing prior research is critical in arguments written for research audiences, including the arguments contained in grant proposals, research reports, and many types of brief research notes and letters. The National Academy of Sciences describes this practice as follows:

> Citations serve many purposes in a scientific paper. They acknowledge the work of other scientists, direct the reader toward additional sources of infor-

review articles situate research in social discourse community mation, acknowledge conflicts with other results, and provide support for the views expressed in the paper. More broadly, citations place a paper within its scientific context, relating it to the present state of scientific knowledge. (NAS 1995, p 12)

As described in Chapters 3 and 6, reports and proposals always begin by reviewing the current state of the field's knowledge of the phenomenon under study. Many proposal formats require a separate section devoted to reviewing prior research (sometimes referred to as a "review of the literature"). Research reports tend to include a more abbreviated review of research in the introduction section, but this brief review serves a critical role in establishing the context for the study. Citations of prior research also appear in the discussion sections of research reports, where authors tend to make very specific references to previous studies in order to help readers interpret the scope, scale, and significance of the reported findings and to help them understand how the new research extends, refines, or challenges the state of knowledge in the field. Previous studies are frequently cited in methods and results sections as well, often as precedents for specific methodological decisions or data analysis procedures. As we discussed in Chapter 3, decisions about what statements need support or justification, and how much support or justification is appropriate, depends on the writer's awareness of what the audience knows and expects. Knowing when and to what degree to qualify, explain, justify, and cite is one of the ways scientists create a professional *ethos* (Herrington 1985).

In thus acknowledging the work of others, researchers demonstrate not only that their knowledge is up to date but also that they have taken advantage of the best of the field's expertise in designing, carrying out, and interpreting their own work. Researchers who fail to acknowledge the relevant prior research will appear either naïve (if readers are charitable) or arrogant (if they have no reason to be charitable). Pons and Fleischmann again represent a case in point. In their first paper in the *Journal of Electroanalytical Chemistry* (Fleischmann and Pons 1989), they claimed that the energy in their cold fusion experiment was produced by "an hitherto unknown nuclear process or processes" (301). In a critique of this paper, Huizenga (1992) criticized the researchers not only for neglecting to qualify their "risky assumption" but also for failing to "acknowledge the extensive literature on nuclear reactions acquired and the basic principles established over the last half century" (25). As described in Chapter 1, the general disapproval with which this research was met was due as much to the way in which Pons and Fleischmann presented and situated their work as it was to the quality of the research itself.

• •

EXERCISE 4.1   Choose one or more full-length research reports or grant proposals, either from this text or from your own field. In each major section of the paper, find 2 or 3 sentences that include citations of previous research. How are citations used in different sections of the paper? Look for citations that support qualifications, explanations, or justifications made in the text. What other functions do citations serve in the texts you've examined?

• •

. . . .

4.2 Reviewing as a genre: The review article

Before we discuss general strategies for reviewing research in scientific reports or proposals, we will briefly examine the *review article*, a distinct genre in which these strategies are paramount. Many journals publish full-length review articles, often solicited from experts in the field on topics of particular interest to the journal's readers. Review articles tend to be written for a journal's broadest readership, including researchers in related fields (Day 1994), and thus are pitched at a somewhat more general level than research reports. As described in *Nature*, review articles "survey recent developments in a field" (see Figure 2.3). Such surveys serve an important function in their respective fields in that they offer a comprehensive synthesis of the results of a wide and complex set of studies. In so doing, the review may have a substantial influence on how readers perceive the nature and implications of recent developments in a field and thus may influence the direction of subsequent research (Myers 1991).

The review article we've included in Chapter 9 (Blaser 1987) illustrates two primary traits implied above: comprehensiveness (it reviews 123 studies!) and recency or timeliness. Blaser's review, published in *Gastroenterology* in 1987, was occasioned by the renewed interest in "gastric bacteria" sparked by Marshall and Warren's work in the mid-1980s. Like other important reviews, this piece helps the field "take stock" of a rapidly developing research area, as the following excerpt from Blaser's introduction indicates:

> . . . Although the presence of gastric bacteria has long been known, their significance has been uncertain. The development of fiberoptic endoscopy, permitting collection of fresh clinical specimens, has ushered in a new era for the management and investigation of gastroduodenal inflammatory conditions. Gastric bacteria now are being observed with regularity (2–4), and recently, Marshall and Warren (5,6) were able to isolate a spiral bacterium that had never been cultivated before. This organism, which they called *Campylobacter pyloridis*, has since been isolated by many other investigators (7–9). The field has moved quickly, and a review of its current status is appropriate.

The goal of such "status reports" is to describe what the field has learned so far: what, if any, consensus is developing, and what questions remain to be answered? Review articles are especially useful to "research managers," who need to distribute resources (human and financial) among research areas—for example, program officers at funding agencies and administrators in research agencies such as EPA. Similarly, reviews are useful to practitioners, including medical professionals, agricultural producers, and resource managers, who must make daily decisions about courses of treatment, feed composition, management practices, and so forth, and who want to base those decisions on the most current information available. Comparable to the discussion and "management implications" sections contained in reports of individual studies, review articles discuss the implications of the complex set of findings under review. Blaser, for example, helps readers make sense of the many reported observations of gastric bacteria by pointing out trends and patterns across studies. His concluding paragraph emphasizes directions for future research and implications for clinical practice:

> The rediscovery of spiral gastric organisms and the cultivation of *Campylobacter pylori* has opened a new era in gastric microbiology that has great clinical relevance. The next few years may provide answers to the perplexing problem of chronic idiopathic gastritis and possibly provide insights into the pathogenesis of peptic ulcer disease. Studies to clarify the role of GCLOs [gastric *Campylobacter*-like organisms] in the pathogenesis of these conditions should be a high priority. At the present, however, clinicians might wait for more definitive clinical studies before attempting to obtain cultures from affected patients or initiating specific antimicrobial treatment.

In contrast to the trends toward multiple authorship in research reports and proposals, review articles are often (though not always) single-authored. The review thus represents one expert reader's interpretation of the state of knowledge in the field. Blaser bases his advice to clinicians on his interpretation of the field's current understanding of the relationship between these bacteria and particular gastric conditions. For this interpretation to be accepted by readers in the field, it must be carefully and clearly supported, not only with citations of relevant studies but with enough information about those studies to enable readers to see the trends the author has seen. We'll return to this issue later in this chapter.

Notice that research reviews typically present a synthesis of *findings* rather than a synthesis of *views*. Direct quotations are rarely found in research reviews because the primary focus is not on what previous authors have believed or said but on what their studies have demonstrated. Reviewers are interested in researchers' claims only insofar as they are supported by the empirical evidence they present. Even in reviews of theory, common in fields like geology, meteorology, and astrophysics, theories tend to be discussed in the context of the physical observations they seek to explain. In short, the goal of the research review is to help readers make sense of all the available evidence.

Given this goal, it seems clear that the central line of reasoning in the research review depends on the information contained in your sources. This contrasts with some term-paper assignments, in which the writer is encouraged to present his or her own position, supported with references to other authors. In such assignments (a common academic model), the line of argument comes first, and the supporting sources come later. This is the inverse of the research review model, for until you find and read your sources, you will have no argument to make.

· · · · · · · · · · · · · · ·

4.3 Locating the literature

Whether writing stand-alone review articles or citing research in other scientific documents, experienced researchers rarely begin their reviews of research from scratch. By the time they are ready to write the introduction to a research report or the background section of a grant proposal, they have become very familiar with previous work in the area, much of which may be their own. Thus, the scientist author already knows which previous studies are pertinent to the argument at hand. But new scientists, or researchers working in new areas, may need to do a more extensive search of the available literature in order to develop this sort of familiarity with the research base. Therefore, some search strategies are in order.

The easiest way to search for prior research on a topic you are investigating is to begin with a reference in hand. This advice is more helpful than it sounds. We often discover topics for research while reading the work of others. Research reports and reviews typically include some discussion of directions for further research (e.g., see Blaser's conclusion quoted in Section 4.2), and this is an excellent place to find topics.

• •

EXERCISE 4.2   Review the discussion sections of two or more research reports, either those contained in this textbook or, ideally, papers from your field. What kinds of further studies are suggested? List the potential research questions proposed or implied by these authors.

• •

Research reviews focus on primary sources—original reports of individual studies published in professional research journals—as opposed to secondary sources such as textbooks or magazine articles written for non-expert audiences. But secondary sources are excellent places to start in your search for the primary literature. Biddle and Bean (1987) recommend beginning a search with the sources you find at home, e.g., your textbooks and lab manuals, both of which may include lists of works cited or suggestions for further reading. To this we would add other course readings, class discussions, and conferences with your professors as easily accessible starting points. If you have chosen a topic that was mentioned in class, your professor or lab instructor should be able to refer you to a recent paper on the subject. If your topic was suggested in the discussion section of another researcher's report, then you've already identified the first study to include in your review.

The advantages of beginning with such a source are many. First, the paper will contain at least a brief review of relevant research and a discussion of implications, providing you with a quick introduction to the topic. Second, if you have even one study in hand, you'll be able to locate the network of previous work by searching "backwards" through its references (and through the references in those references, and so forth). This will take some trial and error; you will undoubtedly come across studies that are not directly relevant to your specific topic. Streamline your search by using the authors' descriptions of these studies in the text to help you decide which are likely to be most pertinent to your particular research interest.

Lastly, your in-hand source can help you navigate manual or on-line indexes and abstract databases if you need to do some broader searching. Find your initial article in the database (perhaps by searching on the first author's name), and see how it is indexed—that is, see what keywords or identifiers are associated with the paper. Then use those keywords to help you design a search for related articles. If your original source is not contained in the database, check to make sure you're in a system which indexes appropriate journals in your research area.

Print and electronic indexes and abstract services can also be useful if you are starting from scratch, but be prepared to spend some time experimenting with different keywords and combinations. The reference staff in your campus library is familiar with the databases available on your campus and will be able to help you get started on keyword searches. They also can help you identify appropriate indexes and databases for topic searches in your field. All college libraries include print indexes in their holdings, and a rapidly increasing number of institutions offer electronic databases as well. Many indexes, such as the interdisciplinary *Science Citation Index,* are available in both forms.

A broad variety of electronic databases are currently available in the sciences, ranging from general multidisciplinary services such as *Current Contents* and *Science Citation Index;* to domain-specific sources such as *Agricola* (which indexes journals and books in agriculture and related fields), *BIOSIS* (biological and biomedical sciences), and *GEOREF* (geology); to sources focusing on specific research areas within domains, e.g., *Fish and Fisheries Worldwide, Textile Technology Digest,* and *VETCD* (veterinary science and medicine).

To identify the most useful indexes and abstract services in your research area, consult with your professors and/or check the masthead page of the primary journals in your field. This page—which provides information about the journal publisher, copyright notices, subscription procedures, etc.—appears in each issue somewhere near the table of contents. The masthead page often lists the databases in which the journal is indexed. Two sample indexing lists are reprinted in Figure 4.1.

Cerebral Cortex

Indexing: this journal is indexed in *Current Contents/Life Sciences, BIOSIS, CABS (Current Awareness in Biological Sciences), Cambridge Scientific Abstracts: Neurosciences, Index Medicus, MEDLINE, Neuroscience Citation Index, Psychological Abstracts,* PsycINFO database, *Research Alert, Reference Update,* and *SciSearch.*

The forthcoming *Cerebral Cortex* contents are available on the World Wide Web at "http://neuroscience.ucdavis.edu/CC/cc_preview.html"

Nonrenewable Resources

Nonrenewable Resources is indexed by the *Alerts* series, *Aluminum Industry Abstracts,* the *Bibliography and Index of Geology, Ecology Abstracts, Engineered Materials Abstracts,* the *Environmental Periodicals Bibliography, GeoArchive, Geographical Abstracts: Human Geography, Geological Abstracts, Geotitles,* the *IMM Abstracts, International Development Abstracts, Metals Abstracts, Pollution Abstracts,* and *Risk Abstracts.*

FIGURE 4.1 Indexing information from the masthead pages of the journals *Cerebral Cortex* and *Nonrenewable Resources*, both published by Oxford University Press.

• •

EXERCISE 4.3   Find the masthead pages of three to five major journals in your field. List the indexing and abstract services used by these journals. Put a star by those currently available in your university library.

EXERCISE 4.4   Working alone or with a partner, choose a topic in your field that you would like to know more about. It may be a topic you identified in Exercise 4.2; it may be an area you are currently exploring in another course or at work; or it may come from your own reading or browsing through journals in the library. Conduct a keyword or subject search on this topic in at least two different indexes appropriate for your field (print or electronic). Compare the outcomes of your searches. How useful was each database? Did the two systems produce similar sets of sources? What journals were cited in each search? Which database contained more relevant references on your topic? How easy was it to narrow the search in each system? Which provided greater flexibility in defining and combining keywords? Which search was more efficient? In what other ways did these indexes differ? List the advantages and disadvantages of each system you examined. Your instructor may ask you to compare these findings with those of your classmates.

• •

• • • • • • • • • • • • • •

4.4 Reading previous research

In most cases, you will be reviewing research as part of a research report or grant proposal. Thus the goal of your literature search is to determine the context for your research: What does the field already know about this topic? What kinds of studies have been done? What methods have been used, and how useful have they turned out to be? What has been found? What kind of information is still needed? Your answers to such questions will help you design a project which represents a reasonable "next step" for your field. When you write the review itself, you will aim to help your readers see the trends that you have seen in this literature.

In conducting your search, use paper titles and abstracts to help you sort through the sources you've located (on-line abstract services are particularly useful at this sorting stage). As you begin to identify the studies that seem most relevant, read the introduction and discussion sections carefully. Your goal in reading each paper is to understand why the authors conducted this research, what questions they hoped to shed light on, and what conclusions they came to. What were they trying to find out, and how does this relate to what you're trying to find out?

As you begin to compare and contrast the studies to be included in your review, skim the methods and results sections as well to see what kinds of materials were used (or sites or subjects observed), what kinds of measurements were taken or observations made, what kinds of analyses were performed. Locate the major findings of the study.

Recall from our discussion in Chapter 3 that in well-structured results sections major findings are highlighted in the text. Look for the authors' generalizations about their results, often contained in topic sentences (McMillan 1997).

4.5 Identifying trends and patterns

Whether your review is a stand-alone document or part of a report or proposal, readers will expect you to have read widely in the research literature and to have selected the most significant and most relevant studies to include in your review. Your goal is to present an overview of what this research has demonstrated. That is, you will want to sift and synthesize, pointing out similarities and differences in the findings these researchers report and, if pertinent, in the methods they used and the focus of their experiments or observations. Notice that this synthesizing goal will lead you to talk about the studies in groups or clusters, rather than describing each one in isolation. The review should not be a list of article summaries but a discussion of the trends you noticed across studies.

A useful way to identify trends is to construct a grid to help you record distinctive features of the studies as you read them. List the studies down the left-hand side of the page, and mark off several columns across the top. In its simplest form, perhaps early in your reading, the grid might include column headings such as "research question," "methods," and "principal results." As you become more familiar with the literature and the issues raised by these studies, you will want to develop more-specific column headings.

For example, in exploring the clinical research on *Campylobacter pylori,* Blaser (1987) might have used a column headed "conditions associated with *C. pylori.*" After a while, he would notice by reading down this column that some researchers documented the presence of *C. pylori* in patients with gastritis, others found associations with peptic ulcer, others with still other conditions. The grid would thus help him notice clusters or subgroups of studies that could usefully be discussed together in his review and may suggest subheadings he could use to organize the body of the paper. (Skim the subheadings in Blaser's review in Chapter 9 to see where the gastritis and peptic ulcer "clusters" were included.) Blaser could then use a similar "gridding" process to help him notice differences and similarities among studies in each cluster, e.g., the gastritis studies. In fact, the table Blaser created to summarize the results of the gastritis studies (Table 2) looks very much like the sort of grid we have been describing.

EXERCISE 4.5   Read the introduction to the research letter by Reynolds et al. (1994) in Chapter 11. Notice where and why previous studies are cited in the text. If these authors had used a grid to organize their initial reading of these sources, what column headings might they have used?

Whether you use a formal grid or some other note-taking system, your primary goal is to identify trends in this body of research. Are findings consistent across the set of studies? If the phenomenon was studied in different regions, at different times of year, with different methods, or under different conditions, were the findings similar or different? Is there theoretical consensus in the field, or are there differences of opinion? To help readers understand the current state of the field's knowledge on the topic, your review should highlight consistent patterns and points of agreement as well as inconsistencies and issues that are unresolved.

Two examples are presented in Figures 4.2 and 4.3. In the first, Blaser (1987) synthesizes the findings from the gastritis studies he had listed in Table 2 (in Chapter 9). Since the presence of *C. pylori* and the condition of gastritis were strongly associated in

Investigators on four continents have now identified GCLOs in gastric biopsy specimens and have shown an association between the presence of GCLOs and gastritis diagnosed by histology in adults (Table 2). Although methods employed in these studies to document the presence of GCLOs have varied, as have the definitions of gastritis used, it is notable that in all but one study the GCLO detection rate was significantly greater in patients with gastritis than in those without. The exception occurred in a small study in Australia in which nearly equal rates of GCLO detection were found in the two groups (44). Whether idiopathic antral gastritis in children is specifically associated with the presence of GCLOs is not yet settled (53–55). In attempts to answer the important question of whether GCLOs are present in healthy persons, three endoscopic studies of asymptomatic volunteers have been reported. Despite absence of symptoms or risk factors and the young age of the volunteers (mean ages 27–30 yr), 20.4% were found to have histologic gastritis; all of these subjects had GCLO present (Table 2). In contrast, no gastritis was found in 79.6% of the subjects and GCLOs were not detected in any of these cases. In total, the volunteer studies indicate that gastritis may be present in asymptomatic young adults, that this gastritis is associated with the presence of GCLOs, but that in the majority of subjects neither gastritis nor GCLOs are present.

FIGURE 4.2 From Blaser (1987). Gastric *Campylobacter*-like organisms, gastritis, and peptic ulcer disease. p 373. Full text included in Chapter 9.

Roger et al (1988) and Leckband, Spangler, and Cairns (1989) point out that another possible mechanism for producing bipolar structure in shell remnants is a systematic dependence of the efficiency of shock acceleration on the obliquity angle θ_{Bn} between the shock normal and the external magnetic field, if the field is assumed to be fairly well ordered on the scale of the remnant diameter. Shock acceleration theorists are divided on whether quasi-parallel ($\theta_{Bn} \sim 0°$) or quasi-perpendicular ($\theta_{Bn} \sim 90°$) shocks are more efficient in accelerating electrons. The quasi-parallel geometry seems more adapted to classical diffusive shock acceleration, (see reviews such as Drury 1983 or Blandford and Eichler 1987) while quasi-perpendicular geometry allows the so-called shock drift mechanism (Pesses, Decker, and Armstrong 1982; Decker and Vlahos 1985, among others) in which electric fields along the shock front accelerate particles, a process which can be considerably more rapid (Jokipii 1987). Leckband, Spangler, and Cairns (1989) attempted to study this issue by examining the limb-to-center ratios of SNRs, inferring the direction of the external magnetic field for each remnant using a model of the galactic magnetic field, and comparing with model calculations for profiles of SNRs. They could not come to a definite conclusion.

FIGURE 4.3. From Fulbright and Reynolds (1990). Bipolar supernova remnants and the obliquity dependence of shock acceleration. p 592. Full text included in Chapter 11.

all but one of these studies, Blaser highlights this consistent trend. Contrast this emphasis on consensus with the example in Figure 4.3. In this "minireview" excerpted from a research report, Fulbright and Reynolds (1990) describe an unresolved issue in shock acceleration theory, the question of whether quasi-parallel or quasi-perpendicular shocks are more efficient in accelerating electrons. Their review emphasizes the lack of consensus in the field, a gap in the field's knowledge that their study goes on to address.

Notice in both examples that the authors offer conclusions or *generalizations* about the set of studies under review, and they cite the *specific studies* on which their conclusions are based. Notice also that, as discussed in Chapter 3, generalizations tend to be stated in present tense (because they describe the current state of knowledge), while past tense is used to describe the results of specific studies (which were conducted in the past).

• •

EXERCISE 4.6   Read the paragraph in Figure 4.4, from the introduction to Mallin et al. (1995). References to prior research have been deleted from this paragraph. Use asterisks to indicate where you think the authors would have included citations to previous research. Then compare your text with the original article, in Chapter 10.

> The ciliated protozoan *Stylonichia* cf. *putrina* has been observed to consume TFVCs of *P. piscicida* without apparent adverse toxic effects. During prolonged feeding events, however, remaining planozygote stages of the dinoflagellate can transform into large amoebae which in turn, engulf the ciliate as predator becomes prey. Interactions between toxic stages of *P. piscicida* and other potential predators are currently unknown, such as whether mesozooplankton or microzooplankton aside from *Stylonichia* can consume or control it, or whether they are adversely affected by its toxin. Other toxic dinoflagellates are used as food resources by some zooplankters, but they reduce fecundity and survival in other zooplankton species or adversely affect higher trophic levels through toxin bioaccumulation.

FIGURE 4.4 From Mallin et al. (1995). Response of two zooplankton grazers to an ichthyotoxic estuarine dinoflagellate. p 352; citations deleted. Full text included in Chapter 10.

EXERCISE 4.7   Choose a sample paper or proposal in your field. Modify one paragraph of a research review section by stripping out research citations as we did in Figure 4.4. Type up the "research-free" review paragraph, and exchange with another member of the class. Read the paragraph you've been given, and note in the text where you think the authors would need to cite prior research and why. Compare your analysis with the original review.

4.6 Organizing the review

There is no standard organizational format for the research review, for the scope and purpose of reviews vary widely. Whether you are writing a review article or the literature review section of a proposal or report, use basic principles of good writing as your guide:

Format for Review

- Introduce your discussion by establishing the significance of the topic. It is helpful to give a quick preview of the major trends or patterns to be described in the review.
- Organize the body of the review to reflect the clusters or subtopics you have identified, using headings if the review is lengthy.
- Use topic sentences at the start of paragraphs and sections to highlight similarities and differences, points of agreement and disagreement.
- Conclude with an overview of what is known and what is left to explore.

    Notice that the headings in Blaser's (1987) review article identify the major sub-topics covered in his review ("Microbiologic Characteristics of *Campylobacter pylori* and Related Organisms"; "Pathological Associations With Gastric *Campylobacter*-like Organism Infection"). These *topical* headings are quite different from the *functional* headings of the IMRAD format followed in research reports. The IMRAD headings—introduction, methods, results, and discussion—let readers know what function each section of the report serves: the introduction introduces, the methods section describes methods, and so forth. Functional headings are useful because they enable readers who are familiar with a standard format to quickly locate the kinds of information they expect the document to include. But unlike the research report, the review article is not subdivided by function and therefore follows no standard functional format. All sections of the review have the same goal or function: to review research. Each section does, however, describe a distinctive trend or feature of that research, and headings can be used to highlight those trends or features. Thus in a research review, headings signal shifts in the focus or content of the discussion.

    Though review articles originally tended to survey historical trends (Day 1994), today they are more likely to concentrate on recent history, as indicated by *Nature's* emphasis on "recent developments in the field." On some topics, however, the significant trends are chronological trends—i.e., changes in the field's understanding or methodological approaches over time—in which case a chronological structure for the review may be warranted. On other topics, a brief discussion of past history may be necessary to contextualize more recent advances. Notice that Blaser (1987) includes a section titled "Historical Developments" at the start of his review of research on *C. pylori*. This section is particularly appropriate in this case because one of the particularly striking features of Marshall and Warren's discovery was the fact that gastric bacteria had been observed for decades but had been largely ignored. Now that the relationship between these bacteria and gastrointestinal conditions is becoming clearer, this "old" evidence is suddenly interesting. As a general rule, however, do not organize your review chronologically unless there are compelling reasons to do so. Your colleagues will be more interested in learning about the trends you have observed in the most recent literature in the field.

    Writers of research reviews often point out limitations in the scope or methods of a study in order to assess the significance of its findings, but by now it should be clear that *review* in this context means to synthesize or characterize a body of information, not simply to look for flaws in individual studies (as is often the case, for example, in movie reviews). Literature reviews, like all analyses, are interpretive. We choose which studies are important and relevant, we decide which of the reported results to highlight and how to describe them, and we present our own summary of what these various results add up to. Though there is clearly an evaluative component to the review, the overall goal is descriptive.[1]

---

[1] In reviewing and citing, scientists indicate what work they consider valuable as well as what work remains to be done. In the rhetoric of science, this is the special function of "epidiectic" discourse, one of three types of discourse that we will talk more about in Chapter 6. In classical rhetoric, epidiectic discourse was used to praise and blame. Today, in reviewing the contributions and limitations of particular studies to assess progress in the field and situate their own project, scientists not only acknowledge but also circumspectly evaluate and judge research in the field. But the purpose of their doing so is to demonstrate continuity between the old and new, between previous research and their own (Sullivan 1991), as well as to subtly exhort a research program to move in a new direction (see Halloran 1984).

4.7. Citing sources in the text

Two primary citation systems are used in scientific journals: the name-year system and the citation-sequence system (CBE 1994). Check the journals in your field to see which system they require authors to use. This information will appear in the "instructions to authors," published along with editorial goals and guidelines in every issue of some journals or in one special issue each year. Full descriptions of these systems can be found in style manuals, such as those published by the Council of Biology Editors (CBE 1994) or the American Institute of Physics (AIP 1990). Some journals will simply refer you to the appropriate style manual for guidelines on handling citations. Others will include samples of their required citation and reference formats in their instructions to authors.

The two citation systems are illustrated in Figure 4.5. Under the **name-year system,** which we are using in this textbook, any sources referred to in the text are identified by the author's last name and the date of publication. These sources are then listed alphabetically in the list of references or works cited at the end of the text. This system is followed in such journals as the *Journal of Plankton Research,* the *Astrophysical Journal,* and *Monthly Notices of the Royal Astronomical Society* (see sample articles in Chapters 10 and 11).

In the **citation-sequence system,** cited sources are numbered in the order in which they are mentioned in the text. These numbers are used as identifiers for citations in the text, and the full references are listed in this order in the list of works cited. Numbers may be in superscript (above the line) or in parentheses, depending on the capacities of your word processor. Of the journals represented in this textbook, the citation-sequence system is used in *Nature* and in the three medical journals, *Annals of Internal Medicine,* the *Lancet,* and *Gastroenterology* (see sample articles in Chapters 9 and 10).

Some journals (e.g., the *Journal of Bacteriology*) use a less common numerical system which combines features of both the systems described above: all cited sources are listed alphabetically at the end of the text and then numbered in this order. These numbers are then used as identifiers for citations in the text. You'll be able to recognize this variation quickly by checking the list of works cited to see whether or not sources are listed alphabetically.

EXERCISE 4.8   Compare a research report or proposal using the name-year citation system with one using the citation-sequence system. What do you notice about the effects of these different systems on your reading? List the advantages and disadvantages of each system.

In addition to the basic form of the citation, the examples in Figure 4.5 illustrate a number of other variations in citation practices, as described below.

Direct versus indirect citations. Authors may be identified directly in the text, as in examples A and C of Figure 4.5, or they may be cited indirectly, using parenthetical

Name-Year System

Example A
In more general terms, Dickel *et al.* (1989) propose a model for radio emission from young SNRs in which interior turbulence resulting from the shock overrunning small inhomogeneities generates magnetic field.

Example B
Other toxic dinoflagellates are used as food resources by some zooplankters (Watras *et al.,* 1985; Turner and Tester, 1989), but they reduce fecundity and survival in other zooplankton species or adversely affect higher trophic levels through toxic accumulation (White, 1980, 1981; Huntley *et al.,* 1986).

Citation-Sequence System

Example C
In 1940 Freedburg and Barron stated that "spirochaetes" could be found in up to 37% of gastrectomy specimens,[4] but examination of gastric suction biopsy material failed to confirm these findings.[5]

Example D
Recent studies have suggested that the eradication of *Helicobacter pylori* infection affects the natural history of duodenal ulcer disease such that the rate of recurrence decreases markedly (2–6).

Example E

A worldwide increase in toxic phytoplankton blooms over the past 20 years[1,2] has coincided with increasing reports of fish diseases and deaths of unknown cause.[3]

FIGURE 4.5 **Sample citations from the following sources included in Chapters 9 through 11:** *A.* **Fulbright and Reynolds (1990);** *B.* **Mallin et al. (1995);** *C.* **Marshall and Warren (1984);** *D.* **Graham et al. (1992);** *E.* **Burkholder et al. (1992).**

or numerical identifiers as in examples B, D, and E. Swales (1984) notes that direct citation focuses discussion on the researcher(s) who did the work; in contrast, indirect citation features the research claim or finding as the subject of the sentence and thus focuses discussion on the research itself. Reviewers use these different "reporting formulae" not only for variety but also to emphasize the contributions of individual researchers or to highlight trends in the research. In fact, the use of reporting formulae can become a strategic dimension of a scientific argument. In the introduction to the report by Marshall and Warren (1984) contained in Chapter 9, for example, note how the authors use direct citation when highlighting the fact that earlier researchers had observed bacilli in the stomach lining of ulcer victims, a finding which supports Marshall and Warren's own claim. But they use indirect citation when reporting the fact that other researchers failed to confirm those observations or recognize their importance.

The effect of indirect citation here is to downplay the negative findings of these other researchers.

In direct citations it is conventional to refer to authors by last name only; first names and titles are not included. Once you have included the authors' names in the text, it is unnecessary (in fact, redundant) to include them in the parenthetical citation. If you are using the name-year system, insert only the date of the source in parentheses immediately after any author mentions in your text, as in example A. In the citation-sequence system, the numerical identifier remains the same in direct and indirect citations.

Placement of indirect citations. Indirect citations appear most frequently at the end of a sentence, as in examples B through E of Figure 4.5, but they are also commonly found at the ends of clauses (examples B and C) or phrases (example E). A parenthetical or numerical citation should be inserted immediately after the statement, word, or phrase to which it is directly relevant (CBE 1994), so that it is clear to readers which part of your claim or observation is based on that particular source.

Citing work by multiple authors. Whether you are citing directly or indirectly, always acknowledge all the authors of a work. If the cited work has two authors, include both names, as in examples B and C of Figure 4.5. For works with more than two authors, use *et al.,* as in examples A and B. An alternative form preferred by the Council of Biology Editors (1994) is to replace the Latin abbreviation with the English equivalent: "Dickel and others (1989)."

· ·

EXERCISE 4.9   Obtain copies of the instructions to authors for three major journals in your field. What citation systems do these journals follow? What style guidelines do they offer? Which style manuals are recommended? Where can these manuals be found on your campus?

· ·

· · · · · · · · · · · · · ·

4.8 Preparing the list of works cited

Any sources cited in your text should be included in a reference list, sometimes called a list of works cited, at the end of the paper. The general format for references is the same in the two systems, except for the placement of the year of publication. In the name-year system, the year of publication must appear directly after the authors' names, so that references can be easily recognized from their name-year citations in the text (CBE 1994). In the citation-sequence system, the year appears later in the reference. The basic format of the two types of references, as prescribed by the Council of Biology Editors (CBE 1994, 634), is as follows:

Name-Year System
Author(s). Year. Article title. Journal title volume number(issue number):inclusive pages.

Citation-Sequence System
Author(s). Article title. Journal title year month;volume number(issue number):inclusive pages.

Sample references are listed in Figure 4.6. You'll notice that though the general ordering of reference components is relatively consistent, some formatting details vary from journal to journal. For example, none of the journals represented in this text include issue numbers in their references, though this is recommended in the generic format. In comparing the sample references in Figure 4.6, you'll find that some journals require first and last page numbers; others allow only the starting page number. Some put dates in parentheses; others do not. Some will ask you to list all authors; others will limit the number of names. (*Annals of Internal Medicine* only allows the first six authors' names to be included. Others must be indicated with *et al.*) Some reference formats omit article titles altogether. Most journals do not use quotation marks around article titles in references (but some do!). Punctuation around authors' initials and journal title abbreviations varies across journals, as does capitalization and the use of italics and boldface. The format of in-text citations varies as well: some journals omit the comma between name and year in parenthetical name-year citations.

Though these differences may seem arbitrary and unnecessary, in most cases the modifications have been adopted in an effort to save space or to make the typesetting

---

Name-Year System

Example A
Dickel, J.R., Eilek, J.A., Jones, E.M., and Reynolds, S.P. 1989, *Ap.J.Suppl.,* **70**, 497.

Example B
Watras, C.J., Garcon, V.C., Olson, R.J., Chisholm, S.W. and Anderson, D.M. (1985) The effect of zooplankton grazing on estuarine blooms of the toxic dinoflagellate *Gonyaulax tamarensis. J. Plankton Res.,* **7**, 891-908.

Citation-Sequence System

Example C
4. Freedburg AS, Barron LE. The presence of spirochaetes in human gastric mucosa. *Am J Dig Dis* 1940;7:443–45.

Example D
2. **Coghlan JG, Gilligan D, Humphries H, McKenna D, Dooley C, Sweeney E, et al.** Camplyobacter pylori and recurrence of duodenal ulcers—a 12-month follow-up study. Lancet. 1987;2:1109–11.

FIGURE 4.6 Sample references from the following sources included in Chapters 9 through 11: *A.* Fulbright and Reynolds (1990); *B.* Mallin et al. (1995); *C.* Marshall and Warren (1984); *D.* Graham et al. (1992).

process more efficient. Unfortunately, journal staffs have tended to experiment with these modifications independently, making a common set of rules difficult to maintain. For this reason, even if you're following the general formatting guidelines listed above, it is important to read the journal's instructions to authors and to use a sample reference list from the journal you're targeting as a model.

The generic format above covers basic references from journals only. For references to other types of sources—such as conference proceedings, technical reports, monographs, and chapters in books—you'll need to consult the appropriate style manual in your field or find an example in the reference list you're using as a model. Style manuals will also provide guidelines for citing multiple articles by the same author, citing authors with the same last name, citing organizations as authors, citing unpublished work, and many other special cases.

· · · · · · · · · · · · · ·

4.9 The research review abstract

Finally, if you are writing a review article—as opposed to a review section embedded in a research report or grant proposal—you will also need to prepare an abstract summarizing the focus and scope of the project. As we noted in Chapter 3, the abstract is a contingent genre; that is, its form is contingent upon the form of the paper it is intended to represent. Your review abstract may be either informative (summarizing the trends observed in the literature) or descriptive (informing readers of the topics to be discussed). Check the journals in your field to see which is preferred. In either case, the abstract should preview the major topics under which you have organized the review itself.

· ·

EXERCISE 4.10 Read the abstract for Blaser's (1987) review article, included in Chapter 9. Is this an informative or a descriptive abstract? Write an alternative abstract in the other mode.

EXERCISE 4.11 Compare Blaser's (1987) abstract with the "Summary" section of Marshall and Warren's (1984) research report, or with another research report abstract of your choosing. Describe any differences in content, organization, language, etc. How do the content and structure of Blaser's abstract reflect the purpose and content of the research review?

· ·

ACTIVITIES AND ASSIGNMENTS

1. Choose a research report on a topic in your field. Read the introduction carefully. Find three to five of the works cited in this section. Write a one-paragraph summary of each cited paper, and a one-sentence explanation of why it is cited in the review.

2. Read the first five paragraphs of the introduction to the article by Mallin et al. (1995) in Chapter 10. For the purposes of this exercise, consider this section as a stand-alone research review article. Write an abstract for this review article.

3. Conduct a manual or electronic search on a topic in your field (or on a topic assigned by your instructor). You may want to limit your search to the 10 most recent publications, excluding monographs or technical reports. Find and read as many of the sources as you can. Construct a grid in which you summarize the critical features of these studies.

4. Write a two- or three-page review of recent research on a topic in your field. Follow a citation format appropriate for a primary journal in your field.

5

Preparing Conference
Presentations

....
5.1 The role of research conferences in the sciences

Although this book is primarily concerned with writing, much, if not most, day-to-day communication in science takes place orally in informal one-on-one discussions, lab meetings, and participation in professional conferences. It is in these oral, more personal modes of interacting that much of the work of science gets done. Thus, the ability to interact and orally communicate with other scientists is essential for scientific progress. For an individual to participate in the knowledge-sharing and consensus-building business of science, publication is not enough (Rief-Lehrer 1990).

A scientist's personality and interpersonal skills come into play in oral modes of communication, perhaps more directly than in writing. We saw in the case of Barry Marshall in Chapter 1 that a scientist's personality and style are an important dimension of conference presentations and of scientific communication generally (see Chazin 1993; Monmaney 1993; SerVaas 1994). As Peter Raven, Home Secretary of the National Academy of Sciences, states: "In contrast to published manuscripts, oral presentations establish personal contact with an audience. The favorable or unfavorable impression created through this contact often permanently shapes a scientist's reputation—a reputation that can either help or haunt the scientist for the rest of his or her career" (Anholt 1994, ix). Of all the oral modes of communication, the conference presentation is the most formal and structured, and therefore it is more explicitly governed by generic conventions that can be studied than other modes.

You discovered in working through Chapter 2 how important professional conferences are in your field. Most scientists present their research at several conferences in their field per year. They attend conferences to share information, get feedback on new ideas or ongoing research, catch up with the latest developments in the field, and create professional relationships—to "network." Conferences provide a more immediate means for scientists to share new research or research in progress—often before it is even published.

Probably the best way to learn about conference presentations is to go to a few professional conferences yourself. If you have not already done so, make plans to attend a meeting in your field with your professors and/or peers. Observe the format and conventions of talks or poster sessions. Do presenters speak from notes, from visual aids, or from prepared texts? What kinds of visuals are common (tables, graphs, photographs), and what kinds of equipment are used (overhead projectors, slide projectors, VCRs, computer stations)? What is the time limit for presentations, and is it strictly enforced? How are the presentations organized? How are they different from written research reports?

We will discuss these and other features of conference presentations in this chapter. We'll focus on two presentation modes common in the sciences: the oral presentation, often referred to as the conference "talk" or "paper"; and the research poster. The poster is an increasingly common form of presentation in many scientific fields. Posters offer several advantages over the traditional oral presentation, especially for conference organizers. Large organizations typically receive proposals for far more conference papers than can be accommodated in the time available, despite the practice of scheduling concurrent sessions. Because many posters can be displayed simultaneously (limited only by the size of the room or hall available), conference organizers are able to accept a much larger proportion of proposals than would be possible if each presenter needed a 10- or 20-minute presentation period. This practice enabled the program committee for the 1990 Annual Meeting of the American Society for Microbiology to promise to include *all* acceptable proposals on the conference program, 75 percent of which were to be scheduled as poster sessions (see the "Poster/Slide Session Assignment" section of Figure 5.1).

As with oral presentations, preparing and presenting research posters requires both visual and verbal skills, as well as detailed knowledge of the expectations of conference goers in your field. We offer here some basic principles that will help you investigate the conventions for presenting research in either mode at conferences in your field.

5.2 Writing conference proposal abstracts

We discussed the writing of abstracts for research papers in Chapter 3. You'll recall that the generic abstract summarizes the major points of the research report in four moves: purpose, method, results, implications. The same is generally true for conference proposal abstracts. Read the "Call for Abstracts" from the American Society for Microbiology, presented in Figure 5.1.

Notice (under "Content of Abstract") that the ASM requires informative abstracts much like the report abstracts described in Chapter 3. Also note the detailed attention

Call for Abstracts

90th Annual Meeting of the American Society for Microbiology

Anaheim, Calif.   13–17 May 1990

READ ALL INSTRUCTIONS BEFORE WRITING ABSTRACT
DEADLINE FOR RECEIPT OF ABSTRACTS: *15 DECEMBER 1989*

Rules Governing Submission and Acceptance of Abstracts

Submission of abstracts for the Annual Meeting is governed by rules established by the Annual Meeting Program Committee as listed below.

1. You *must* be a full or student member to sign an abstract.
2. You may sign only *one* abstract.
3. You may be an author on only *three* abstracts.
4. You may not submit an abstract if you are not one of the authors.
5. You must obtain the approval of *all* coauthors *before* placing their names on the abstract.
6. No author may be delinquent for dues.
7. Abstracts must not be submitted if the work will have been published or presented at another national meeting by 13 May 1990.

Abstracts must be submitted on an **original Official Abstract Form (p. vii).** Any abstract submitted on a photocopy of the form will be retyped at headquarters, and the submitter will be charged a $25 retyping fee. *Once a paper has been accepted by the Program Committee, it cannot be withdrawn.* ASM will not make any changes in abstracts or other program copy after they have been submitted to the Program Committee. Authors submitting abstracts containing possible information relating to patent matters especially should make note of this point.

Preparation of Abstracts

Content of Abstract

The main factors in acceptance by the Program Committee are the quality of research, the importance of the topic to the membership, as shown in the abstract, and the content of the abstract. An abstract should contain a concise statement of (i) the problem under investigation, (ii) the experimental method used, (iii) the essential results obtained in summary form, and (iv) a conclusion. *Do not state,* for example, "The results will be discussed." An abstract is not a formal publication. Therefore, abstracts should not contain figures, literature references, grant acknowledgments, and the like.

Directions for Completing Abstract Form

Practice typing the abstract in a rectangle 5 inches wide by 4¾ inches long on plain paper before using the official form. Consult the example on pg. v for proper format.

Abstracts of the Annual Meeting is prepared by direct photographic reproduction. Abstracts will not be edited in any way. Thus, every error that appears in the submitted abstract will appear in the printed abstract.

Editorial style. The editorial style of submitted abstracts should conform to standard ASM style. Consult the *Instructions to Authors* in the January 1989 issue of any ASM journal.

Title. Use a concise title that indicates the content of the abstract. Capitalize the first letter of each word except prepositions, articles, and species names. *Underline or italicize scientific names of organisms.*

Authors and institutions. Type authors' names in CAPITAL letters. *Place an asterisk after the name of the author presenting the paper.* List each author by institution, city, and state, omitting department, division, etc. All addresses should follow the last author's name. In addresses, use standard abbreviations such as those listed on p. vi.

Directions for typing abstract. The following recommendations will assist you in typing the abstract on the Official Abstract Form.

1. Type the abstract, using a good-quality ribbon. Daisy wheel or laser printing is acceptable, but dot matrix is not. Because of the direct reproduction process and size reduction, the type must be completely legible.
2. Single space all typing on the form. *All text, including title and author-ship, must be within the box.* See example on pg. v for appropriate format. Use ASM style (see Editorial Style, above).
3. Any special symbols, such as Greek letters, that are not typed or printed must be carefully drawn by hand in BLACK ink. No other write-ins are permitted.
4. DO NOT ERASE. Your abstract will be published exactly as you submit it, with any erasure smudges, errors, misspellings, poor hyphenations, and deviations from good usage glaringly apparent.

NOTE: Any poorly prepared abstract unsuitable for direct reproduction will be retyped, and the author will be charged $25.

(continued)

FIGURE 5.1 Call for abstracts: 90th Annual Meeting of the American Society for Microbiology (1990).

SAMPLE

Official Abstract Form

(Read all instructions before typing)

Start → Interruption of Adherence of <u>Bordetella pertussis</u>
to Human Respiratory Cilia by Monoclonal Antibodies
to Eukaryotic Glycoconjugates. E. TUOMANEN*, H. TOWBIN, G.
ROSENFELDER, D. BRAUN, and A. TOMASZ. The Rockefeller Univ.
New York, N.Y. and CIBA GEIGY Ltd., Basel, Switzerland.

Whooping cough is a noninvasive infection in which adherence of <u>Bordetella</u> <u>pertussis</u> to respiratory cilia is an ongoing process essential for the maintenance of the organism at the site of disease. Adherence is highly specific
and involves 2 bacterial proteins and a carbohydrate-con-
taining host cell receptor. We sought to determine if antibodies (Ab) to human cell surface glycoconjugates could
interrupt adherence by blocking the receptor for bordetella.
Monoclonal Ab of known specificity for glycosphingoli-
pids were tested in 3 assays: 1) immunofluorescence to detect Ab binding and its topography on ciliated cells, 2)
Western blot of glycolipid extracts of cilia or ciliated
cell bodies, and 3) adherence assay for virulent bordetella.
Ab recognizing fucosylated galactose-n-acetylglucosamine
moieties were found to 1) bind to cilia but not to cell
bodies, 2) bind to specific glycolipids present in extracts
of cilia but not cell bodies, and 3) completely block adherence of bordetella to ciliary tufts.
We suggest that the bordetella receptor on cilia is related to the Lewis <u>a</u> blood group differentiation antigen. Ab
recognizing this glycoconjugate can prevent adherence in
vitro and could potentially interrupt natural disease.

Poster/Slide Session Assignment

Because of the flexibility in programming afforded by poster sessions, the Program Committee will attempt to schedule *all abstracts* which (i) are considered by elected divisional officers to be of acceptable quality and (ii) conform to rules established by the Program Committee. **The decision of whether an abstract is scheduled in a slide or a poster session will be made by the elected Program Committee.** Approximately 75% of the abstracts will be scheduled in poster sessions. By submitting an abstract, the author agrees that the paper will be presented as scheduled.

FIGURE 5.1 Call for abstracts: 90th Annual Meeting of the American Society for Microbiology (1990).

given in the ASM call to the preparation of the abstract: the deadline for submission; the content, style, typing, and formatting of the abstract; and the submission procedure itself. As described in the "Directions for Completing Abstract Form," and illustrated in Figures 5.2 and 5.3, abstracts are often reproduced in conference proceedings exactly as they are submitted. While not all calls for abstracts provide as much detail as the ASM, any guidelines that are provided should be adhered to with great care. Confer-ence proposals are often reviewed under a tight schedule. Abstracts that exceed the stip-

SESSION AC: NUCLEAR STRUCTURE I: $A < 80$
Thursday afternoon, 12 October 1989
13:30
F. Ajzenberg-Selove (Penn), presiding

13:30
AC 1   Deuteron-α angular-correlation measurements in $^{13}C(^6Li,d\alpha)^{13}C(gs)$ A. H. WUOSMAA and R. W. ZURMÜHLE, University of Pennsylvania[†]. We have used d-α angular-correlation measurements to examine α-unbound levels in ^{17}O populated in the $^{13}C(^6Li,d)^{17}O^*(\alpha)$ reaction. Results of recent measurements[1] have suggested that this technique may be used to obtain spin assignments for states in ^{17}O. That study indicates that structure in the d-α angular correlation for a group at $E_x(^{17}O)=13.6$ MeV, with the deuteron detected at $0°$, $8°$ and $10°$ arises from the interference of two closely spaced, interfering levels. We wished to study the $(^6Li,d\alpha)$ reaction on ^{13}C in more detail, for a large range of deuteron and α-particle angles. In our experiment we bombarded $30\mu g/cm^2$ ^{13}C targets with a 70 pna 28 MeV ^6Li beam. We used a position-sensitive $\Delta E-E$ telescope to identify deuterons in a range of $\theta_d=8°$ to $32°$ in the lab. Coincident α particles were detected using segmented position sensitive detectors spanning and angular range of $30° \leq \theta_\alpha \leq 170°$. An analysis of the d-$\alpha$ angular correlation in terms of a DWBA framework will be presented.
†Work supported by the National Science Foundation.
[1] G. Cardella, *et al.*, Phys. Rev. **C36**, 2304 (1987).

13:54
AC 3   MASS DETERMINATIONS OF EXOTIC NEUTRON-RICH RECOILS IN THE Z=14-26 REGION[*] X.L. Tu, X.G. Zhou, V.G. Lind, Utah St. U., D.J. Vieira, J.M. Wouters, Los Alamos Nat. Lab., Z.Y. Zhou, Nanjing U. & Los Alamos Nat. Lab., H.L. Seifert, U. Giessen & Los Alamos Nat. Lab. An extensive set of mass measurement data for neutron-rich nuclei in the Z=14-26 region were collected using the TOFI spectrometer at LAMPF. Since a gas-ionization Bragg-curve spectrometer replaced the $\Delta E-E$ Si detector telescope in the experimental system[(1)], it is required the development of a new Z determination method. In the end, the best Z resolution was obtained from a Q-specific two-dimensional table that relates Bragg peak energy and the range of the recoil in the CH_4 gas. Masses are extracted from Z and Q gated mass-to-charge spectra using moments analysis. A mass-to-charge ratio calibration is obtained for each run by fitting the measured centroids of the known mass lines. The final mass for each nuclide is determined by taking a weighted average of masses from (a) 23 runs and (b) all significantly populated charge states for a given nuclide. The resulting mass determinations will be compared to nuclear mass models.
[(1)] Wouters et al. Z. Phys. A331, 229 (1988)
* Work supported by U.S. Department of Energy.

FIGURE 5.2 Promissory abstracts from the American Physical Society's Division of Nuclear Physics, 1989.

ulated length or that fail to provide required information are likely to be excluded from the review pool altogether.

We have included this section on abstracts at the beginning rather than the end of this chapter because, while abstracts for research reports are almost always written last, abstracts for conference papers are often written "*before* the paper is written, indeed sometimes even before the work itself is done" (Olsen and Huckin 1991, 369). Olsen and Huckin describe this type of abstract as "promissory." In promissory abstracts, the research purpose and methodology are described, but results and implications may be projected, or omitted entirely. In some cases, an overview of preliminary results is provided. In Figure 5.2 we have reprinted two sample promissory abstracts from the program of the 1989 Annual Meeting of the American Physical Society's Division of Nuclear Physics.

You'll see in Figure 5.2 that instead of discussing the implications of the findings (which are at this point unknown), these promissory abstracts announce the context or theoretical framework in which the results will be discussed *in the presentation*. Notice also that the conference proposal abstracts in Figure 5.2 cite references and include explicit acknowledgment of funding sources, neither of which is necessary in a research report abstract (because this information appears elsewhere in a printed article). Remember, though, that the conference abstract is often the only written version of your paper that conference attendees will see; thus it is conventional in many organizations to include these critical acknowledgments.

It is important to point out that conference organizers, like journal editors, establish their own guidelines. As a consequence, the critical features of conference proposals vary across and within fields, as a comparison of Figures 5.2 and 5.1 will reveal. In contrast to the APS Division of Nuclear Physics, the American Society for Microbiology bars any mention of references or grant acknowledgments and explicitly prohibits the promissory abstract (see "Content of Abstract") by requiring a statement of results. Always check the organization's call for proposals before submitting!

In proposal abstracts, as in all the work you do as a scientist, hypotheses, methods, results, and implications must be logical and reasonable to professionals in your field, as well as appropriate to the conference. Since the abstract is the document conference organizers use to select the presentations to be given, it almost goes without saying that these abstracts must be exceedingly well written. The guidelines from the American Society for Microbiology make this absolutely clear.

· ·

EXERCISE 5.1   In Figure 5.3 we have reprinted some abstracts from the American Physical Society's 1989 spring meeting; the abstracts were submitted for a special session on cold fusion, no doubt prompted by the controversy surrounding Pons and Fleischmann's press announcement the previous fall.

1. Examine the content and structure of these abstracts. Do you see any "promissory" moves?

2. Note that these abstracts differ in amount of detail from the abstracts in Figure 5.2. Speculate on some possible reasons for this.

American Physical Society 1989 Spring Meeting Special Sessions on Cold Fusion
Baltimore, Maryland; 1–2 May 1989

Abstracts of talks submitted to the Special Session on Cold Fusion. For further information on a particular talk, please contact the author(s) directly. Videotapes of a selected number of the talks are available for purchase from The American Physical Society.

SPECIAL SESSION ON COLD FUSION
Monday evening, 1 May 1989 at 7:30 P.M.; Exhibit Hall E, Baltimore Convention Center; E. F. Redish, presiding

CF 1 Cold Nuclear Fusion: Recent Results and Open Questions. S. E. JONES, *Brigham Young University.*

```
We have shown that nuclear fusion between hydrogen isotopes can be
induced by binding the nuclei closely together for a sufficiently long
time, without the need for hig-temperature plasmas. For example,
muon-catalyzed fusion occurs rapidly when negative muons are added to
liquid deuterium-tritium mixtures, forming small muon-bound d-t
molecules that fuse in picoseconds. Recent experimental results
illuminate the rich tapedtry of processes that constitute the muon
catalysis cycle, while a number of question remain yet unresolved [1]. We
have also accumulated considerable evidence for a new form of cold
nuclear fusion which occurs when hydrogen isotopes are loaded into
various materials, notable crystalline solids (without muons).
Implications of these findings on geophuysics and fusion research
will be considered.

Supported by the U.S. Department of Energy, Advanced Energy Products
Division
[1] S.E.Jones, J. Rafelski, H.J.Monkhorst, eds. "Muon Catalyzed
Fusion 1988", AIP Publication 181, pp.1-469 (1989).
```

CF 2 Cold Fusion: Can it be True? A Theoretical Point of View. J. RAFELSKI, *University of Arizona, Tucson.*

```
It is shown that the fusion rates observed by the BYU team of S.E. Jones
during electrolytic infusion of hydrogen into Pd and Ti cathodes can
readily be explained by combination of standard nuclear physics data
and WKB penetration integrals in the metal lattice environment. A specific
mechanism for the process invoking formation of Bose macroscopic state
(drop) of deuterium ions neutralised by an electron cloud will be
described.
State of the attempts to skew the branching ratios of nuclear reactions
by 12 orders of magnitude towards processes not involving production of
neutrals (neutrons, gammas) will be given. This would be needed
to account for production of heat without penetrating radiation in a
nuclear
process, as suggested by the press release of the University of Utah.
```

CF 3 Theoretical Issues and Problems Raised by Cold Fusion Experiments.* S. E. KOONIN, *Institute for Theoretical Physics, UCSB.[†]*

I will discuss several challenges to our current understanding posed by recent cold fusion experiments. In particular I will review calculations of the rates for various hydrogen fusion reactions in molecular and condensed matter systems. I will also discuss the potentially large effect of lattice fluctuations on fusion rates in solids. Finally, I will review the shortcomings of various proposals to "hide" the radiation produced in d + d and p + d fusion.

*Supported by the National Science Foundation, grants PHY86-04197 and PHY88-17296.
[†]On leave from the California Institue of Technology.

FIGURE 5.3 Abstracts of talks submitted to a special session on cold fusion at the American Physical Society 1989 spring meeting.

EXERCISE 5.2   Return to the list of conferences you developed while exploring your field in Chapter 2, and select one you'd be interested in attending. Obtain the conference guidelines and copies of sample abstracts from a researcher in your field or from the World Wide Web.

Read over the conference instructions, and look at the logic and structure of the sample abstracts. Do these abstracts tend to include "promissory" statements, or do such statements seem to be prohibited? Describe the distinctive features of these abstracts, such as length, format, use of references, and funding-source acknowledgments.

• •

• • • • • • • • • • • • •

5.3 Organizing the research talk

It is important to recognize that the conference presentation, like its written counterpart the journal article, is primarily an act of persuasion, not simply of fact dissemination. As we noted in Chapter 3, when presenting results to other scientists, researchers seek to convince their audiences that their research questions are important, their methods sensibly chosen and carefully carried out, and their interpretations of results sound—in short, that their work represents a valid contribution to the developing knowledge of the field. Like the author of a journal article, a scientist presenting his or her research at a conference must build a convincing case for the project. The primary difference is that it must be done quickly—often within 10 minutes—and without the benefit of a formal written document in the "reader's" hand.

Because of the strict time constraints, scientists carefully plan oral presentations to ensure that the essential parts of the research argument are included. As in the written report, you will need to report your research questions, methods, results, and conclusions; but the amount of attention devoted to each component is somewhat different in the oral presentation. You will not have time for an extensive review of literature or a detailed description of your experimental methods. Instead, you will want to devote as much of your brief time as possible to reporting your new results and discussing their implications.

Thus, the first two components of the IMRAD format are rather abbreviated in the conference presentation. State the purpose and rationale for your study quickly but clearly at the start. The standard four-part strategy described in Chapter 3 provides a useful framework for conference paper introductions as well: Announce the topic, briefly overview trends in previous research, explain the gap or question raised by that research, and state the purpose of your study. Though the first three of these moves will be abridged in the interests of time, the final move is often expanded in an oral presentation to more fully outline for the listening audience the research to be presented or the major ideas to be discussed. Some written reports also provide this sort of preview for readers (see Figure 3.5), but it is particularly important in the conference setting because listeners cannot "turn back" later to the first page to see what your main point is. Thus, in addition to announcing the purpose of the study, introductions to oral presentations have the added burden of orienting the audience to the organization and scope of the presentation itself.

Carl Blackman, a biophysicist at the EPA's National Health and Environmental Effects Research Laboratory, told us an interesting story that highlights the importance

mention this story

of the orientation function of the introduction. Early in his career, Blackman discovered while preparing a conference talk that his data would take less time to describe than he had expected, leaving his presentation rather short. To fill the time, he finished his introduction by elaborating on the structure of the presentation itself, telling his listeners in some detail what he was going to talk about. After talking about the data, he also summarized what he had said. To his surprise, the president of the society hosting the conference and others attending his session told him that this presentation was one of the clearest presentations they had heard, and one of his best. Taking time to orient the audience can be well worth the effort.

After your introductory remarks, briefly outline your research design, identifying who or what was studied under what conditions (or where), but unless your research focuses on methodological issues, save the details for later. A slide or overhead transparency outlining the basic features of your design can be very effective in communicating this basic information quickly. A conference audience is not going to replicate an experiment on the basis of a talk alone, so listeners do not need the comprehensive description of materials and methods required in the written report. Interested listeners will ask for further information during the question-and-answer period after the session.

Spend most of your allotted time describing the results of your study, illustrating your presentation with carefully prepared visuals that clearly highlight the major trends in your results. If you are presenting a series of findings, it may be helpful to spend some time discussing the implications of each finding while the visual is still on the screen, rather than saving all your discussion comments for later. In any case, you will want to prepare your concluding comments with great care, as this is the part of your talk that is most likely to be remembered. Attention typically wanders during the course of a presentation (distractions, fatigue, and cogitation can all affect the attention span of your audience), but it returns when listeners sense the speaker is wrapping up (Olsen and Huckin 1991). Since the introduction and concluding sections of your talk are when you have the most attention of your audience, extra care should be taken in preparing these sections, and they should be delivered slowly and clearly.

mention conference fatigue

· · · · · · · · · · · · · ·
5.4 Methods of oral presentation

Scientists employ a variety of oral presentation techniques, depending not only on the customary practices of their fields but also on such variables as the type of topic and its reliance on visual as opposed to verbal explication, the length of time allowed for the presentation, and the speaker's own experience, comfort level, and familiarity with the material. Physicist Stephen Reynolds, some of whose work is contained in Chapter 11, reports that he never writes out a talk. Instead, he practices his talks several times beforehand, giving careful thought to how the presentation will begin and end, and then he works from his overhead transparencies when presenting.

Researchers use prepared text and notes to varying degrees when presenting. Carl Blackman prepared scripts for his talks early in his career. We've included an excerpt from one of his scripts in Figure 5.4. Notice that Blackman writes out his introduction, indicates which slides he will present in which order, and scripts some of his commen-

Introduction

For a number of years, we have studied the influence of electromagnetic fields, on the efflux of calcium ions from brain tissue, in vitro. Our aim has been to define all the characteristics of the field, that are essential for the effect to occur.

Today, I'll report on additional responses we have observed, that provide more clues, but do not in themselves provide any clear indication of a mechanism of action.

Background

[slide #1] Procedure (Expose vs. Sham Expose)
[slide #2] Crawford Cell with Function Generator.
            Sample orientation & load producing EM field.
[slide #3] Overview-Freq vs Int. (White dots vs. Blue dots)
            Tested only a few of the many possible combinations.
            Have extended the frequency response at 15 Vrms/m to 510 Hz

Frequency Data

[slide #4] Differences 1–510 Hz
            features: 1. includes previous data 1–120 Hz
                       2. repeats at 165, 180, & 405 Hz
                       3. basically, no apparent pattern

At this point, I will digress from a formal data presentation to a more intuitive viewing of the data. I don't claim that this view is rigorously correct, I'm just using it to help develop some crude hypotheses. Specifically, in order to distill some pattern in this frequency-response data, I've replotted it using a function that reflects both the magnitude of the difference between the exposed and the sham tissues, and also the variance of each group.

[slide #5] Log P vs Freq. (Note lines drawn at P = 0.05 and p = 0.01)
            features: 1. one group of data at $p<0.01$
                       2. other data at $p<0.05$, which includes 60 and 180
                           Hz and 405 Hz

Remember, we must exercise caution, because I've imposed an interprettion on the data that was not established before the experiments began. These interpretations should only be used for hypothesis generation. Any hypothesis still must be tested. In addition, the step size in frequency is 15 Hz. A smaller step size might reveal more fine structure or in some way change this picture.

FIGURE 5.4 Excerpt from script of 1985 presentation for the Bioelectromagnetics Society (BEMS) by Blackman, Benane, and House.

tary on the slides. He believes that such scripts might be especially useful for young scientists, particularly those who do not teach and thus do not regularly practice speaking to groups.

Scripting sections of your talk beforehand can help you stay within the time limit and ensure the accuracy of your remarks. Blackman suggests that if your text is written out beforehand and rehearsed, your presentation may be better organized and more accurate, subtle, and polished in wording and argument, even if you don't read from the written script when you present. He cautions against simply reading a paper to an audience, however. When notes and texts are used, they serve as an aid or a prompt for the presenter, not as a substitute for careful practice and preparation. Blackman now prepares his talks by drafting detailed outlines, which he then condenses and incorporates into overheads or slides, which then serve as notes for a more extemporaneous presentation.

· ·

EXERCISE 5.3   In Figures 5.5 and 5.6 we have reprinted an outline Carl Blackman prepared for a 1994 presentation at a meeting sponsored by the Ohlendorf Foundation, an organization which supports biomedical research on the use of electric and magnetic fields, as well as the opening set of slides from which he presented. Compare Blackman's outline and overhead transparencies. How has material in the outline been reworked into slides (that is, what has been changed, reworded, added)? Which version has more detail? Why? How difficult would it be to deliver the presentation without the slides? What is the value of preparing an outline to begin with?

partial set

How would you do this differently?

· ·

Title:
"The interaction of static and time-varying magnetic fields with biological systems"

I. Introduction
   A. Objective
      Identify field conditions responsible for changes in biological endpoints, NOT to examine physiological significance of changes, nor whether their health implications.
   B. Question
      Can EMF cause changes in biological systems that were not due to heating? If so, how did this happen?

II. Research History
   A. Identification of Non-Thermal Radiofrequency Field Effects
      1. Biological preparation—excised chick hemisphere in test tube
      2. Frequency tuning curve, "window"—Bawin 1975

slide 7
slide 8

(continued)

FIGURE 5.5 Outline of talk prepared by Carl Blackman for a 1994 conference on Scientific and Technical Foundations for Therapeutic MRI, sponsored by the Ohlendorf Foundation.

      3. Intensity window—Blackman 1979
      4. Predictions, different carrier frequencies—Blackman & Joines 1980
   B. Influence of Various Exposure Parameters. To simplify emphasize ELF.
      1. Intensity
         a. One intensity window, but few points—Bawin 1976
         b. Two intensity windows—Blackman 1982
         c. RF shows multiple intensity windows—Blackman 1989
         d. Low intensity ELF—Blackman ~1989
      2. Frequency
         a. Frequency series (1–510 Hz)—Blackman 1988
         b. Islands of intensity and frequency—Blackman 1985a
      3. Static (DC) Magnetic Field
         a. Bdc assigns effective frequency—Blackman 1985b
         b. Orientation dependence of Bac/Bdc—Blackman 1990
      4. Other Salient Factors
         a. Prior embryo exposure to E field affects tissue test—Blackman 1988
         b. Temperature history of tissue—Blackman 1991

III. Hypotheses—Mechanistic Basis for Response
      1. Initial transduction of EM signal into physical chemical change
         a. NMR/EPR-like signature in data (405; 15–315 Hz)
         b. Cyclotron resonance also possible (60, 90 & 180 Hz)
            —Liboff developed ICR for frequency dependence
            —Lednev added concept of intensity dependence
            —Models incomplete
      2. Amplification of initial effect by biological system
         a. Biological systems poised at instability point, like high diver
      3. Expression
         a. Observable—perhaps far removed from initial transduction

IV. New Biological System
      1. Requirements
         a. more direct physiological relevance
         b. standard, well-established and widely used assay
      2. Essential features
         a. NGF ->NO
      3. Results
         a. Dose response to B—Blackman 1993
         b. Induced E field not involved—Blackman 1993
         c. Frequency dependence—just like chick brain (in review)
         d. DC magnetic field influences results (unpublished)
      4. Consequences—collaboration w/ JPB

[handwritten annotations: "Slide 11" at left; "comparison end — Tell students ✱" at right]

FIGURE 5.5 Outline of talk prepared by Carl Blackman for a 1994 conference on Scientific and Technical Foundations for Therapeutic MRI, sponsored by the Ohlendorf Foundation.

(1)

**Interaction of Static and Time-Varying
Magnetic Fields with Biological Systems**

Carl F. Blackman
Health Effects Research Laboratory
US Environmental Protection Agency

Presented at

"Scientific and Technical Foundations for Therapeutic MRI"
27-29 January 1994
Margarita Island, Venezuela

(2)

Opinions Voiced

during this talk are my own

and do not necessarily represent those of the

US Environmental Protection Agency

(3)

Outline

Objectives of Research

Results with Model System

Theoretical Considerations

New Biological System

(4)

Objective of Research

Identify & Characterize
EMF field conditions that cause biological change

NOT
to determine physiological significance, or
examine health implications

(5)

Question 1

Is there a non-thermally based
biological response to RFR exposure?

(6)

(continued)

FIGURE 5.6 Slides prepared by Carl Blackman for the 1994 Ohlendorf Foundation meeting.

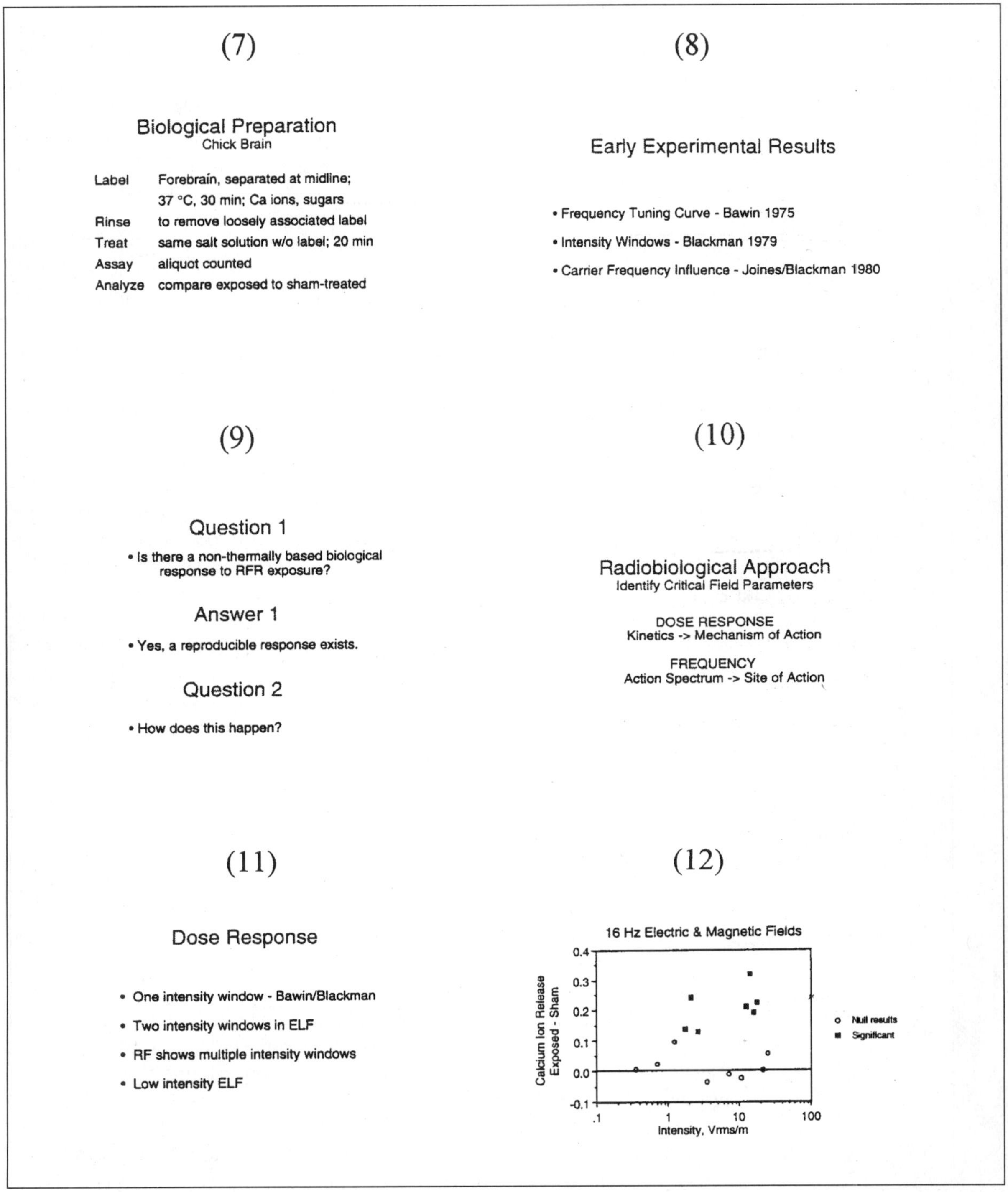
(7)
Biological Preparation
Chick Brain
Label Forebrain, separated at midline;
37 °C, 30 min; Ca ions, sugars
Rinse to remove loosely associated label
Treat same salt solution w/o label; 20 min
Assay aliquot counted
Analyze compare exposed to sham-treated
(8)
Early Experimental Results
• Frequency Tuning Curve - Bawin 1975
• Intensity Windows - Blackman 1979
• Carrier Frequency Influence - Joines/Blackman 1980
(9)
Question 1
• Is there a non-thermally based biological response to RFR exposure?
Answer 1
• Yes, a reproducible response exists.
Question 2
• How does this happen?
(10)
Radiobiological Approach
Identify Critical Field Parameters
DOSE RESPONSE
Kinetics -> Mechanism of Action
FREQUENCY
Action Spectrum -> Site of Action
(11)
Dose Response
• One intensity window - Bawin/Blackman
• Two intensity windows in ELF
• RF shows multiple intensity windows
• Low intensity ELF
(12)
16 Hz Electric & Magnetic Fields
Calcium Ion Release
Exposed - Sham
0.4
0.3
0.2
0.1
0.0
-0.1
.1
1
10
100
Intensity, Vrms/m
Null results
Significant

FIGURE 5.6 Slides prepared by Carl Blackman for the 1994 Ohlendorf Foundation meeting.

Both Blackman and Reynolds, then, work out their presentations ahead of time. Blackman develops a written outline; Reynolds rehearses from his slides. And both present from overheads or slides. The degree to which one uses these strategies depends on the personality and confidence of the presenter, according to Blackman, as well as on the audience, purpose, and conventions of the discipline. However, Blackman adds, "Students should not feel something is lacking if they can't do a presentation from overheads or slides; it is a skill that needs to be developed." Relying on prepared scripts can be an effective interim measure.

In actual practice, conference presenters usually combine three presentation methods: memorizing, reading, and speaking extemporaneously (Ehninger et al. 1984).

Memorizing can be useful for *short* presentations for which time limits are severe (Rief-Lehrer 1990): if a speaker sticks to the script and does not ad lib, memorizing will allow him or her to fairly accurately gauge the length of the talk; memorizing also may allow for continuous eye contact with the audience when presenting. However, memorizing takes time, and a memorized delivery is often stilted and stiff; there also is the danger that if a speaker forgets one part, the rest may unravel as well. Presenters rarely work entirely from memory, using it instead to help deliver short pieces of their talk or to make transitions from one part of a presentation to another.

Reading from a full text is rare at scientific conferences, as noted earlier, but speakers may prepare and read particularly critical or intricate parts of their presentation to facilitate clarity, coherence, and timing. The danger with the reading strategy is that it may hinder looking at the audience. There is also the danger when reading a prepared text that the speaker's delivery will be flat and that the presentation will lack energy and variety. Scientists reading parts of their presentation must be very aware of these tendencies and make a special effort to maintain eye contact with the audience. Experienced speakers look up between sentences or paragraphs, as well as at important or strategic points that they want to emphasize in the text. Good speakers, even when reading, also vary their rate of speed, pitch, and intonation pattern to enliven their speech and hold listeners' attention.

Speaking extemporaneously from notes or slides is the most frequently used method of presenting at scientific conferences. Speaking extemporaneously is not the same as speaking impromptu or "off the cuff," which is often required after a formal presentation during the question-and-answer period. To "extemporize" means to elaborate, usually on written notes or graphics. As the examples of both Blackman and Reynolds demonstrate, professional scientists spend time thinking about and preparing visuals and notes to elaborate on during their presentations. (In fact, Blackman goes so far as to prepare multiple slides to handle questions that may arise after his talk. Blackman anticipates his audience's reaction, considers his answers and arguments, and creates "rebuttal graphics" that will provide backup illustrations and evidence; he does not present this material unless he is asked, at which time he can extemporize rather than speak impromptu because he is better prepared.)

As we have seen, extemporaneous presenters may choose to prepare a formal introduction and conclusion, but they will usually speak only from slides or overhead transparencies or from an outline of the main points and supporting subpoints. Notecards, outlines, text, or graphics can be used as "cue cards" to organize a talk and keep the speaker from rambling or straying too far from the subject. Thus, the extemporaneous presentation is a combination of reading, speaking impromptu, and perhaps to some extent memorizing, and it shares some of their advantages and disadvantages. In an extemporaneous presentation important parts may be read so that they can be worded carefully; the discussion of the visuals may be part impromptu and part memorized, while the general structure and movement of the presentation may be both in outline form and in memory. The result is a presentation that appears spontaneous but is actually based on a basic structure worked out beforehand. Extemporaneous talks thus have all the energy of a spontaneous talk but are clearer, better organized, and fit within the time allotted. To speak extemporaneously, Blackman adds, "you have to reach a comfort zone; to get there, you might need notes, crib sheets, etc."

No matter what method of presentation is used, both Blackman and Reynolds advise that speakers time themselves. Practice your entire presentation several times at the speed you will deliver it.

5.5 Delivering conference presentations

Although much about presenting orally is innate and physical and is influenced by such factors as the speaker's personality, bodily voice, and physical appearance, presentation skills can be developed and improved through practice. In a recent book on oral scientific presentations, zoologist Robert Anholt asserts that a scientific presentation "stands or falls with delivery. . . . Delivery is important in establishing the impact a scientific presentation makes on an audience, and speaking skills can be determining factors in scientific careers" (1994, 152). In addition to the content of the presentation, delivery speed, body language, eye contact, voice, and gestures all influence how effectively the speaker communicates and how professional an *ethos* is created.

Because papers are presented orally to listeners who do not have a text and are more distant than they would be in conversation, speakers need to deliver their presentation more slowly than they would normally speak if it is to be comprehensible. This is especially true if parts of the presentation are read. The normal rate of speech in conversation is probably between 200 and 250 words per minute; public speaking experts recommend that presentations to large audiences average 120 to 150 words per minute (Ehninger et al. 1984, 89). Thus, for example, a 250-word typed page should take about two minutes to read. (To establish a good presentation speed, you can practice this yourself.) It is important to try to observe time limits: it is unfair and infuriating to other speakers to have someone use up a part of their time. (Imagine trying to adjust your carefully planned 20-minute presentation to fit the 12 minutes left to you by the previous speaker[s].)

Remember, presenting orally is a different experience from reading to yourself. Whether using notes, visuals, or text, it is much easier for you to lose your place when addressing a live audience. In addition to any nervousness you may experience, you may lose your place as you extemporize or look up from reading. To make your notes, visuals, or text easier to "perform," double- or even triple-space them (as newscasters do). Pick a readable typeface. Underline words in your notes or text that you want to emphasize, and "underline" them vocally with a change of volume or intonation when you deliver your presentation.

Unlike written presentations, oral presentations are very physical—very much rooted in the body. It is the person (not a piece of paper) who presents, the person who is the focus of attention. Thus, in delivering a conference paper, the speaker becomes a part of the presentation; and except for situations involving broadcast, the audience is live and present too. A good speaker is aware of the reactions and needs of his or her audience throughout the presentation.

One of the most important ways of doing this is to maintain eye contact with the audience. Blackman notes that, in his experience, eye contact does not occur naturally but rather comes with verbal flexibility, must be cultivated, and requires practice. There are at least four reasons to maintain eye contact with the audience during a presentation (especially when reading from notes, visuals, or text). The first is to monitor the audience: Can they hear you? Can they see the graphics? Are they following you? Do they agree with you? Do they need a point explained? The second reason is more subtle: when speakers periodically make direct eye contact with individual members of the audience, they are less likely to let their attention wander; psychologically, they know the speaker is watching, and so will want to pay closer attention. Third, by looking at people directly, speakers establish a good rapport with them—a personal relationship of sorts, which is very important in scientific oral presentations (Anholt 1994). Finally, eye contact helps establish the speaker's sincerity and belief in what he or she is saying (Can the speaker "look them in the eye" and say it?). In other words, eye contact helps establish the credibility of the speaker, an important component of the scientist's *ethos* or professional character.

Likewise, in an oral presentation your voice is the medium of communication. Even if you use visual aids, the bulk of the presentation will be carried by your voice. Just as the typeface of a manuscript has to be dark enough and free of errors to render the text as readable as possible, so your voice has to be loud enough and clear to render your words audible. There are several variables of voice that can be improved with practice (Ehninger et al. 1984; Anholt 1994); we will discuss these briefly.

Volume is crucial even with electronic amplification, for a microphone can only pick up what's coming out of the speaker's mouth. The worst presentation is a mumbled one, essentially because there is no presentation; no one can hear what is being said, even if it's brilliant. You must speak loudly enough so that those at the very back of the room can hear you. You can do a sound check before you begin your presentation by asking people in the back of the room if they can hear you.

Voice is physical, and so it is difficult to change. But it can be improved. If you are someone who naturally speaks softly, try opening up your entire vocal passage and work on pushing air up from your diaphragm as you speak; think of your whole body as a hollow instrument, and let your voice resonate in it. Sometimes people speak softly or choke up because they are nervous; we all have noticed how people get louder as a presentation goes on and nerves subside and they gather more confidence. Breathing helps. So does a glass of water. Try to channel and use nervous energy to increase your volume and your presence, rather than letting it stiffen or choke you up. If you are nervous, look into the eyes of audience members in different parts of the room. No one wants you to fail—for they are more or less a captive audience and will have to suffer with you.

In addition to volume, the **rate** or speed of delivery is important in oral presentations. Again, since a speaker's voice is the text, the rate of speed at which a talk is delivered should not be so fast that listeners will not be able to follow or keep up with it. It also should not be so slow that listeners will fall asleep. An intelligible, but lively pace (150 words per minute), with some variation, is best. Some speakers tend to speed up when they're nervous. Taking a deep breath before every sentence will help you both slow down and calm down—and will lend a clarity and dignity to your presentation that will do much for your confidence if you are nervous!

Pitch, the musical tone of your voice, also is important in an oral presentation (Anholt 1994). Sometimes people start speaking at a higher pitch than they normally talk because they are nervous, and then find they are uncomfortable with that pitch but are too embarrassed to change it; find a starting pitch that is natural or comfortable for you. Also pay particular attention to your intonation pattern, the musical scale, of your voice. Variety of pitch (as opposed to a droning monotone) keeps your presentation pleasant, lively, interesting, and intelligible. Intonation can also be used to highlight your key points. In fact, just as volume can be used to underline key words, phrases, or concepts, so too pitch can be used for emphasis.

* * *

EXERCISE 5.4   For this exercise you can use the introduction from Blackman's 1985 BEMS script (Figure 5.4), the Huyghe (1993) article in Chapter 10, or something that you have written.

1. Read the first few paragraphs to yourself, listening to the volume, rate of delivery, and intonation pattern in your mind. Now read the paragraph out loud. Do the volume, speed of delivery, and intonation pattern best capture and express the nuances of meaning in this paragraph? Try delivering the paragraph again, varying the volume, the rate of delivery, and intonation pattern as needed, until your delivery conveys most accurately the nuances of meaning. How does the voice help communicate meaning? Deliver this paragraph out loud a few times to practice volume, rate, and pitch.

2. Now listen to someone else read the same passage, and compare your presentations. Where did he or she alter pitch, volume, or intonation? What different decisions did you make in presenting the same passage?

* * *

If in an oral presentation the voice is a verbal text, the body is a visual element. Body stance and movement, facial expressions, and hand gestures—whether intentional or not—all become part of the message being communicated and in fact often reveal a speaker's feelings toward his or her subject and audience (Ehninger et al. 1984). How do you feel about someone when they can't look you in the eye? If they hang their head and shoulders as they speak? If they keep jingling change in their pockets? If they keep twiddling their thumbs? If they have a pained expression on their face as they talk? In addition to being distracting, this kind of body language reveals much about a speaker's attitude and so becomes a part of his or her message—becomes part of the speaker's *ethos.*

However, body language, like voice, can also be used to subtly communicate parts of the talk (Ehninger et al. 1984). Purposeful body movements can visually communicate a change in the focus or direction of a discussion, as when a speaker shifts weight or changes position as they move to a new topic. Hand gestures too can help a speaker emphasize key points in a scientific presentation. While hand gestures are generally toned down in scientific presentations, they can be useful, when used in moderation, to reinforce important points.

Your body is a natural visual, and it is always with you. Your body is part of your professional character. There's no getting around it: an oral presentation is a self-presentation.

5.6 The use of graphics in oral presentations

It is generally assumed that "two channels" working together can be more effective than one channel in getting a message across and retained (Hsia 1977). Therefore most speakers try to complement the spoken word with graphics. According to Anholt (1994), graphics are the heart of a scientific presentation. In Chapter 3 we discussed some of the more conventional graphics used in research papers; in Chapter 7 we will talk about those used by scientists to communicate with general audiences. In this section we will focus on conventions for using graphics in conference presentations.

The kinds of graphics used to represent trends or summarize results in written reports are also used in conference presentations. The difference between the two lies mainly in the amount of detail that can be accommodated in each medium. While many scientists simply reproduce the visuals from their text for use on an overhead or slide projector at a conference, visuals that are too "busy" are distracting and at worst confusing (Day 1994). If you believe this might be the case with one of your graphics, consider simplifying or redesigning it explicitly for the purpose and audience of your oral presentation. For example, turn to Figure 1 in the research report by Mallin et al. (1995) in Chapter 10. Readers of this report can distinguish among the four curves on the graph by examining the figure closely, but the curves may not be as easily distinguishable when projected on a screen. If Mallin were preparing a slide from this figure, he could use a number of devices to clarify this figure for an oral presentation. He could eliminate the standard-error bars or deemphasize them by putting them in a fainter color. He could draw the curves of the graph in different colors to make the distinctions among groups more obvious. He could indicate group means with symbols that are more easily distinguished from each other at a distance than are those in the printed graph.

In Chapter 3, we also discussed how graphics must be both independent (clearly labeled) and interdependent (referenced in the text). The same applies to a graphic in an oral presentation. A graphic without a label will be less comprehensible, especially if someone in the audience misses your verbal reference to it. Labeling all your graphics and making the letters large enough to be readable from the back of the room can solve this problem.

At the same time, be sure your visuals and your talk are interdependent. Even a labeled graphic should not be displayed without explanation. The time and attention the audience spends trying to figure out how the visual relates to what you're saying is time and attention taken away from your presentation. Take advantage of the opportunity to interact with your visuals, an opportunity unique to the oral presentation mode. Present your visual at the appropriate time in your discussion, refer to it directly ("Here are the results from the two trials"), and as you proceed, point to significant features with a pointer or a pen ("As you can see in the final column, temperatures in the second trial were substantially higher than in the first"). If the overhead screen is high enough and the projector is located near the speakers' table, you can work right off the projector without getting in the way; otherwise stand clear, and point to the screen with a pointer or light pen, always trying to face the audience.

The effective use of graphics is an adventure with many hidden pitfalls. But like conference presentations themselves, graphics can be improved by an awareness of the pitfalls and with practice. If they are designed well and are appropriate, graphics can make complex information visually comprehensible at a glance.

5.7. Preparing research posters

As noted at the start of this chapter, the research poster is an increasingly popular mode of presentation in the sciences. In addition to the logistical advantages for conference organizers, poster sessions offer presenters a chance to meet and talk one-on-one with scientists who are specifically interested in their research topic. In a poster session, presenters can field questions and provide clarification to their audience directly and immediately in a less formal setting than in the oral presentation format. The size of the audience and the effectiveness of poster sessions depends to a large extent on how the sessions are designed by conference organizers. If posters are displayed in an out-of-way room or hallway, attendance can be sparse, but if they are prominently displayed and easily accessible between sessions, poster sessions can offer broad exposure. Steve Reynolds notes that at his physics and astronomy conferences posters are often displayed for one or two days in a central location; presenters are thus able to reach a much larger fraction of the attendees at a given meeting than are likely to be reached through a 10-minute talk in a small, remote meeting room.

Poster titles are listed in conference programs, and authors typically stand by their posters to be available for questions and discussion with conference attendees who have come to find out about their work. At some conferences it is conventional to provide a written abstract or summary for "listeners" to take away with them, often including important figures and tables and always including contact information to facilitate further dialogue with the research team.

The size and format of posters varies across conferences and is constrained by the space and facilities available at the conference site. Always check for guidelines before you begin designing your poster. Most organizations provide easels or bulletin-board space and specify dimensions for the poster itself. Steve Reynolds reports that in the astronomical community each presenter receives about 1 square meter of bulletin-board space. The Society for Neuroscience allows a bit more room, specifying that materials be mounted on a poster board that is 5 feet, 8 inches (1.75 m) wide and 3 feet, 8 inches (1.1 m) high (Neuroscience 1994). In Figure 5.7 we have reprinted the sample poster that this organization developed as a model for presenters at their 1994 annual meeting.

In addition to guidelines provided by conference organizers, you'll want to adhere to general principles of effective page design when preparing your poster. As with any type of presentation, an aesthetically pleasing poster is more likely to communicate effectively (and to attract an audience in the first place) than one that is put together without regard to readers' needs. The sample poster in Figure 5.7 illustrates several important design principles:

- Arrange the presentation in columns within the poster or allotted bulletin-board space. A columnar arrangement is congruent with natural reading and viewing patterns, which proceed from top to bottom and from left to right

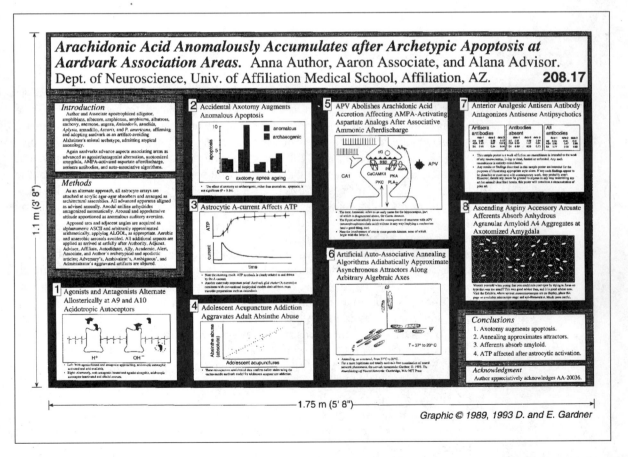

FIGURE 5.7 Model poster developed by D. and E. Gardner for the Society for Neuroscience (1994) as a guide for presenters.

(Dondis 1973). Use ample space between "blocks" and be sure the vertical and horizontal margins clearly indicate the path readers are intended to take (Day 1994). The Society for Neuroscience recommends placing an introduction in the top-left position and a conclusion in the bottom-right portion of the poster, taking further advantage of natural reading preferences.

- Use a large, readable typeface for titles and headings. Day (1994) suggests the title should be readable from a distance of 10 feet; the Society for Neuroscience (1994) requires letters at least 1-inch high for the main title, authors' names, and authors' affiliations. Highly visible titles ensure that conference goers can easily discern whether the project is relevant to their own research interests and thus worth a closer look.

- Clearly label figures, tables, and other graphics. Figures should, of course, be uncluttered and easy to read, but they will not need to be simplified for the poster as much as is needed in the oral presentation, for poster audiences can take as much time as they want to examine figures closely.

- Avoid large blocks of uninterrupted text. Use bulleted lists where appropriate, e.g., in listing major conclusions.

- If previous research is cited on the poster, include a list of works cited at the end.

- Finally, color and variety are appropriate and desirable in posters (Day 1994), but don't get carried away. The poster should be readable and informative, not purely decorative. Substance and clarity impress professional readers more than glitz. The Society for Neuroscience guidelines recommend the use of background colors or shades as organizing elements to underscore connections between closely related portions of the presentation and to distinguish among major sections.

As with all forms of presentation, you will want to notice the distinctive features of research posters in your field. Look around your department for sample posters that faculty and graduate students may have on display in offices or hallways. What types of visuals are typical? What kinds of information are included in captions? How much color is used? How much variation is there in poster size and shape? What is the ratio of visual to verbal information?

The poster presentation is in effect a hybrid form of communication, combining features of both oral and written modes. The poster session offers the immediacy of one-to-one interaction between the researcher and his or her audience—as listeners—but also provides a carefully structured formal presentation which that same audience—as readers—can peruse at their own pace.

ACTIVITIES AND ASSIGNMENTS

1. Imagine it is 1987. You have been working as a research assistant for Allan Brody and Michael Pelton at the University of Tennessee, who were at that time engaged in the research on black bears that they later reported in the *Canadian Journal of Zoology* (Brody

and Pelton 1988). You've been asked to write a conference proposal abstract for this research, to be submitted to the International Conference on Bear Research and Management. Their 1988 journal article was reprinted in Chapter 3, Figure 3.11. Though the article would not yet have been published at the time of your employ, use it to help you understand the goals, methods, and outcomes of that research. Write a promissory abstract for the project.

2. A. Prepare and deliver to the class an oral presentation of a set of research findings in your field. You may choose to present the results of a study you have conducted in another class or at work. Or your instructor may suggest you practice by developing a presentation based on published research conducted by another research team. (In this case, identify the original researchers, and present yourself as a "representative" of that team.) Choose an appropriate professional conference for your presentation, and identify the target audience at the start of your presentation. Use either the extemporaneous or the reading method of presentation, whichever you are more comfortable with. Supplement and illustrate your presentation with carefully prepared visuals, using overhead transparencies or other media available to you.

Rank your response to each question from 1 (low) to 5 (high). Beside each, write a brief statement explaining the reason for your ranking and/or the effect of that element on the presentation.

1. Was the type of presentation appropriate and effective? ___
2. Was the length of the presentation appropriate? Were time limits adhered to? ___
3. Was the rate (speed) of delivery effective? ___
4. Did the introduction establish an adequate context? ___
5. Was the presentation well organized? ___
6. Was the conclusion effective in emphasizing the major point of the presentation? ___
7. Was there sufficient (and direct) eye contact? ___
8. Did the speaker:
    • Speak loudly enough? ___
    • Vary pitch (intonation)? ___
    • Use voice to emphasize important points? ___
9. Was body language purposeful and appropriate? ___
10. Were hand gestures meaningful and appropriate? ___
11. Were visuals:
    • Appropriate (to the content, purpose, and method of presentation)? ___
    • Independent (adequately labeled)? ___
    • Interdependent (referenced and used in the presentation)? ___
    • Well designed and professional in appearance? ___
    • Informative? ___
12. Was the medium for the visuals (overhead, etc.) used effectively? ___

FIGURE 5.8 Criteria sheet for evaluating conference presentations.

B. As part of this assignment, you may be asked to serve as a respondent for one of the scheduled talks in your area of expertise. Your assigned speaker will give you a copy of his or her conference paper ahead of time so you can read it and prepare two or three follow-up questions to ask during a question-and-answer period following the presentation.

3. Use the criteria in Figure 5.8 to evaluate the effectiveness of in-class presentations. Even if you do not follow all the technical detail in presentations outside your research area, you should be able to tell from the discussion in this chapter which presentations were effective and why.

4. Prepare a poster presentation of a set of research findings in your field, following the topic guidelines outlined in Activity 2. As in that activity, identify an appropriate conference and audience for the poster session.

5. Develop a list of evaluative criteria to use in assessing the effectiveness of the poster presentations in your class. Using Figure 5.8 as a starting point, add and delete criteria as needed in order to prepare an effective and usable evaluation instrument. Your instructor may ask you to do this project in pairs or groups.

6. Use the criteria in Figure 5.8 and the poster criteria you developed in Activity 4 to evaluate presentations given at a professional conference that you attend. Which speakers gave the best presentations? Which posters were most readable? Which generated the most interest? Why were these presentations more effective than others you observed?

6

Writing Research Proposals

6.1 The role of the proposal in science

In this book we have taken the position that all scientific texts are persuasive documents, not simply presentations of facts. This persuasive dimension of scientific discourse is nowhere more obvious than in the research proposal. In writing reports and journal articles, we present and defend a particular interpretation of the prior research, of our new findings, and of the relationship between the two; the research report aims to convince readers that our work is valid and important. When we submit a grant proposal to a potential funding agency, however, we must go a step further. Now we must convince readers not only that the work will be valid and important but also that they should pay for it!

In a classical rhetorical framework, the proposal may be classified primarily as a *deliberative argument*. Aristotle distinguished three types of argument: forensic, epideictic, and deliberative, which can be described as arguments of fact, of value, and of policy. Most types of writing contain elements of more than one argument type. In research reports the overriding goal is forensic: we try to convince readers to accept a set of "facts"—the results of our study. However, because the introduction and discussion sections assess the current state of knowledge, and argue at least implicitly about the value and limitations of that knowledge, the report has an epideictic dimension as well. And the recommendations for future research or policy included in discussion sections are deliberative. The research review serves a forensic function in that it presents a summary of the current state of the field's knowledge; but in assessing the outcomes and limitations of individual studies, it is also epideictic. The proposal is distinguished from both these genres in that, although it too contains forensic and epideictic elements, its purpose is primarily deliberative: the proposal argues for a specific research plan as well as a general research direction.

forensic = fact
epideictic = value
deliberative = policy

116

The stakes become clear when we think about the number of scientific questions that could potentially be explored; the amount of time, effort, and money that would be needed to support all these explorations; and the resources actually available for scientific research. The National Science Foundation funds roughly one-fourth of the 40,000 proposals it receives each year (NSF 1994). The National Institutes of Health receives some 20,000 proposals each year, of which 6093 were funded in 1992, a "success rate" of 29.7 percent (NIH 1993). The proposal document plays a critical role in how scientific resources are allocated.

Research proposals are written for a variety of audiences and purposes in science: they are submitted to funding agencies to solicit financial support for new research, to academic departments to request approval of dissertation projects, to research facilities to gain access to equipment and resources. Though this chapter will focus primarily on the proposal for research funding, the goal in all these cases is to convince readers that a particular problem is significant enough to justify the time and effort needed to explore it. The ideal research proposal, then, presents a well-reasoned and carefully documented argument.

The US Department of Energy's grant application guidelines describe the proposal as "the document intended to persuade the staff of the DOE and other qualified members of the scientific and engineering community who review and advise on the proposed work, that the project represents a worthwhile approach to the investigation of an important, timely problem" (DOE 1991, 8). This definition highlights the two primary persuasive goals of the research proposal:

- To convince your scientific audience that the problem you propose to investigate is important and worth exploring
- To convince them you will explore the problem in a sensible way

Thus your proposal must convince readers not only that the research problem is significant but that your research approach is likely to succeed. Proposal reviewers will need to ascertain whether the methods you propose to use represent the most efficient and most worthwhile use of their agency's resources. They will also need to determine whether you are well qualified for the job. Funding agencies will request a description of your track record as part of the grant application, usually in the form of a *curriculum vitae,* an "academic résumé" that lists your research training and experience. But this ancillary material comes later in the application package; it is not the primary focus of readers' attention as they read and evaluate your project description. The text of the proposal itself must also demonstrate that you know what you're doing.

Recall that a text creates a professional *ethos;* it projects a character. Readers will form an impression of you based on the competence revealed in your proposal: Does your review of research demonstrate a thorough knowledge of the field? Have you exercised good judgment in the design of the study and the choice of materials and procedures (and are you therefore likely to exercise good judgment in other phases of the study, such as in recording and interpreting data)? Does your description of methods demonstrate the technical competence needed to carry out the project effectively? Thus, while your proposal contains an explicit argument for the importance and validity of the study and its design, it also contains an implicit argument for your own research competence (Myers 1985). It is important to note that even superficial features of your presentation may enhance or detract from the professional *ethos* created in the proposal. The NIH cautions that "sloppy" applications, with typographical or grammatical errors, create a negative impression of the author's competence and attention to detail; proposal writers are advised to leave ample time for proofreading and editing (NIH 1993).

· ·

EXERCISE 6.1   In Chapter 11 we have included a proposal that Reynolds, Borkowski, and Blondin (1994) submitted to NASA's Astrophysics Theory Program. Read this proposal, paying special attention to the *ethos* it projects. What clues are there to the authors' experience and expertise? List these. What have you learned about the researchers themselves from reading their proposal?

· ·

· · · · · · · · · · · · · ·

6.2 Multiple audiences of the proposal

In the Department of Energy guidelines excerpted above, the audience for the research proposal is identified as "the staff of the DOE and other qualified members of the scientific and engineering community who review and advise on the proposed work." Like other federal agencies, DOE uses a peer review mechanism which is similar in principle to, though typically more elaborate than, the peer review systems used by journal editors. Because most agencies fund research in many different areas and must allocate the agency's resources among these areas, the typical proposal document is reviewed by in-house "generalists" as well as by outside specialists and thus must be comprehensible and persuasive to a broader range of readers than is typical for journal articles. In-house reviewers are program officers and other agency staff who read proposals from a number of related topic areas; though educated scientists, readers at this level are not likely to be specialists in the particular topic area of a given proposal. The in-house readers will solicit reviews from specialists in the scientific community, as many as 3 to 10 in the case of the National Science Foundation (NSF 1995). The number of outside reviews requested may depend on the topic of the proposal and the expertise of the program staff, as NASA explains in their 1995 instructions for proposers (Figure 6.1).

The logistics of the proposal review process vary from agency to agency and sometimes within agencies as well, as Figure 6.1 illustrates. The NIH uses a "dual review" system (NIH no date) in which grant applications are read first by a standing "study section" consisting of outside experts who evaluate for scientific merit, and then by the appropriate institute's advisory council, composed of about eight scientists and four lay people (Seiken 1992) who evaluate proposals in light of the institute's program goals and priorities. These evaluations are central to a many-layered decision process, illustrated in Figure 6.2.

Instead of sending proposals "out" to experts in the field, some agencies assemble panels of experts who serve for longer periods of time and meet periodically during the

Evaluation Techniques

Selection decisions will be made following peer and/or scientific review of the proposals. Several evaluation techniques are regularly used within NASA. In all cases proposals are subject to scientific review by discipline specialists in the area of the proposal. Some proposals are reviewed entirely in-house, others are evaluated by a combination of in-house and selected external reviewers, while yet others are subject to the full external peer review technique (with due regard for conflict-of-interest and protection of proposal information), such as by mail or through assembled panels. The final decisions are made by a NASA selecting official. A proposal which is scientifically and programmatically meritorious, but not selected for award during its initial review, may be included in subsequent reviews unless the proposer requests otherwise.

FIGURE 6.1 Summary of NASA's proposal evaluation techniques. From "Instructions for responding to NASA research announcements," (NFSD 89–19) July 31, 1995.

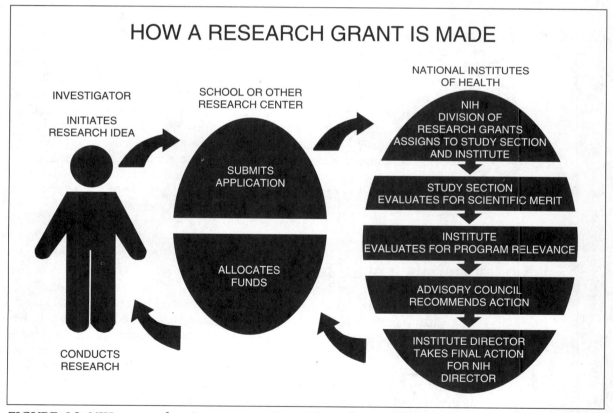

FIGURE 6.2 NIH proposal review process. From "Peer review of NIH research grant applications," NIH Division of Research Grants (no date, 12).

year (as in the case of the NIH study sections) or who are convened for the purpose of evaluating proposals received in response to a specific request for proposals (RFP) issued by the agency. Study sections at the NIH include representation from a fairly wide range of medical fields, which further illustrates the importance of pitching a proposal at a more general level than is necessary for journal articles and other technical reports. As a case in point, the NIH lists the following areas of competence among members of a single study section: biochemistry, general surgery, nutrition, immunology and transplantation, pulmonary embolism, clinical anesthesiology, pharmacology, and seven others.

     In sum, research proposals will be read and evaluated by readers with varying types and degrees of experience and expertise. The audience for a proposal typically includes members of your immediate research community as well as members of the scientific community at large, some of whom will know more about your research topic than others. This multiplicity of audiences clearly complicates the writing process, for the proposal must address all these readers at once. Given the variety of audiences to be addressed and the multiple agendas to be accomplished, writing the research proposal is one of the most challenging—and most important—tasks scientists engage in.

EXERCISE 6.2   Choose a research topic in your field and identify a funding agency likely to support research in this area. (Review Section 4.3 on identifying topics.) Obtain a copy of the agency's proposal guidelines. Researchers in your department will have guidelines and RFPs, but many agencies now make these materials available on-line as well. (For example, visit the NSF home page at *http://www.nsf.gov/* or the NASA home page at *http://www.nasa.gov/.*) Once you've located the guidelines, find the section(s) that describe the review or evaluation process. Based on this information, write a brief paragraph describing the mixed audience for a proposal on your topic, your readers' likely areas of expertise, and ways in which you might adapt your topic for those readers.

6.3 Logic and organization in the research proposal

Whether you are responding to an RFP or submitting an unsolicited proposal to a general program, your proposal must meet the specific guidelines established by your target funding agency. In addition to examining the main argument or project description, funding agencies will request an abstract or summary, a table of contents, a complete budget, biographical information about the investigator(s), and other supporting material pertinent to the type of research being proposed—for instance, information about the treatment of laboratory animals, the protection of human subjects' rights and confidentiality, or the handling of hazardous materials. The scope, form, and ordering of these ancillary materials will vary across research fields and funding agencies.

The components of the main argument in a proposal are, however, fairly consistent. All proposals are designed to do the following:

- Introduce the purpose, significance, and specific objectives of the proposed research.
- Explain the background and rationale for the project, by surveying previous research, summarizing the current state of the field's knowledge on the topic, and showing how the proposed project will further that knowledge.
- Describe the methodology to be used in the proposed study, and explain the rationale behind these methodological choices.

Where these goals are accomplished in the proposal document will differ from one proposal to another. The arrangement and labeling of sections and subsections may depend as much on the nature of your research topic or on your own writing preferences as it does on the funder's guidelines. For example, the federal agencies NSF and NASA use essentially the same language in their instructions for writing the main body of the proposal (Figure 6.3). But in responding to these instructions, the Burkholder team and the Reynolds team developed very different organizational plans. Figure 6.4

Instructions for Project Description, NASA 1993

The main body of the proposal shall be a detailed statement of the work to be undertaken and should include objectives and expected significance, relation to the present state of knowledge in the field, and relation to previous work done on the project and to related work in progress elsewhere. The statement should outline the general plan of work, including the broad design of experiments to be undertaken and an adequate description of experimental methods and procedures. The project description should be prepared in a manner that addresses the evaluation factors in these instructions and any additional specific factors in the NRA.

Instructions for Project Description, NSF 1990

The main body of the proposal should be a detailed statement of the work to be undertaken and should include: objectives for the period of the proposed work and expected significance; relation to longer-term goals of the investigator's project; and relation to the present state of knowledge in the field, to work in progress by the investigator under other support, and to work in progress elsewhere. The statement should outline the general plan of work, including the broad design of activities to be undertaken, an adequate description of experimental methods and procedures, and, if appropriate, plans for preservation, documentation, and sharing of data, samples, physical collections, and other related research products.

FIGURE 6.3 Instructions for project description from NASA's Astrophysics Theory Program (NRA 93-OSS-06, B-4) and from the NSF Grant Proposal Guide (NSF 90-77, 3).

TABLE OF CONTENTS

Page

I. Project Summary ii
II. Project Description 1
  A. Introduction 1
  B. General Objective and Hypotheses 3
  C. Uncharted Waters: Ambush Predator Dinoflagellates in Eutrophic Estuaries 4
  D. Research Design 6
    (i) Culture and Experimental Conditions 6
    (ii) Major Objectives 8
      Objective A: Saprotrophic Response to Fish Metabolites 8
      Objective B: Phagotrophic Nutrition of Flagellated and Amoeboid Stages 10
      Objective C: Interactions Between Flagellated / Amoeboid Stages and Microinvertebrate Predators 12
    (iii) Complementary Field Studies 14
  E. Overall Significance of the Proposed Research 15
  F. References 16
III. Funding Request with Supporting Explanation
IV. Curricula Vitae of the Principal Investigators
V. Cooperator on the Microfaunal Component
VI. Related Information: Available Facilities
  Current and Pending Support

FIGURE 6.4 Table of contents from proposal submitted to the NSF by JoAnn Burkholder and Alan Lewitus in 1994: "Trophic interactions of ambush predator dinoflagellates in estuarine microbial food webs." Burkholder and Lewitus used the NSF 1990 instructions (see Figure 6.3) as a guide.

contains the table of contents for a proposal by Burkholder and Lewitus, written in response to the NSF instructions. Chapter 11 contains a proposal by Reynolds, Borkowski, and Blondin, written in response to the NASA guidelines.

EXERCISE 6.3 Compare the table of contents (TOC) for the Burkholder and Lewitus (1994) NSF proposal, reproduced in Figure 6.4, with that for the NASA proposal by Reynolds, Borkowski, and Blondin (1994), included in Chapter 11. Aside from the obvious differences in level of detail (Reynolds's TOC omits subheadings; you'll have to find these in the text itself), how do these two organizational plans differ? Identify the distinctive features of each proposal in a list or brief paragraph. What do these distinctive features tell us about the nature of research in each area?

The three basic components of the proposal—purpose, background, methods—may appear in distinct sections or may be combined, subdivided, or redistributed, as both the Burkholder and the Reynolds proposals illustrate. In Sections 6.4 through 6.6, we examine these components in some detail.

6.4 Introducing the research problem and objectives

In some respects, the proposal introduction is similar in structure and content to the introductions found in most research reports. The four introduction moves Swales (1984) observed in research reports provide a useful heuristic for proposal introductions as well (see Figure 3.6). In both the report and the proposal, your goals are to introduce the research topic; summarize the current state of the field's knowledge on the topic; identify the gap, question, or problem that motivates the study; and announce the purpose of the study.

But within this general framework, emphasis and development may differ considerably in the two genres, due to the different audiences these texts address and the purposes for which they are written. In the research report, the introduction serves to quickly establish the context for the study and announce its purpose. Unless you are writing for a multidisciplinary journal such as *Nature* or *Science,* the journal audience consists of specialists in your field, and the stance is one of expert to expert; you are reminding fellow marine biologists or physicists or medical pathologists of the state of the field's knowledge on your topic, showing them which specific issues or previous findings you consider most pertinent to your study, and highlighting the gap your study responds to. In other words, you are guiding them through a review of information that is at least generally familiar to them, refocusing the discussion to situate the new results you are reporting. Because most of your research report audience shares your specialized knowledge to some extent, this review can be accomplished rather quickly, sometimes in a single paragraph.

The proposal, however, carries a greater burden of proof and must therefore provide a more fully elaborated introduction to the project at hand. We noted in Section 6.2 that proposal readers are a diverse group, including specialists in your field and "generalists" from other areas of expertise. The significance of your line of research will not be obvious to all the members of this group; therefore you will need to make a stronger case than would be necessary in a journal article. Some of your proposal readers will also need to be educated about the topic before they can make an informed judgment about the merits of your project. Lastly, keep in mind that you are asking more of your proposal readers than of a journal audience—you're asking them not just to entertain your ideas but to invest in them.

Thus, one of the primary goals of the proposal is to convince this audience of the significance of the proposed work. It will not be enough to assert that a problem exists, or that a question has not been answered. The researcher must convince proposal reviewers that the problem is important enough to spend money on, that answering the question will be worthwhile. What readers consider "worthwhile" will be largely a function of the goals and priorities of the agencies they represent. Just as research journals differ with respect to the topics and types of research they are interested in publishing (see Chapter 3), funding agencies also have distinctive research agendas and preferences. The first step in writing a successful proposal is to choose an appropriate agency to submit it to.

A funding agency's general priorities are well known to experienced researchers in the field, but they are also explicitly stated in the agency's proposal guidelines or requests for proposals (RFPs). Most funders accept proposals in their areas of interest on an ongoing basis but also issue RFPs to solicit proposals on particular topics for which they have set aside special funds. (The NIH uses the term *RFA: request for applications;* NASA uses *NRA: NASA Research Announcement.*) RFPs typically announce a new initiative in a carefully delimited research area. Researchers responding to RFPs must demonstrate that their proposed research fits within these parameters and will significantly further the announced goals. For example, the Reynolds team's proposal in Chapter 11 was submitted in response to an NRA soliciting proposals for "Theory in Space Astrophysics." The NRA was intended to encourage basic theoretical work needed for NASA's Space Astrophysics Program. Accordingly, the program description stipulates that "the proposed studies should facilitate the interpretation of existing data from Space Astrophysics missions or should lead to predictions which can be tested with Space Astrophysics observations" (NASA 1993, A-1). This goal is implied throughout Reynolds's proposal and is directly addressed in a summary paragraph toward the end of the introduction:

> The proposed research is timely—it is new observations of SNRs with the ASCA satellite, and to a lesser degree with the ROSAT satellite, which have been the primary cause of the current renaissance of SNR studies. We bring to this field the only tool capable of addressing the spectral imaging data being collected by the ASCA satellite, and one well able to meet the theoretical challenge of the AXAF mission. (Reynolds, Borkowski, and Blondin 1994, 5)

EXERCISE 6.4   Read the Reynolds team's (1994) proposal to NASA and the accompanying program description, both in Chapter 11. Examine the other criteria mentioned in the program description. How and where are each of these issues addressed in the proposal?

As illustrated in the example just quoted, successful proposal authors leave nothing to chance. In discussing the significance of their research, they carefully highlight the specific ways in which the proposed work will further the goals and interests of their target funding agency. In so doing, the researchers are appealing to the agency's *values.* The appeal to values is one type of classical rhetorical argument. Aristotle described three types of rhetorical appeals: those based on the logic of the subject matter *(logos),* those based on the character of the speaker *(ethos),* and those based on the emotions

logos
ethos
pathos

and values of the audience (*pathos*). Most arguments contain all three types of appeal, and each can clearly be seen in the proposal genre: <u>successful proposals present a logical, well-supported line of reasoning; project a professional *ethos* of competence and knowledgeability; and clearly address the values and concerns of the funding agency.</u>

These values are both articulated and implied in the RFP or proposal guide. Each agency has a fairly well-defined research domain which gives it a distinctive character or identity. For instance, the NSF is committed to supporting research that will "contribute to better understanding or improvement of the quality, distribution or effectiveness of the Nation's scientific and engineering research, education and manpower base" (NSF 1995, 13). This concern with the nation's science infrastructure and education is revealed in a number of places in the NSF Grant Proposal Guide. Notice in Figure 6.5, for example, that the NSF's description of "who may submit" clearly highlights this concern. The categories themselves, the ordering of the categories, and the descriptions

---

WHO MAY SUBMIT

Scientists, engineers and educators usually initiate proposals which are officially submitted by their employing organization. Before formal submission, the proposal may be discussed with appropriate NSF program staff. Graduate students are not encouraged to submit research proposals, but should arrange to serve as research assistants to faculty members. Some NSF divisions accept proposals for Doctoral Dissertation Improvement Research Grants when submitted by a faculty member on behalf of the graduate student. The Foundation also provides support specifically for women and minority scientists and engineers, scientists and engineers with disabilities, and faculty at primarily undergraduate academic institutions. (See Chapter V. for information about Special Programs.)

Categories of Proposers

1. *Universities and colleges:* U.S. universities and two- and four-year colleges (including community colleges) acting on behalf of their faculty members.
2. *Non-profit, non-academic organizations:* Independent museums, observatories, research laboratories, professional societies and similar organizations in the U.S. that are directly associated with educational or research activities.
3. *For-profit organizations:* U.S. commercial organizations, especially small businesses with strong capabilities in scientific or engineering research or education. (See Section V.K. for specific information on the Small Business Innovation Research (SBIR) program.) An unsolicited proposal from a commercial organization may be funded when the project is of special concern from a national point of view, special resources are available for the work or the proposed project is especially meritorious. NSF is interested in supporting projects that couple industrial research resources and perspectives with those of universities. Therefore, it especially welcomes proposals for cooperative projects involving both universities and the private sector.
4. *State and Local Governments:* State educational offices or organizations and local school districts may submit proposals intended

to broaden the impact, accelerate the pace and increase the effectiveness of improvements in science, mathematics and engineering education in both K-12 and post-secondary levels.
5. *Unaffiliated Individuals:* Scientists, engineers or educators in the U.S. and U.S. citizens may be eligible for support, provided that the individual is not employed by or affiliated with an organization and:
    • the proposed project is sufficiently meritorious and otherwise complies with the conditions of any relevant program announcement/solicitation;
    • the proposer has demonstrated the capability and has access to any necessary facilities to carry out the project; and
    • the proposer agrees to fiscal arrangements which, in the opinion of the NSF Grants Officer, ensure responsible management of Federal funds.
Unaffiliated individuals should contact the appropriate program before preparing a proposal for submission.
6. *Foreign organizations:* NSF rarely provides support to foreign organizations. NSF will consider proposals for cooperative projects involving U.S. and foreign organizations, provided support is requested only for the U.S. portion of the collaborative effort. (For further information, contact the Division of International Programs, Appendix A.)
7. *Other Federal agencies:* NSF does not normally support research or education activities by scientists, engineers or educators employed by other Federal agencies or Federally Funded Research and Development Centers (FFRDCs). However, a scientist, engineer or educator who has a joint appointment with a university and a Federal agency, such as a Veterans Administration Hospital, or with a university and an FFRDC, may submit proposals through the university and may receive support if he/she is a bona fide faculty member of the university, although part of his/her salary may be provided by the Federal agency. Under unusual circumstances, other Federal agencies and FFRDCs may submit proposals directly to NSF. Preliminary inquiry should be made to the appropriate program before preparing a proposal for submission.

FIGURE 6.5 Categories of proposers, NSF Grant Proposal Guide, 1995, 1–2.

within them—all seem to point to the NSF's overarching concern with the infrastructure of American science and education. According to this list, proposals that do not
appear to further these goals (e.g., research at foreign institutions or proposals from professional societies *not* "directly associated with educational or research activities") are
rarely funded. Funding agencies spend time, money, and effort, carefully wording their
proposal guides in order to clearly indicate the kinds of research they will and will not
consider. These guidelines can be an invaluable resource in selecting an appropriate
agency for your proposal and in developing the significance argument for that audience.

Read   ✗

EXERCISE 6.5   Read the RFPs we have included from the National Sea Grant College Program (Chapter 10) and from NASA (Chapter 11). Describe the values and goals of each agency as
revealed in these documents. Now compare and contrast the values and goals of these
two programs. What do we learn about the identity and priorities of these agencies by
reading their proposal guidelines?

The introduction does not have to build the case for significance all on its own,
of course; this is the purpose of the full proposal. In most cases the introduction serves
as a preview of issues to be discussed further in other sections of the document
(Pechenik 1987). The introduction identifies the problem, but specific research objectives are often elaborated elsewhere, sometimes in the background or methods sections
(as in the Burkholder and Lewitus proposal, Figure 6.4), and sometimes in a separate
section altogether, as in the Burkholder and Rublee proposal, Chapter 10. Similarly, the
introduction overviews previous research and the questions that motivated the study,
but a more extensive discussion of prior research and explanation of the research problem is generally presented in a background section. The introduction also identifies the
methodological approach to be used in the study, but a detailed description of methods
is saved for the methods section. In some cases, the significance argument may be elaborated in a separate section as well, as in the proposals by Burkholder and Rublee
(labeled "Expected Results," Chapter 10) and Burkholder and Lewitus (Figure 6.4).
These sections, appearing in both cases at the end of the document, provide a neat conclusion to the proposal argument as a whole, leaving readers with a strong sense of the
importance and potential value of the proposed work.

In short, the introduction orients readers to the topic, purpose, and significance
of the research, providing a framework or scaffold for other sections of the proposal to
build on.

EXERCISE 6.6   The Reynolds team's proposal (Chapter 11) does not contain a special section or sub-
heading for research objectives. Where are the specific goals of this study introduced?
In one or two paragraphs, identify the objectives of the study, and explain the rationale
behind them.

EXERCISE 6.7   The Burkholder and Rublee proposal (Chapter 10) combines the introduction and background sections. As you read this section, look for the four standard introductory moves (announce topic, review prior research, identify gap, introduce new research). Are these moves recognizable in this elaborate introduction, and, if so, where is each move made? In one or two paragraphs, describe the structure and logic of this section.

6.5 Providing background

Most proposals contain a separate background section or sections where the authors can present a more extensive explanation of the research problem, grounded in a thorough review of previous research. This section serves different purposes for the two segments of your audience. Your discussion should bring program officers and other generalists up to speed on the nature of the problem and the reasoning behind your specialized project. At the same time, it provides an opportunity for in-field readers to judge how familiar you are with the current state of knowledge in the field and how well you understand the issues and constraints involved in conducting research of this sort. This second purpose is clearly evident in the advice the NIH offers on reviewing prior research:

> Refer to the literature thoroughly and thoughtfully. Explain what gaps in the literature would be filled by your project. In the past, research proposals have not been funded when applicants seemed to be unaware of relevant published work or when the proposed research or study design had already been tried and judged inadequate. (NIH 1993, 5)

As this quotation indicates, the way in which you review the literature in the background section, as in the introduction, significantly influences the professional *ethos* projected by your text. Readers will want to know not only that you understand what other researchers have done but that you appreciate the contributions their studies have made to the developing knowledge in your research area. Situating your work in this context is essentially a cooperative as opposed to a competitive gesture. While oversights and methodological limitations must be taken into account in interpreting the results of individual studies, the primary goal in a review is to highlight what has been learned by the field so far (see Chapter 4). In the proposal, the review of research shows how far the previous research has gone, and where it still needs to go. Once this groundwork is established, you will be in a position to explain how the proposed study will take the field forward.

Chapter 4 provides some general guidelines for structuring research reviews and citing sources. You'll see in the sample proposals in Chapters 10 and 11 that subheadings are a useful device for organizing the background section and are usually needed due to the length and complexity of these discussions. Notice that, as in the research report, in the proposal the major headings are *functional* headings (introduction, background, methods). But subheadings within sections are *topical*; that is, they identify the topic to be discussed in the section rather than simply announcing the function the section serves. Reynolds's (1994) background section contains three subdivisions: "X-Ray Emission of SNRs," "Radio Emission of SNRs," and "Dynamics of SNRs." Topical subheadings provide important cues to help readers navigate your background argument, and they are particularly effective when the organizational scheme is introduced at the start of the section, as in the Reynolds proposal.

As you develop the background section, keep in mind the overall purpose of your proposal. You are reviewing research in order to introduce your study and show how it will further the field's knowledge and the agency's goals. You'll want to be sure to tie the background research clearly to the proposed research, especially as you bring this section to a close. Reynolds, Borkowski, and Blondin (1994) finish the background section with a review of the broad need their study addresses: "We shall work to correct this major gap between theory and observations, so that the observations can in fact serve as a useful check on the theoretical work." This statement brings the discussion back to the proposed research, clearly recalling the program description's emphasis on applying theory to data.

Another way to connect the background argument to the proposed study is to follow this section with a list of specific research objectives (Pechenik 1987), as in the Burkholder and Rublee (1994) proposal in Chapter 10. Once readers have read the background section, they are well prepared to understand and appreciate these specific goals of your study. Another advantage of placing the specific objectives here (rather than earlier in the proposal) is that they can serve as a preview of the methods section, enhancing the coherence of the overall proposal argument.

EXERCISE 6.8   Choose a sample research topic in your field, perhaps one suggested in the discussion section of a research report you've read recently. List the research areas that would need to be reviewed in the background section of a proposal for this project. Design sample subheadings for this section.

6.6 Describing proposed methods

The methods section of a research proposal is distinguished from that of a journal article in two respects: the proposal generally contains fewer details but more explanation of rationale. Fewer details are to be expected, given that the research being proposed has not yet been conducted, but the need for rationale is more a function of the deliberative purpose of the proposal. This section must do more than describe how the study will be carried out; it must explain why this approach, as opposed to others, was chosen. The researcher must articulate and defend the methodological decisions he or she has made in such a way that the diverse readers in the proposal audience will be able to understand and appreciate those decisions. Remember that your target agency is being asked to pay for these activities. Its program boards must be convinced that this approach represents the best possible use of their limited funds.

As a consequence, the description of methods in the proposal tends to be heavily documented. In effect, it is an extension of the background discussion and serves a similar dual purpose. This extended explanation of materials and procedures helps your generalist readers understand what's needed to accomplish this research and allows your in-field readers to determine whether *you* understand what's needed for this research. The NIH cautions:

> While you may safely assume the reviewers are experts in the field and familiar with current methodology, they will not make the same assumption about you. . . .
>
> Since the reviewers are experienced research scientists, they will undoubtedly be aware of possible problem areas, even if you don't include them in your research plan. But they have no way of knowing that you too have considered these problem areas unless you fully discuss any potential pitfalls and alternative approaches. (NIH 1993, 6)

EXERCISE 6.9   In either the Burkholder and Rublee (1994) proposal or the Reynolds team's (1994) proposal, look for places where the authors include a brief or extended rationale for a methodological choice. What kinds of decisions have they chosen to explain? Why do these issues warrant further explanation? What role do citations play in these explanations? Look for examples of each of the three levels of procedural explanation described in Chapter 3: routine procedures, procedures established in previous studies, and new procedures or substantial modifications (see Section 3.4).

Of course you will describe your proposed methods in future tense rather than past, but otherwise the basic guidelines presented in Chapter 3 (Section 3.4) can be followed in developing this section. See Section 3.5 for advice on incorporating figures and tables. As in the background section, subheadings are common in this section of the proposal. Headings and subheadings help readers keep track of the basic components of your methodology. The Burkholder and Reynolds teams both use topical subheadings within their methods sections. Notice that Reynolds et al. begin this section with an overview of topics to be covered, as they did in the background section. In some proposals, the research objectives introduced earlier become subheadings for organizing the methods section. In the Burkholder and Lewitus (1994) proposal outlined in Figure 6.4, for example, the research hypotheses, first introduced in section B of "Project Description," reappear as subheadings under section D, "Research Design," creating a clear link between the goals and methods of the study.

In sum, the proposal must convince readers that the project is significant and that it will be conducted expertly. Each section of the proposal document plays a role in building this case.

· · · · · · · · · · · · · ·

6.7 The research proposal abstract

The proposal abstract or summary is likely to serve a number of purposes and reach a number of audiences: it may be used at the start of the review process to help program officers sort proposals and select appropriate reviewers; it will be used by reviewers during the process as a preview of the larger document; and it may be used in reporting or publicizing an agency's funding decisions after those decisions are made. Thus, the NSF's proposal guidelines specify that the proposal summary should be "suitable for publication" and "informative to other persons working in the same or related fields and, insofar as possible, understandable to a scientifically or technically literate lay reader" (NSF 1995, 5).

The general shape of the proposal abstract reflects the shape of the proposal itself and thus includes a synopsis of the research problem, goals, and methods. Specifications for abstracts are variable. A limit of 200 to 300 words is common, but some agencies will accept more elaborate summaries, as illustrated in the Burkholder and Rublee (1994) proposal, Chapter 10. The NSF now requests a "proposal summary" of up to a page in length. A proposal summary written for the Burkholder and Lewitus project is presented in Figure 6.6.

· ·

EXERCISE 6.10   Read the project summary in Figure 6.6, from the Burkholder and Lewitus (1994) proposal to the NSF. Outline the "major moves" in this summary. Compare this summary with the table of contents for this proposal (Figure 6.4). What components of the proposal are highlighted in the summary? In what ways is this summary "understandable to a scientifically or technically literate lay reader"?

· ·

Whether brief or elaborated, the overview of methods in the abstract or summary must of course be "promissory" in that it describes what you *will* do. But the discussion of the research problem and goals is expected to be informative rather than descriptive (see Section 3.7); it must provide a clear, concise summary of what the project is about. Most, if not all, of your readers will read the abstract; some will read only the abstract (Olsen and Huckin 1991).

· ·

EXERCISE 6.11   The research announcement for NASA's Astrophysics Theory Program asks authors to include a 200- to 300-word abstract "describing the objective of the proposed effort and the method of approach" (NASA 1993, B-4). Write such an abstract for the Reynolds team's proposal, Chapter 11.

EXERCISE 6.12   Using their one-page Project Summary as a starting point, write a 200-word abstract for the Burkholder and Rublee (1994) proposal, Chapter 10. What parts of the summary did you use, and what parts didn't you use? Why?

· ·

Trophic Interactions of Ambush Predator Dinoflagellates
in Estuarine Microbial Food Webs

I. PROJECT SUMMARY

The diverse heterotrophic dinoflagellates include free-living estuarine species that demonstrate "ambush predator" behavior toward algal, protozoan, or fish prey. This behavioral pattern is widespread; thus far, it has been reported from the Mediterranean Sea, the Gulf of Mexico, and the western Atlantic. Within the past two years, we have discovered a new genus of ambush predator dinoflagellate, *Pfiesteria* (nov. gen.), with known species having a complex life cycle that includes at least 15 stages. Among these stages are persistent amoeboid heterotrophs and ephemeral "phantom-like" flagellates that are highly toxic to fish. *Pfiesteria* (nov. gen. et sp.) and its close relatives are common in eutrophic mid-Atlantic and Gulf Coast estuaries, and likely are widespread throughout warm temperate and subtropical waters. The ubiquitous occurrence and abundance of these stages in the water column and sediments, their voracious phagotrophy on bacterial, algal, and microfaunal prey, and their lethality to fish point to a major role of ambush predator dinoflagellates in the structure and function of estuarine food webs. In the proposed research, we plan to experimentally examine the nutritional ecology and trophic interactions of abundant stages of *Pfiesteria piscimorte* (nov. gen. et sp.) as a representative toxic ambush predator, toward determining the role of these organisms—in their many forms—within estuarine food webs. In a three-pronged approach, we will (1) examine saprotrophic nutrition of the biflagellate and amoeboid forms on stimulatory secretions of finfish prey; (2) characterize phagotrophic interactions between *Pfiesteria* (nov. gen.) and bacterial, algal and microfaunal prey; and (3) determine the influence of the biflagellate and amoeboid stages on representative microbial predators. The insights gained from this research will alter general paradigms about the role of dinoflagellates in the structure and function of food webs in eutrophic warm temperate estuaries.

FIGURE 6.6 Project summary from Burkholder and Lewitus (1994) proposal to the NSF.

• • • • • • • • • • • • • •

6.8 How scientists write research proposals

As discussed in Chapter 3, research scientists, like all writers, demonstrate a wide range of writing processes and preferences. This variation applies equally to the processes of writing and revising research proposals. Individual scientists may prefer one drafting method over another; research teams may develop preferred patterns of collaboration that enable them to produce and revise documents efficiently. Authors may vary their methods of composing and collaborating from one occasion to the next. The one constant in all this variation is an overriding concern with audience in developing the proposal. A critical dimension of the proposal writing process is the assessment of the potential funding agency's interests, values, and goals.

We have noted that agency goals are revealed in RFPs and are part of an agency's reputation, but many researchers also actively seek out new information about their target agency and establish a dialogue with program staff well before they submit proposals. Funding agencies encourage this sort of early contact and information exchange. Program staff are readily available by phone and via electronic mail for such consultations. It is in the agency's interests to provide this guidance in advance to ensure that the proposals it receives fall within its program guidelines and include all the information needed in the review process. In a recent study of the funding process, Mehlenbacher (1994) found this give-and-take among researchers and program staff to be quite common. The prominent researchers Mehlenbacher interviewed consistently described the proposal process as a long-term, interactive process.

If a proposal is not accepted on its first submission, the dialogue generally continues. Most agencies will send copies of reviewers' comments to the proposer and will continue to consult with the researcher as he or she revises the text for resubmission. In an extensive case study of this revision process, Myers (1985) found that proposal arguments changed significantly as researchers evaluated and responded to reviewers' concerns. Changes were effected on several levels, from the shape of the argument, to the amount of explanation, to the tone and *ethos* established in the text. But significantly, basic content remained relatively unchanged.

In sum, the proposal writing process is a complex, long-term endeavor that involves the participation of many. Given the high degree of interaction among researchers, program staff, and reviewers—and the continuing interaction among members of a research team (some of whom may be geographically separated)—it is no wonder that researchers in Mehlenbacher's (1994) study cited management and organizational skills as essential components of the research process.

• •

EXERCISE 6.13   Reread the transcript at the end of Chapter 1 (Figure 1.2), from a dialogue between two scientists collaborating on a research proposal. What assumptions are these two proposal authors making about their target audience's knowledge, interests, and information needs?

• •

• • • • • • • • • • • • •

6.9 How reviewers evaluate research proposals

The proposal review process exerts a powerful influence on the direction of research in scientific fields. In issuing special RFPs on specific topics, funding agencies encourage research in some areas and not others; proposal guidelines ensure that researchers who work in these areas address the agency's goals and priorities; and in responding to reviewers' comments in the revision process, researchers may further tailor their research to fit within the parameters established by the targeted research program.

This degree of control makes some scientists nervous. As with the journal article review system, there are always worries about potential abuses such as favoritism, censorship, breaches of confidentiality, or misappropriation of ideas—concerns that may be enhanced by the political context in which these organizations operate; federal funding agencies such as the NIH and the NSF depend on congressional approval of their budgets and thus may be subject to the economic tug of political priorities. An additional concern in the granting process is the role that nonspecialists play in the review process, as, for example, in NIH study sections (Cohen 1996). In response to such concerns the NSF and other agencies have initiated changes in the review process designed to give proposers more "say" in the choice of reviewers and to provide authors of unsuccessful proposals more complete information about the reasons for the rejection (NSF 1995). Like the peer review process used by journal editors, the proposal review process may perhaps best be described as an imperfect system that nevertheless provides careful scrutiny of thousands of proposals every year, of which roughly a third are funded.

Proposal guidelines and RFPs routinely include a list of the criteria reviewers will use in evaluating new applications. The review criteria for the Sea Grant program to which Burkholder and Rublee (1994) submitted their proposal are presented in Figure 6.7. NASA's review criteria, in effect for the Reynolds team's (1994) proposal, are presented in Figure 6.8. Notice that these criteria clearly reflect the different goals of the two programs (see Chapters 10 and 11).

CRITERIA FOR EVALUATION OF PROPOSALS

The following criteria will be used to evaluate the proposals.

1. *Rationale*—the degree to which the proposed activity addresses an important issue, problem, or opportunity in marine biotechnology, and how the results will contribute to the solution of the problem.

2. *Scientific Merit*—the degree to which that activity will advance the state of the science or discipline through use and extension of state-of-the-art methods. *+ Valid Science*

3. *User Relationship*—the degree to which users or potential users of the results of the proposed activity have been brought into the execution of the activity, will be brought into the execution of the activity, or will be kept apprised of progress and results.

4. *Innovativeness*—the degree to which new approaches (including biotechnological ones) to solving problems and exploiting opportunities will be employed or, alternatively, the degree to which the activity will focus on new types of important or potentially important issues.

5. *Programmatic Justification*—the degree to which the proposed activity will contribute an essential or complementary unit to other projects, or the degree to which it addresses the needs of important state, regional, or national issues.

6. *Relationship to Priorities*—the degree to which the proposed activity relates to guidance priorities in this document.

7. *Qualifications and Past Record of Investigators*—the degree to which investigators are qualified by education, training, and/or experience to execute the proposed activity and past record of achievement with previous funding.

FIGURE 6.7 Review criteria, National Sea Grant College Program. "Statement of Opportunity for Funding: Marine Biotechnology." 1994. No page number.

13. Evaluation Factors

a. Unless otherwise specified in the NRA, the principal elements (of approximately equal weight) considered in evaluating a proposal are its relevance to NASA's objectives, intrinsic merit, and cost.

b. Evaluation of a proposal's relevance to NASA's objectives includes the consideration of the potential contribution of the effort to NASA's mission.

c. Evaluation of its intrinsic merit includes the consideration of the following factors, none of which is more important than any other:

> (1) Overall scientific or technical merit of the proposal or unique and innovative methods, approaches, or concepts demonstrated by the proposal;

> (2) The offeror's capabilities, related experience, facilities, techniques, or unique contributions of these which are integral factors for achieving the proposal objectives;

> (3) The qualifications, capabilities, and experience of the proposed principal investigator, team leader, or key personnel who are critical in achieving the proposals objectives;

> (4) Overall standing among similar proposals available for evaluation and/or evaluation against the known state-of-the-art.

d. Evaluation of the cost of a proposed effort includes the consideration of the realism and reasonableness of the proposed cost and the relationship of the proposed cost to available funds.

FIGURE 6.8 Review criteria, NASA Astrophysics Theory Program. 1993. p B-7.

A comparison of Figures 6.7 and 6.8 will also reveal some common concerns which reflect the generic goals of the research proposal. Both evaluation guides highlight the relevance of the study to the agency's goals or mission, the scientific merit of the proposed work, and the capabilities of the investigators. This chapter has argued that while there are specific places in the proposal document to establish the relevance of the study and the scientific merit of the design, the professional competence of the researcher is indirectly demonstrated throughout. As these review criteria indicate, the researcher's professional *ethos* plays a critical role in the evaluation of his or her work.

• •

EXERCISE 6.14   The NIH has identified the 10 most common reasons given by reviewers for rejecting a proposal. We have reprinted these in Figure 6.9. Consider this list carefully. Which of these reasons are related to the researcher's competence or professional *ethos*?

• •

**Most Common Reasons for the Rating
"Not Recommended for Further Consideration"**

- Lack of new or original ideas
- Diffuse, superficial, or unfocused research plan
- Lack of knowledge of published relevant work
- Lack of experience in the essential methodology
- Uncertainty concerning the future directions
- Questionable reasoning in experimental approach
- Absence of an acceptable scientific rationale
- Unrealistically large amount of work
- Lack of sufficient experimental detail
- Uncritical approach

FIGURE 6.9 Most common reasons cited by NIH reviewers for rejection of proposal submissions. (NIH no date, 58).

ACTIVITIES AND ASSIGNMENTS

1. In a two- or three-page essay, discuss the significance argument developed by Burkholder and Rublee in their proposal to the National Sea Grant College Program. After reading the proposal and excerpts from Sea Grant's RFP, both in Chapter 10, describe the primary goals of the marine biotechnology initiative, the primary goals of Burkholder and Rublee's proposed project, and the relationship between them. Explain where and how in the proposal document Burkholder and Rublee demonstrate that their research meets the initiative's goals and fits the specified parameters for the program.

2. Obtain a copy of an RFP from an appropriate agency in your field. Imagine that you and a colleague at another institution plan to submit a proposal to this agency for a collaborative project you have in mind. Based on the information contained in the RFP, write a letter to your colleague describing what you think are the most important goals and constraints the proposal will have to meet. Identify the intended audiences and their interests, needs, and knowledge. Summarize the program goals, the evaluation criteria that will be used, and any special considerations you will have to keep in mind in preparing the proposal.

3. Obtain a copy of a successful research proposal on a topic in your field, along with the appropriate RFP and/or proposal guidelines. Write a descriptive analysis in which you begin by examining the organization of the proposal as revealed in the headings and subheadings of sections. Comment on any distinctive features of this organizational plan that reflect distinctive characteristics of research in this area (as in Exercise 6.3). Then examine the program goals and criteria listed in the RFP. Describe how and where each of these issues is addressed in the proposal.

4. Write a two- or three-page analysis of the rhetorical appeals in a research proposal from this text or one from your field. In your essay, examine how logical, ethical, and emotional appeals (*logos, ethos, pathos*) are made or implied in the text. That is, you should describe the shape of the logical argument being presented (the major claims made in each section and the kind of evidence offered in support of those claims); describe the character or *ethos* projected by the author(s) and the ways in which that *ethos* is created in the text; and describe any appeals to readers' interests, needs, and broader social concerns. In which section(s) of the report is each type of appeal found?

5. We noted in Section 6.1 that most texts contain elements of all three types of classical argument: arguments of fact (forensic), of value (epideictic), and of policy (deliberative). Choose a proposal from this text or from your field and look for instances of each type of argument. In which sections of the proposal does each type appear? In which sections does one or the other type predominate? Write a two- or three-page analysis outlining the role each type of argument plays in the proposal.

6. Write a five- or six-page (1250 to 1500 words) proposal for a new research project on a topic in your field. The project should build on, and represent a logical "next step" in, a current line of research reported in the literature. (Review Chapter 4 on identifying topics and sources.) Target a specific funding agency that is likely to support work in this area, and obtain the agency's proposal guidelines or an RFP. Structure your text in accordance with the generic form described in this chapter (introduction, background, proposed methods), unless your guidelines include more specific instructions for organizing the proposal. Use an appropriate documentation format in citing sources and in preparing your reference list. You may be asked to include a table of contents, a budget, biographies of investigators, and other ancillary information pertinent to your specific research focus.

7. In this activity you and your classmates will be organized into groups. Your group will role-play a peer review committee for a research proposal written by a team in your class. The team will provide copies of the RFP, proposal guidelines, or other agency material that they used to write their proposal, as well as the proposal itself. First, evaluate the proposal on your own: Is it well thought out, well argued, and well written? Use the criteria discussed in this chapter to help you evaluate the merits and limitations of the proposal: Does the proposal adhere to the agency's guidelines for organizing and formatting proposals (or follow the generic form discussed in this chapter if the agency does not specify one)? Does the introduction demonstrate the merits of the project based on the goals, priorities, and values of the funding agency? Does the review of literature seem to demonstrate a knowledge of the subject and the field? Does the design of the study and choice of methods seem to reflect good judgment? Does the description of methods seem to demonstrate the technical competence needed to carry out the project? Is the proposal free of surface-level errors in spelling that would detract from the professional *ethos* of the writer? In short, are you persuaded that this project is worth funding?

  After you have made your own assessment, discuss the merits and limitations of the proposal with your fellow reviewers, and arrive at a consensus concerning which of the following actions should be taken on this proposal. Would your committee:

1. Recommend this proposal for funding?
2. Reconsider with revision?
3. Reject?

Write a statement to the research team explaining the reasons for your decision. Your instructor may ask review teams to review their findings with the proposers.

7

Communicating
with Public Audiences

Define public audience

See p. 138 for list of all the kinds of people

7.1 Why do scientists communicate with public audiences?

In Chapter 1 we noted that scientific disciplines are not only communities in themselves but also parts of the larger society in which these communities are situated and in which scientists live. In previous chapters we have been exploring important genres and conventions that professional scientists use to communicate with each other. In this chapter we will explore some conventions that scientists use to communicate with public audiences in the larger society.

The term *public audience* is a rather loose one and is meant here to imply a wide range of listeners and readers with a variety of educational backgrounds, interests, and needs. It may be a group or a professional person (even with a Ph.D. in another field) in need of information about a particular scientific topic, or a general reader or listener simply curious about science. Public audiences for science thus can include but are not limited to those who regularly use the results of scientific research in the course of their daily work (such as agricultural producers, fish and game managers, medical professionals); administrators, local government agencies, and other public officials who need scientific information to make decisions about issues such as waste management, industrial and environmental regulations, and road construction; clubs, classes, and other educational or special-interest groups that want to learn about science; private citizens who use natural resources for hunting, fishing, hiking, and other recreational purposes; and the public at large, which has a vested interest in science insofar as they support it financially through government funding and must live with its consequences.

Thus, in addition to addressing diverse audiences *within* the scientific community (e.g., readers of journals such as *Science* and *Nature,* or the mixed audience of the research proposal), scientists may choose to write articles for the much broader audiences that read science-oriented journals such as *Scientific American* and *Discover,* general-interest publications such as *Time* and *Newsweek,* or the feature section of a newspaper. They may also make presentations to public audiences, participate in question-and-answer sessions, take part in public policy debates affected by developments in their research areas, and give press releases and interviews on important discoveries or issues in their field.

● ●

EXERCISE 7.1　　Choose a topic in your field (perhaps derived from a research report you have read), and identify some specific public audiences you might address on this topic. Who outside your field might read or listen to what you as an expert have to say about this topic? Why would they be reading or listening? How do the needs and interests of these groups differ from those of experts in your field? What do they want or need from your presentation, and how much would they already know? (Are there any public audiences with whom experts in your field might communicate but currently don't?)

● ●

There are three major reasons why scientists communicate with the general public: moral, economic, and political. The National Academy of Sciences asserts that scientists have a moral responsibility to understand and explain the effect of the work that they do on the society in which they live:

> The occurence and consequences of discoveries in basic research are virtually impossible to foresee. Nevertheless, the scientific community must recognize the potential for such discoveries and be prepared to address the questions that they raise. If scientists do find that their discoveries have implications for some important aspect of public affairs, they have a responsibility to call attention to the public issues involved. (NAS 1995, 20)

Genetic engineering and nuclear power are two obvious cases in which scientific discoveries are morally controversial or have complicated, long-term implications for society (Associated Press 1996). Other controversial areas of research include environmental protection and wildlife preservation, the use of laboratory animals in medical research, the development of drugs, the application of medical technology to prolong life, and biological studies of race and gender.

A second reason why scientists communicate with the general public is economic and hinges on the practical question: Who funds science? In addition to private labs, corporations, and universities, it is the government, and therefore the public, that funds science through tax dollars and so indirectly chooses which projects to support. Public financial support of science takes two forms: the funding of governmental agencies that conduct scientific research and the funding of government grant programs that support research by scientists at other institutions. In 1993, the federal government spent $8,629,000,000 for research and development at NASA, $6,731,000,000 for the Department of Energy, $1,337,000,000 for the Department of Agriculture, and $2,247,000,000 for the National Science Foundation (DOC 1994). But many government-run programs have either been scaled back because of budget cuts (NASA's planned space station and the Mars expedition, for example) or eliminated altogether (the Superconducting Super Collider). "Big science" projects in which the federal government plays a major role, such as the Hubble Telescope, the Space Station, the human genome project, require costly equipment and the coordination of efforts of scientists around the world and can be prohibitively expensive. In a time of budget deficits and economic belt tightening, scientists must be able to convince not only their peers but also the public and its official representatives in government of the worthiness of scientific projects. Responsibility for garnering public understanding of, enthusiasm for, and goodwill toward science ultimately rests with scientists.

The third reason scientists should learn to communicate with the general public is related to the politics of a democracy. A democratic society requires that its citizens (both electorate and elected) be informed about the issues that confront it. Since science is a major cultural force in our democracy, many of the policy decisions we make are about or based on science. As the National Academy of Sciences states:

> [S]cience and technology have become such integral parts of society that scientists can no longer isolate themselves from societal concerns. Nearly half of the bills that come before Congress have a significant scientific or technological component. Scientists are increasingly called upon to contribute to public policy and to the public understanding of science. They play an important role in educating nonscientists about the content and processes of science. (NAS 1995, 21)

Thus, a scientific democracy such as ours needs a scientifically informed citizenry to arrive at good decisions about what research to support and how to interpret and apply its results. In recent survey of the American public, the National Science Foundation found that while 72 percent of participants considered scientific research valuable, only 25 percent performed well on a basic test in science and economics (Associated Press 1996). Scientists themselves can help citizens understand and participate in our scientific democracy by becoming aware of the various genres through which the public gets its information and by learning to use the conventions of those genres to effectively communicate with the public. According to John Wilkes (1990), director of the Science Communication Program at the University of Southern California–Santa Cruz,

> U.S. citizens get up to 90% of their information about science from newspapers, magazines, and, to a lesser extent, television. As producers of scientific knowledge, scientists are in the best position to use the media to teach the public what it wants—and needs—to know about developments in medicine, science, and technology. (15)

While most scientists are not professionally trained in speaking or writing to general audiences, scientists are increasingly recognizing that the general public is an important audience to reach. Books and articles by such famous scientists as physicist Stephen Hawking, paleontologist Stephen Jay Gould, marine biologist Rachel Carson, research physician Lewis Thomas, and anthropologist Richard Leakey—just to name a few—as well as a host of popular television programs about science (*National Geographic Explores, Nature,* and *NOVA*) suggest that rather than heading to ivory-tower labs and leaving the communication of science to journalists, scientists are taking to the public pages and airwaves to explain their work and pique an interest in their science.

7.2 Understanding "general" audiences

Public audiences can be as diverse as the general population in their knowledge, interests, and needs. Listeners and readers in these audiences will possess varying types and degrees of scientific knowledge. What these general audiences have in common is a presumed lack of knowledge about *your* topic.

Just as scientists must understand the conventions governing communication with their peers to be successful professionals, so too must they understand the conventions of communicating with public audiences in order to do so successfully. In this book we cannot describe in detail the many forms public communication may take. But we can discuss some general strategies for adapting scientific information to meet the needs and interests of nonscientists in a wide range of situations. These strategies can then be applied to specific audiences, genres, and mediums as the occasion demands.

EXERCISE 7.2   In Figure 7.1 we have reproduced the introductions to five different articles on the same topic. You'll see that they are clearly intended for different audiences. As you read, think about what kind of publication these pieces would have appeared in. (This exercise has been adapted from Bradford and Whitburn 1982.)

1. Read the five introductions, and categorize them according to the level of specialized knowledge assumed on the part of the audience. Use a scale of **1** (general audience) to 5 (most-specialized audience). Speculate about where each piece might have been published.

2. After you've categorized the texts, reflect on what criteria you used to do so. In what ways do these introductions vary? What made you decide that one article is intended for a more general audience than another? Be sure to consider all dimensions of the text, including such features as content and organization, terminology and phrasing, formatting and visual presentation, tone and point of view. List the many ways in which these texts vary. Illustrate each of these features with a pair of contrasting examples from the passages.

3. Your instructor may ask you to work in small groups to develop a consensus ordering of the five texts, a consensus list of the features on which they vary, and a set of contrasting examples to illustrate each feature.

[handwritten annotations: "PASSIVE", "Distant", "highly specialized audience", "from Physiological Zoology", "MOST TECHNICAL"]

INTRODUCTION A

RECENT studies have provided reasons to postulate that the primary timer for long-cycle biological rhythms that are closely similar in period to the natural geophysical ones and that persist in so-called constant conditions is, in fact, one of organismic response to subtle geophysical fluctuations which pervade ordinary constant conditions in the laboratory (Brown, 1959, 1960). In such constant laboratory conditions a wide variety of organisms have been demonstrated to display, nearly equally conspicuously, metabolic periodicities of both solar-day and lunar-day frequencies, with their interference derivative, the 29.5-day synodic month, and in some instances even the year. These metabolic cycles exhibit day-by-day irregularities and distortions which have been established to be highly significantly correlated with aperiodic meteorological and other geophysical changes. These correlations provide strong evidence for the exogenous origin of these biological periodisms themselves, since cycles exist in these meteorological and geophysical factors.

In addition to possessing these basic metabolic periodisms, many organisms exhibit also overt periodisms of numerous phenomena which in the laboratory in artificially controlled conditions of constancy of illumination and temperature may depart from a natural period. The literature contains many accounts, for a wide spectrum of kinds of plants and animals, of regular rhythmic periods ranging from about 20 to about 30 hours. The extent of the departure from 24 hours is generally a function of the level of the illumination and temperature. It has been commonly assumed, without any direct supporting evidence, that the phase- and frequency-labile periodisms persisting in constant conditions reflect inherited periods of fully autonomous internal oscillations. However, the relationships of period-length to the ambient illumination and temperature levels suggest that it is not the period-length itself which is inherited but rather the characteristics of some response mechanism which participates in the derivation of the periods in a reaction with the environment (Webb and Brown, 1959).

It has recently been alternatively postulated that the timing mechanism responsible for the periods of rhythms differing from a natural one involves, jointly, use of both the exogenous natural periodisms and a phenomenon of regular resetting, or "autophasing," of the phase-labile, 24-hour cycles in reaction of the rhythmic organisms to the ambient light and temperature (Brown, Shriner and Ralph, 1956; Webb and Brown, 1959; Brown, 1959). It is thus postulated that the exogenous metabolic periodisms function critically as temporal frames of reference for biological rhythms of approximately the same frequencies.

(continued)

FIGURE 7.1 Five introductions. Bradford and Whitburn. "Analysis of the Same Subject in Diverse Periodicals." Technical Writing Teacher 9 (Winter 1982).*

The basic principle of audience adaptation is that we build a discussion or argument on the knowledge, goals, values, and experience of the audience. This principle is the basis of *all* successful communication and teaching, including that between professional scientists. It is also essential for the scientist communicating with public audiences. In adapting scientific information for public audiences, the writer or speaker introduces new knowledge by trying to relate it to what the audience knows or values; this new knowledge, grounded in what the audience already knows, then becomes the foundation for more new knowledge and so forth.

When you write for other scientists in your area of specialization, as in the research report, you are able to assume your readers have some degree of familiarity with the topic to begin with, as well as some degree of interest. When you write for public audiences, however, you need to be more cautious. Instead of assuming knowledge on the part of your readers, you must help them develop that knowledge; instead of

*Original formatting and typeface are reproduced as closely as possible. Sources for these introductions are listed in the Bibliography and identified in the Instructor's Guide.

[handwritten annotations: "Friendly", "notice 'hook'", "very general audience", "ACTIVE"]

INTRODUCTION B

[handwritten: "Saturday Evening Post — Least Technical", "Least Technical"]

One of the greatest riddles of the universe is the uncanny ability of living things to carry out their normal activities with clocklike precision at a particular time of the day, month and year. Why do oysters plucked from a Connecticut bay and shipped to a Midwest laboratory continue to time their lives to ocean tides 800 miles away? How do potatoes in hermetically sealed containers predict atmospheric pressure trends two days in advance? What effects do the lunar and solar rhythms have on the life habits of man? Living things clearly possess powerful adaptive capacities—but the explanation of whatever strange and permeative forces are concerned continues to challenge science. Let us consider the phenomena more closely.

[handwritten: "question to begin"]

Over the course of millions of years living organisms have evolved under complex environmental conditions, some obvious and some so subtle that we are only now beginning to understand their influence. One important factor of the environment is its rhythmicality. Contributing to this rhythmicality are movements of the earth relative to the sun and moon.

The earth's rotation relative to the sun gives us our 24-hour day; relative to the moon this rotation, together with the moon's revolution about the earth, gives us our lunar day of 24 hours and 50 minutes. The lunar day is the time from moonrise to moonrise.

[handwritten: "defined"]

The moon's arrival every 29.5 days at the same relative position between the earth and the sun marks what is called the synodical month. The earth with its tilted axis revolves about the sun every 365 days, 5 hours and 48 minutes, yielding the year and its seasons.

[handwritten: "known to unknown"]

The daily and annual rhythms related to the sun are associated with the changes in light and temperature. The 24.8-hour lunar day and the 29.5-day synodical month are associated most obviously with the moon-dominated ocean tides and with changes in nighttime illumination. But all four types of rhythms include changes in forces such as gravity, barometric pressure, high energy radiation, and magnetic and electrical fields.

Considering the rhythmic daily changes in light and temperature, it is not surprising that living creatures display daily patterns in their activities. Cockroaches, earthworms and owls are nocturnal; songbirds and butterflies are diurnal; and still other creatures are crepuscular, like the crowing cock at daybreak and the serenading frogs on a springtime evening. Many plants show daily sleep movements of their leaves and flowers. Man himself exhibits daily rhythms in degrees of wakefulness, body temperature and blood-sugar level.

[handwritten: "example", "Crepuscular = pertaining to twilight"]

We take for granted the annual rhythms of growth and reproduction of animals and plants, and we now know that the migration periods of birds and the flowering periods of plants are determined by the seasonal changes in the lengths of day and night.

In a similar fashion creatures living on the seashore exhibit a rhythmic behavior corresponding to the lunar day. Oyster and clams open their shells for feeding only after the rising tide has covered them. Fiddler crabs and shore birds scour the beach for food exposed at ebb tide and retreat to rest at high tide.

(continued)

FIGURE 7.1 Five introductions. Bradford and Whitburn. "Analysis of the Same Subject in Diverse Periodicals." Technical Writing Teacher 9 (Winter 1982).

ACTIVE & PASSIVE

Well educated general science audience

INTRODUCTION C

Science   *MODERATE (middle)*

passive

not defined

Figure sci. and. not defined

Familiar to all are the rhythmic changes in innumerable processes of animals and plants in nature. Examples of phenomena geared to the 24-hour solar day produced by rotation of the earth relative to the sun are sleep movements of plant leaves and petals, spontaneous activity in numerous animals, emergence of flies from their pupal cases, color changes of the skin in crabs, and wakefulness in man. Sample patterns of daily fluctuations, each interpretable as adaptive for the species, are illustrated in Fig. 1. Rhythmic phenomena linked to the 24-hour and 50-minute lunar-day period of rotation of the earth relative to the moon are most conspicuous among intertidal organisms whose lives are dominated by the ebb and flow of the ocean tides. Fiddler crabs forage on the beaches exposed at low tide; oysters feed when covered by water. "Noons" of sun- and moon-related days come into synchrony with an average interval of 29½ days, the synodic month; quite precisely of this average interval are such diverse phenomena as the menstrual cycle of the human being and the breeding rhythms of numerous marine organisms, the latter timed to specific phases of the moon and critical for assuring union of reproductive elements. Examples of annual biological rhythms, whose 365¼-day periods are produced by the orbiting about the sun of the earth with its tilted axis, are so well known as scarcely to require mention. These periodisms of animals and plants, which adapt them so nicely to their geophysical environment with its rhythmic fluctuations in light, temperature, and ocean tides, appear at first glance to be exclusively simple responses of the organisms

to these physical factors. However, it is now known that rhythms of all these natural frequencies may persist in living things even after the organisms have been sealed in under conditions constant with respect to every factor biologists have conceded to be of influence. The presence of such persistent rhythms clearly indicates that organisms possess some means of timing these periods which does not depend directly upon the obvious environmental physical rhythms. The means has come to be termed "living clocks."

Autonomous-Clock Hypothesis

From the earliest intensive studies of solar-day rhythmicality during the first decade of this century by Pfeffer (*1*), with bean seedlings, certain very interesting properties of this rhythm became clearly evident. Pfeffer found that when his plants were reared from the seed in continuous darkness, they displayed no daily sleep movements of their leaves. He could easily induce such a movement, however, by exposing the plants to a brief period of illumination. Returned to darkness, the plants possessed a persisting daily sleep rhythm. The time of day when the leaves were elevated in the daily rhythm was set by the time of day when the single experimental light period commenced. It was apparent that the daily rhythmic mechanism possessed the capacity for synchronization with the outside daylight cycles while having its cyclic phases experimentally altered by appropriate light changes made to occur at any desired time of day. These alterations would then persist under constant

conditions. Since Pfeffer's time, this property has been abundantly confirmed for numerous other plants and animals. The daily rhythms, therefore, exhibit the capacity for synchronization with external, physical cycles while having freely labile phase relations.

A second discovery, also made by Pfeffer, was that the daily recurring changes under constant conditions could occur earlier, or later, day by day, to yield regular periods deviating a little from the natural solar-day ones. Periods have now been reported ranging from about 19 to 29 hours. The occurrence of persisting rhythmic changes under constant conditions, with regular periods of other than precisely 24 hours, clearly indicated that these observed rhythmic periods could not be a simple direct consequence of any known or unknown geophysical fluctuation of the organism's physical environment.

A third fundamental contribution to the properties of the daily rhythms was made by Kleinhoonte (*2*). While confirming, in essentials, all of Pfeffer's findings, she discovered that the daily sleep movements of plants could be induced to "follow" artificial cycles of alternating light and dark ranging from about 18-hour "days" to about 30-hour "days." When the "days" deviated further than these limits from the natural solar-day period the plants "broke away" to reveal their normal daily periodicity, despite the continuing unnatural light cycles. This observation clearly emphasized the very deep-seated character of the organismic daily rhythm.

passive

passive

(continued)

FIGURE 7.1 Five introductions. Bradford and Whitburn. "Analysis of the Same Subject in Diverse Periodicals." Technical Writing Teacher 9 (Winter 1982).

Biology, modality, specialized science audience (handwritten)

PASSIVE (handwritten)

INTRODUCTION D

Biological Bulletin (2nd most technical) (handwritten)

connection to audience (handwritten)

A deep-seated, persistent, rhythmic nature, with periods identical with or close to the major natural geophysical ones, appears increasingly to be a universal biological property. Striking published correlations of activity of hermetically sealed organisms with unpredictable weather-associated atmospheric temperature and pressure changes, and with day to day irregularities in the variations in primary cosmic and general background radiations, compel the conclusion that some, normally uncontrolled, subtle pervasive forces must be effective for living systems. The earth's natural electrostatic field may be one contributing factor.

A number of reports have been published over the years advancing evidence that organisms are sensitive to electrostatic fields and their fluctuations. More recently Edwards (1960) has found that activity of flies was reduced by sudden exposures to experimental atmospheric gradients of 10 to 62 volts/cm., and that prolonged activity reduction resulted from gradient alternation with a five-minute period. In 1961, Edwards reported a small delay in moth development in a constant vertical field of 180 volts/cm., but less delay when the field was alternated. The moths tended to deposit eggs outside the experimental field, whether constant or alternating, in contrast to egg distribution of controls. Maw (1961), studying rate of oviposition in hymenopterans, found significantly higher rates in the insects shielded from the natural field fluctuations, whether or not provided instead with a constant 1.2 volts/cm. gradient, than were found in either the natural fluctuating field, or in a field shielded from the natural one and subjected to simulated weather-system passages in the form of a fluctuating field of 0.8 volts/cm.

A study in our laboratory early in 1959 (unpublished) by the late Kenneth R. Penhale on the rate of locomotion in *Dugesin* suggested strongly that the rate was influenced by the difference in charge of expansive copper plates placed horizontally in the air about six inches above and closely below a long horizontal glass tube of water containing the worms. Locomotory rates in fields of 15 volts/cm. (+ beneath the worms) were compared with those in fields between equipotential plates. The fields were obtained with a Kepco Laboratories, voltage-regulated power supply. A comparable study with the marine snail, *Nassarius*, by Webb, Brown and Brett (1959), employing a Packard Instrument Co., high-voltage power supply, confirmed the occurrence of such responsiveness to vertical fields of 15 to 45 volts/cm., and advanced evidence that the response of the snails displayed a daily rhythm.

(continued)

examples from biology (handwritten, right margin)

FIGURE 7.1 Five introductions. Bradford and Whitburn. "Analysis of the Same Subject in Diverse Periodicals." Technical Writing Teacher 9 (Winter 1982).

assuming interest in the topic, you must generate interest. The way to begin is by assessing your audience's knowledge, needs, and goals.

Consider your subject *from the point of view of your readers or listeners*. What expectations does your audience have about the subject? About you? How is your audience likely to see and/or understand the topic or issue? Are there conflicting beliefs or concepts that will have to be dealt with, and how will you deal with them? Are familiar explanations trite or boring? Does the audience have firsthand experiences that you can draw on to illustrate points in your discussion? (Campbell 1982).

To answer these questions, you will need to find out as much as possible about the background, areas of expertise, probable beliefs, values, and general interests of members of your audience. One way to learn about your audience is to ask the organization sponsoring your talk about who will be in attendance and why. Or read the magazine you are writing for and look at the kinds of text features you noted in Exercise 7.2.; these features can often provide hints about how to interpret and adapt to a particular audience.

ACTIVE

educated audience, not necessarily in science, people interested in science

Scientific American (2nd least technical)

INTRODUCTION E

question {

Everyone knows that there are individuals who are able to awaken morning after morning at the same time to within a few minutes. Are they awakened by sensory cues received unconsciously, or is there some "biological clock" that keeps accurate account of the passage of time? Students of the behavior of animals in relation to their environment have long been interested in the biological clock question.

Most animals show a rhythmic behavior pattern of one sort or another. For instance, many animals that live along the ocean shores have behavior cycles which are repeated with the ebb and flow of the tides, each cycle averaging about 12½ hours in length. Intertidal animals, particularly those that live so far up on the beaches that they are usually submerged only by the very high semimonthly tides when the moon's pull upon the ocean waters is reinforced by the sun's, have cycles of behavior timed to those 15-day intervals. Great numbers of lower animals living in the seas have semilunar or lunar breeding cycles. As a result, all the members of a species within any given region carry on their breeding activities synchronously; this insures a high likelihood of fertilization of eggs and maintenance of the species. The Atlantic fireworm offers a very good example of how precise this timing can be. Each month during the summer for three or four evenings at a particular phase of the moon these luminescing animals swarm in the waters about Bermuda a few minutes after the official time of sunset. After an hour or two only occasional stragglers are in evidence. Perhaps even more spectacular is the case of the small surface fish, the grunion, of the U.S. Pacific coast. On the nights of the highest semilunar tides the male and female grunion swarm in from the sea just as the tide has reached its highest point. They are tossed by the waves onto the sandy beaches, quickly deposit their reproductive cells in the sand and then flip back into the water and are off to sea again. The fertilized eggs develop in the moist sand. At the time of the next high tide when the spot is again submerged by waves, the young leave the nest for the open sea.

Almost every species of animal is dependent upon an ability to carry out some activity at precisely the correct moment. One way to test whether these activities are set off by an internal biological clock, rather than by factors or signals in the environment, is to find out whether the organisms can anticipate the environmental events. The first well-controlled experimental evidence on the question was furnished by the Polish biologist J. S. Szymanski. In experiments conducted from 1914 to 1918 he found that animals exhibited a 24-hour activity cycle even when all external factors known to influence them, such as light and temperature, were kept constant. During the succeeding 20 years various investigators, especially Orlando Park of Northwestern University, J. H. Welsh of Harvard University and Maynard Johnson (currently in the U.S. Navy), demonstrated that comparable rhythmic processes persisted in many insects, in crustaceans and in mice. Persistent daily rhythmicity has been found in animals ranging from one-celled protozoa to mammals. And the Austrian biologist Carl von Frisch, using a slightly different approach, discovered that bees could be trained to come to a feeding station at the same time on successive days but not at different times—a finding which suggested that bees have an internal daily cycle.

FIGURE 7.1 Five introductions. Bradford and Whitburn. "Analysis of the Same Subject in Diverse Periodicals." *Technical Writing Teacher* 9 (Winter 1982).

EXERCISE 7.3   Consider this scenario. Jane, a premed student working with lasers, wanted to show her friend Mike, a zoology major, how a new laser in her lab worked. "Come on over to my lab and I'll give you a demonstration." Mike had never studied or worked with lasers, but from what he had heard, they seemed fascinating; so one day he took Jane up on her offer. When Mike got to the lab, Jane escorted him through a maze of machines to the lab table where her laser was set up, and she proceeded to take the laser apart and explain each major component. Mike quickly lost interest and wandered away to look at the other machines, while Jane continued to discuss at length the technical details she had learned about lasers, oblivious to the fact that he was no longer listening. What do you think happened?

1. Using the information given about Mike, and your own common sense and empathy (both necessary in audience adaptation), what do you think Mike expected when he walked into Jane's lab? What would Mike like to see and learn about lasers?

2. Now, if Mike were an administrator of the lab—say in charge of personnel and finance—why would he be interested in the laser? What would he want to see and learn?

3. If Mike were a parent whose child's school was about to purchase a laser for use in science classes, what would he have wanted to know about it?

• •

To extend a distinction drawn by Flower (1993), Jane was "speaker-based" rather than "listener-based": She was more concerned with what she as the speaker was interested in and had to say about lasers than in what Mike as the listener was interested in and may have wanted to hear about lasers. The story of Jane and Mike is a microcosm of large- and small-scale breakdowns in communication that occur every day in our scientific society. In fact, communication breakdown has been cited as a major contributing factor in a number of serious technological accidents. In the case of the Three Mile Island nuclear reactor meltdown, for example, faulty assumptions about how to communicate with the public hampered the efforts of officials to find the best way to inform them and control the emergency and led to widespread panic and social disarray (Farrell and Goodnight 1981). The inability of engineers and managers to understand each others' values has been directly implicated as a major cause of the Challenger shuttle explosion (Herndl et al. 1991). And the inability of experts and government officials to consider the values and emotions of public audiences has been a factor in unsuccessful attempts to site low-level radioactive waste facilities all around the country (Katz and Miller 1996).

In all these situations, communicators in one area of expertise seriously misunderstood the audience outside their fields, with dire consequences. Unlike the expert colleague, who has both knowledge of the subject and an intrinsic interest in it, public audiences have different perspectives on and interests in the subject, and thus different expectations and needs that must be appealed to. While experts are interested in theory and technical details, in methods and results, public audiences are generally interested in what things "do" and their effect on public safety, health, and welfare.

In addition to the three general modes of appeal (*logos, pathos, ethos*) discussed in Chapter 6, two special appeals often come into play in the accommodation of scientific knowledge to public audiences (Fahnestock 1986). The first is the *wonder appeal,* which emphasizes the sense of surprise and joy and awe people (both generalists and specialists!) often feel when confronted with an exciting scientific discovery. The second is the *application appeal,* which emphasizes the practical benefits of a scientific concept or discovery for a particular audience, society at large, or humankind. This appeal is especially effective with administrators and public officials.[1]

Most of us are fascinated by the accomplishments and spectacle of science, and interested in what things do from the point of view of common experience or daily life. Both the wonder appeal and the application appeal work well with general audiences. In the scenario above, Mike may have secretly wanted to ask: "Can we see the laser burn a cool hole in something?" (wonder). "How can lasers be used to shoot down missiles in outer space and also perform delicate eye surgery?" (wonder and application). "How could I use lasers in my zoology major?" (application). He never got a chance.

• •

EXERCISE 7.4   Assume you're a member of Stephen Reynolds's research team. You've been invited to speak to a local amateur astronomy club on the topic of supernova remnants. Using the audience analysis questions on pages 145–146 as your starting point, write a paragraph describing your audience. Write a second paragraph describing what kinds of appeals you would use with this audience.

• •

In the remainder of this chapter, we will explore several different strategies writers can use to adapt scientific and technical discussions for general audiences.

• • • • • • • • • • • • • •

7.3 Adapting through narration

a story

Narration is a powerful means of audience adapation. Research has shown that stories are basic to the formation of human identity and knowledge—even scientific knowledge (Fisher 1987; Polkinghorne 1988). By creating a human story through which an audience can identify with a scientific subject, narration can make science accessible and acceptable to general audiences (Katz 1992a). Unlike the research report—with its hypothesis, methods, results, and conclusions sections—science, when presented to the general public, often takes the form of a dramatic story, with characters, plot, conflict, and resolution. When the *New Yorker* reported the discovery of *H. pylori,* for example, its story was titled "Marshall's Hunch" (Monmaney 1993) and focused on Marshall himself, tracking his rise to prominence from relative obscurity and emphasizing his iconoclastic style. Marshall the rebel, not *H. pylori,* was the hero of this story for a general audience.

Parts of the "story of science" often take the form of a *history,* a brief chronology of events. It may be a recounting of steps leading to a discovery or the development of

---

[1] The administrator is interested in what things do from the point of view of cost, production, public health, environmental safety, and resource management: How can this new piece of equipment be used in the lab? How much does it cost to purchase and maintain? What products (or discoveries) are likely to arise from it, and will they be profitable for the lab or company? Is the piece of equipment or product cost effective? Safe? Efficient?

a phenomenon, concept, project, or field. Historical narrative can be especially effective if the story can be accommodated to the audience's interest and/or general (cultural) experience. Such narratives are usually given dramatic flair, as in the following "scene" in Huyghe's description of the discovery of the killer algae (contained in Chapter 10): "The story properly begins one night early in 1988, with a massacre in Edward Noga's laboratory at North Carolina State University in Raleigh" (1993, 72).

· ·

EXERCISE 7.5   Read the Huyghe (1993) article from *Discover* magazine about the Burkholder team's research on toxic dinoflagellates (included in Chapter 10). Look for instances of narrative storytelling and character development. What parts of this project are dramatized in this article written for a lay audience? Who are the main "characters"? What is the "plot"? What is the "conflict"? How is it "resolved"?

EXERCISE 7.6   Watch an episode of a science program such as *Nature* or *Nova* on television. Is the scientific information adapted to a general viewing audience through the use of narrative storytelling? What parts of the science are being dramatized? Who are the main "characters"? What is the "plot"? What is the "conflict"? How is it "resolved," and how does the resolution support scientific research?

· ·

· · · · · · · · · · · · · ·

7.4 Adapting through examples

Examples play a prominent role in audience adaptation and are essential for comprehension. As we see in Figure 7.2, examples provide a more particular instance of a general concept or class. Because examples are particular instances of a general concept or

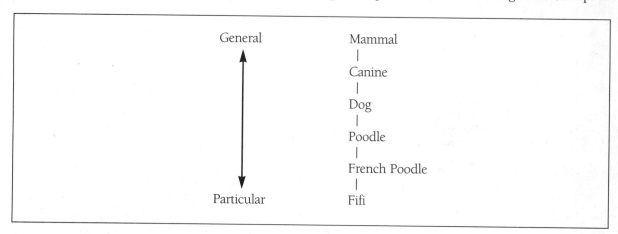

FIGURE 7.2 The use of examples: Movement from general classes to particular instances.

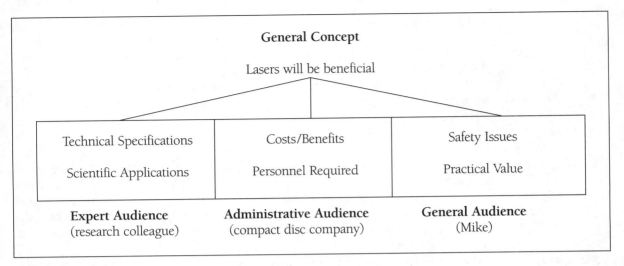

FIGURE 7.3 Topics for different audiences.

class, they tend to be more concrete, and so they tend to make the general concept or class easier to grasp. Red tide, which many coastal visitors have seen firsthand, is an example of the class of biological phenomenon called *algae blooms*. In this book we have used examples to illustrate our discussion of various genres and conventions. Imagine this book without them! Examples also provide an effective way to relate unfamiliar concepts to what readers or listeners already know, value, and are interested in, if they are drawn from the reader's or listener's experience. As you discovered in Exercise 7.3, different audiences are interested in different issues. Figure 7.3 outlines three sets of issues that might be useful in explaining lasers to three different audiences. In speaking or writing to one of these audiences, you would want to draw examples from that particular audience's area of interest.

Thus, there are two general rules concerning the use of examples as a strategy for audience adaptation: (1) examples must *logically* support your point by representing the general class of items they're meant to illustrate; (2) examples should *rhetorically* support your point by being drawn from the knowledge and experience of your audience.

7.5 Adapting through definition

Several definitional strategies can be used to "unpack," or "translate," scientific terms for a general audience. Before you consider any of these strategies, you will want to ask yourself whether the particular term is needed by the audience at all. Think about the knowledge, needs, and interests of your audience. Is the term absolutely essential for the audience to learn? If you decide it will bog down the reader, the term should be omitted. For example, when you read our descriptions of the appeals to wonder and application in Section 7.2, did you need to know that the technical term for the appeal to application is the *teleological appeal* and that the appeal to wonder is the *deontological appeal* (Fahnestock 1986)? What good does it do you to know this information now? What would have been the effect if we had used these terms instead of the general terms in the earlier discussion? Are there circumstances in which this technical terminology might be appropriate?

If you decide a term is important enough to take the time to explain it to your audience, then you can use one of the following definitional strategies to define it. These strategies are not just different ways of explaining terms but also different ways of thinking about concepts. Each definitional strategy gives you (and thus the audience) a different "angle" on a term, a different perspective on a concept. Thus, your choice of definitional strategy will depend on (1) the term or concept; (2) the audience's knowledge, values, interests, and needs; and (3) what you want to say to this audience (your purpose).

When you define using the *classical (Aristotelian) definition,* you place a term or concept in a general class of things to which it is similar and then delineate how the term or concept is different from other members of this class. This genus-differentia definition is the one we find the most in dictionaries. It depends on classification as the primary intellectual strategy to explain a concept. For example, the term *nova* (used as a matter of course for an expert audience in Reynolds's et al. paper on supernova remnants, in Chapter 11) is defined in an ordinary dictionary as "a star that suddenly becomes thousands of times brighter and then gradually fades" (Random House). The *genus* is "a star," and everything that follows is the *differentia,* that part of the formal definition that distinguishes novas from other stars.

The definitional strategy *etymology* explains a term by examining the history of the word(s). Notice how in the following definition Huyghe (1993, 75) employs the narration strategy to enhance audience appeal as well: "They have proposed calling [the dinoflagellate] *Pfiesteria piscimorte* (the genus name was picked to honor the late dino specialist Lois Pfiester of the University of Oklahoma and also because Burkholder liked the name's echo of both *feast* and *cafeteria;* the species name means 'fish killer')."

7.6 Adapting through analysis

Analysis breaks a whole into constituent elements. Thus, in analysis, division, not classification, is the primary strategy used to unpack a concept. Analysis can be used to discuss the structure of a molecule or the stages of a supernova. Huyghe (1993) uses analysis to divide the life cycle of dinoflagellates into stages to emphasize the scientific oddity of these creatures and their devastating effect:

> "[The new dinoflagellate's] life cycle consists of more than 15 stages. . . . The dino lives most of its life as an amoeba; in one toxic "giant" amoeboid stage it grows to be nearly 20 times the size of the flagellated toxic cells. The stage

when the dino leaves its cyst to attack fish is actually quite ephemeral, appearing only when fish are present. The researchers also found that this stage is the only time the creature sexually reproduces. . . ." (Huyghe 1993, 75)

7.7 Adapting through comparison

Comparison is a particularly useful method for explaining concepts to nonspecialists. Using this strategy, the scientist indicates how a phenomenon is similar to or different from other phenomena the audience is more familiar with: "Dinoflagellates are twilight-zone creatures: half-plant, half-animal. They produce chlorophyll, but they also move about, using their two flagella, or whiplike tails, to swim rapidly through the water" (Huyghe 1993, 72).

The use of synonyms is one type of comparison strategy. *Synonym,* substituting one or more words for another, can be used to define a term in a context the audience is more familiar with, as in the example above: "flagella, or whiplike tails." This sort of definition in context is also known as *parenthetical explanation* (if in parentheses), and *apposition,* if inserted without parentheses.

It is because no two words have exactly the same meaning or connotation that synonym falls under the category of comparison: for the writer or speaker, the use of synonyms really involves a comparison of terms and concepts rather than exact substitution—a search for a similar word that the audience will be more likely to understand and be able to relate to. Given that technical terms are part of a precise vocabulary of a field, the scientist using synonym usually has to trade off some accuracy for the sake of gaining the audience's comprehension.

You've all learned that *simile* is a comparison that uses *like* or *as.* Simile can be used to drive home a point figuratively, as in the following example: "Efforts to describe [the dinoflagellate] have evoked the strangest comparisons: like grass feeding on sheep, said one scientist. And that's not stretching things much" (Huyghe 1993, 72). Unlike simile, *metaphor* does not use *like* or *as.* It is thus a less obvious, but a much more pervasive, method of comparison than we might think. By omitting *like* or *as,* metaphors not only conceal the comparison but also imply an identification of the things compared, as in the following example: "The creature is in fact a tiny plant with a Jekyll and Hyde personality, one that preys on animal life a million times its size" (Huyghe 1993, 72). Notice that the attribution of a specific behavior to the alga through the personality metaphor of "Jekyll and Hyde" is grounded in the audience's cultural knowledge.

Some researchers believe that metaphors actually structure, and to some extent determine, the way we conceptualize the world (see Lakoff and Johnson 1980). It has been argued that metaphors are implicit in scientific models (Turbayne 1970; Boyd 1979; Kuhn 1979; Leary 1990); one immediately thinks of Neils Bohr's early model of the atom as a solar system, but there are many others. In any case, common sense tells us that the metaphors we choose both color and reflect the way we think about phenomena. Metaphors are used not only in communicating with general audiences (e.g., "biological clock" [Brown 1959]; "killer algae" [Huyghe 1993]), but also by scientists themselves in exploring new phenomena (e.g., "genetic code" [Watson and Crick 1953]; "phantom dinoflagellate" [Burkholder et al. 1992]).[2]

In creating metaphors for nonexpert audiences, it is important to remember two things: (1) Find the metaphors in the language, knowledge, and experience of your audience; do not merely switch to other technical metaphors embedded in your field that the audience will not understand. (2) Use metaphors consistently; mixing metaphors confuses readers and may introduce inconsistencies into your discussion.

EXERCISE 7.7   An excerpt from Monmaney's (1993) profile of Barry Marshall in the *New Yorker* is presented in Figure 7.4. Identify and circle the metaphors Monmaney uses to discuss the *H. pylori* bacteria. What are the bacteria being compared to? Why? Are the metaphors related to each other (i.e., consistent)? What do the metaphors communicate? Compare the description of this bacterium in the *New Yorker* with its description in the Marshall and Warren letters (1983) in Chapter 9. Are there any metaphors in the technical letters? What conclusions can you draw about the use of metaphors for scientific and non-scientific audiences? What is the scientific basis for the metaphors in the *New Yorker* excerpt? What is the popular basis of the metaphor?

EXERCISE 7.8   Identify and compare the metaphors on the first page of Huyghe's essay in *Discover* (1993) and the first page of the Burkholder team's (1992) letter in *Nature* (1992), both contained in Chapter 10. You will see that Huyghe and the Burkholder team use similar metaphors to describe microorganisms to specialist and nonspecialist readers. What does this tell you about the role of metaphors in science?

Analogy is an extended comparison which may include similes and metaphors. In fact, an analogy might comprise a series of related metaphors running throughout a text. Like other methods of comparison, the advantage of an analogy is that the writer can begin with what the audience knows and then move back and forth between the known and the unknown as the concept is explained or new terms further defined. For example, in the Hughye piece, the metaphor "killer algae," announced in the title of the article, "surfaces" throughout the text.

EXERCISE 7.9   In the Huyghe (1993) piece, look at definitions, similes, metaphors, and other places where the metaphor "killer algae" is extended in the discussion through analogy. Are there places where the analogy breaks down or is violated? Why? What is the effect? Do

---

[2] Quotation marks are often used around "popular" terminology by scientists writing to general and specialist audiences to distance themselves from it, reflecting an abiding concern for their professional credibility.

Iᴛ takes a kind of cunning for *Helicobacter pylori* to fill its niche, for the adult human stomach is one of nature's most hostile habitats. Each day, the stomach normally produces about half a gallon of gastric juice, whose strong hydrochloric acid and digestive enzymes readily tear meat and microbes apart. Gastric juice is like a binary chemical weapon—so destructive that it's constituted only on the way to the target. As cells in the stomach lining secrete the raw ingredients of gastric juice into the mucus that coats the stomach lining, the ingredients mix into an even more caustic brew, which then oozes into the cavity. There the gastric juice breaks pabulum down chemically while muscles in the stomach wall act to crush it. The viscoelastic mucus, as thick as axle grease, keeps the stomach from digesting itself.

Once *Helicobacter* reaches the stomach, it probably does not linger out in the open cavity—a tossing sea of toxic chemicals. It heads for cover. The bacterium's helical shape seems to have been designed for speedy travel in a dense medium. *Helicobacter* is living torque; a microscopic Roto-Rooter, it corkscrews through the mucus. Then, instead of penetrating the cells of the stomach lining, it settles in the mucus just beyond the lining. More often than not, it settles in the pylorus. No one knows why. Under the microscope, a *Helicobacter* infection looks like a satellite image of an armada gathered off a ragged shore. At one end of the bacterium is a cluster of long, wispy, curving flagella, which may serve as anchors. It's a graceful menace.

Helicobacter possesses a vital defense against stomach acid, and this adaptation, too, is a marvel of evolutionary design. Its coat is studded with enzymes that convert urea—a waste product, virtually unlimited supplies of which can be found in the stomach—directly into carbon dioxide and also into ammonia, a strong alkali. Thus *Helicobacter* ensconces itself in an acid-neutralizing mist. In like fashion, it generates another antacid—bicarbonate, as in Alka-Seltzer.

A *Helicobacter* infection that establishes itself succeeds largely because the immune system can't reach it. In response to a *Helicobacter* invasion, immune-system cells in the bone marrow produce white blood cells, killer cells, and other microbe destroyers, and those float through the bloodstream to the very edge of the stomach lining—and go no farther, because the lining holds them back. The *Helicobacter*, hovering in the mucus, are out of range. And yet the immune system sends reinforcements. Killer cells pile up, gorging the stomach lining; permanently alerted, seldom engaged, the killers become trigger-happy. Some die, fall apart, and spill their microbe-fighting compounds into the host tissue. Friendly fire begets friendly fire. Metabolic hell breaks loose. The lining is now inflamed: acute gastritis. Micronutrients are pumped from the bloodstream to the front lines, to feed the killers, but loads of them seep out of the stomach lining and into the mucus. Offshore, the *Helicobacter* feast; having drawn the immune system into battle, the bacteria now loot the provisions. "I propose that inflammation is good for *Helicobacter*," Blaser says. "That's what it wants."

Chronic gastritis, a standoff between the bacteria and the host's immune system, may persist for years—decades, according to some estimates. As it happens, whatever serious damage is done to stomach or intestinal tissue is apparently done not by the bacteria themselves but by the inflammatory response they provoke. That the host plays such a large role in his own pathology may help explain why *Helicobacter* infection affects different people differently. (Also, scientists recently discovered that there are at least two strains of *Helicobacter pylori*, and that one of them is far more likely to cause a peptic ulcer than any other.) Inevitably, though, for *Helicobacter* to be really successful it has to meet a parasite's final challenge: to start another colony before the host dies.

FIGURE 7.4 Metaphor identification: Excerpt from "Marshall's Hunch" (Monmaney 1993).

you see any similarity between the analogy in the Huyghe piece on dinoflagellates and the analogy in the excerpt from Monmaney's piece on *H. pylori?* What do these analogies tell you about our attitude toward microrganisms?

• •

You should note that *analogy* **is not** *example.* While examples are drawn from the general class of concepts they are a part of, the "examples" in analogies are drawn from similar or parallel classes of concepts; that is, they are based on comparison. Scientists can draw on family and social relationships as well as causal ones in comparing unfamiliar concepts with concepts more familiar to their general audiences (Young et al. 1970). In the example in Figure 7.5, from Zukav's (1979) *The Dancing Wu Li Masters: An Overview of the New Physics,* the population of a city is used not as an *example* of the concept of discontinuity in new physics; people are not members of the class of particles. Rather, the parallel between people and particles is being used to explain the concept of discontinuity in New Physics by *comparison* with something the audience is more familiar with: population statistics.

Usually several strategies are used together to unpack a scientific concept; they may even be embedded in each other, as in the following excerpt: "Under the right conditions, for example, sodium reacts to chlorine (by forming sodium chloride—salt), iron reacts to oxygen (by forming iron oxide—rust), and so on, just as humans react to food when they are hungry and to affection when they are lonely" (Zukav 1979, 71). In this passage, we see that in *comparing* chemical and human reactions the author uses *examples;* these examples are *analyses* of chemical reactions; the results of the reactions are clarified by *synonyms* drawn from common experience. Thus, several layers of explanation revolve around each other, as illustrated in Figure 7.6. Notice that at the center of this complex system are two ordinary concepts, salt and rust.

> What had [Max] Planck discovered that disturbed him so much? Planck had discovered that the basic structure of nature is granular, or, as physicists like to say, discontinuous.
>
> What is meant by "discontinuous"?
>
> If we talk about the population of a city, it is evident that it can fluctuate only by a whole number of people. The least the population of a city can increase or decrease is by one person. It cannot increase by .7 of a person. It can increase or decrease by fifteen people, but not by 15.27 people. In the dialect of physics, a population can change only in discrete increments, or discontinuously. It can get larger or smaller only in jumps, and the smallest jump that it can make is a whole person. In general, this is what Planck discovered about the processes of nature.

FIGURE 7.5 Excerpt from *The Dancing Wu Li Masters: An Overview of the New Physics* (Zukav 1979, 73).

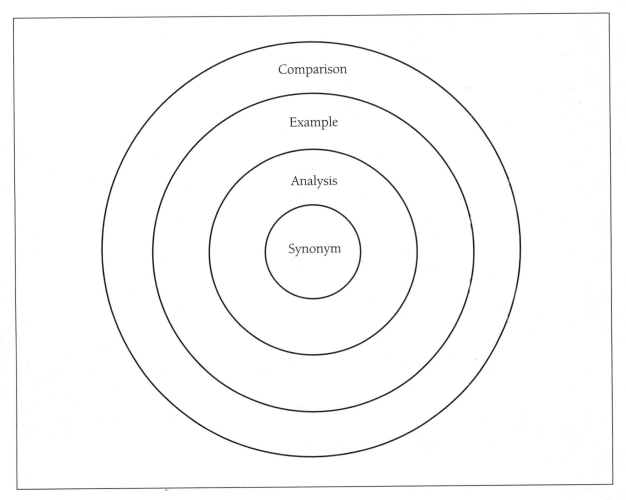

FIGURE 7.6 Orbits of explanation in Zukav's comparison.

· · · · · · · · · · · · · · ·
7.8 Adapting through graphics

Graphics are visual representations of phenomena. As discussed in previous chapters, visual aids such as tables, line graphs, diagrams, and site maps are conventionally used to present information to scientific audiences in research reports and conference presentations. But graphics can also be used to explain concepts to general audiences. Note the use of graphics to explain audience adaptation strategies in this chapter. Like verbal examples, visual examples should be based on your audience's knowledge, interests, and values.

EXERCISE 7.10   If you did Activity 1 at the end of Chapter 1, look at the three different ways you represented the information contained in that exercise. What interests, values, and needs does each (graphic) representation appeal to (or assume) on the part of your audience?

abstract to concrete

For general audiences, photographs, slides, maps, drawings, bar graphs, pie charts, and simple diagrams serve to make the verbally abstract visually concrete. One advantage of these types of pictorial representation is that they can be made colorful, attractive, and thus very appealing. They also can be discussed at any level of generality that is appropriate. A catchy or attractive photograph can even be used to introduce a topic. For example, the opening page/spread of the "Killer Algae" article (Huyghe 1993) catches readers' attention by juxtaposing a "supersized" image of the toxic dinoflagellate with smaller pictures of the fish that become its victims, effectively dramatizing the magnitude of the threat posed by the microscopic algae. In the piece Martin Blaser (1996) wrote for *Scientific American* (Chapter 9), a colorful diagram of a stomach lining inhabited by *H. pylori* decorates the first two pages of the text.

EXERCISE 7.11   Imagine you are an editor working for *Discover* or *Scientific American*. You have assigned one of your staff writers to write an article describing Stephen Reynolds's recent research on supernova remnants. While your writer reviews Reynolds's papers and flies to North Carolina State University to conduct an interview, you call in your design staff to start planning the visual presentation. After reviewing Reynolds's work yourself (see Chapter 11), think about what kind of visual effect you'd like to create. Describe (or better yet, sketch) the first page spread for the article. Think of an effective title for the piece, and incorporate it into your design.

Simplicity & Creativity

The visual medium, size, shape, and color of a visual representation also can be used to communicate symbolically. Line graphs use colors (e.g., black and red) as well as the direction of the line (up and down) to symbolically (and dramatically) show profit and decline *visually*. Thus, a pie chart showing percentages of food consumed by Americans during the day versus the night, for instance, might color the day part of the graph yellow and the night part blue, communicating basic information at a glance through color as well as labels. A graph showing annual deforestation by fire might have bars colored red or orange (known as "hot colors," as opposed to blue or green, known as "cool colors") or even shaped like flames! The use of size, shape, and color in graphics is especially important for nonexpert audiences when knowledge of the appearance of phenomena cannot be assumed but can be communicated quickly through the visual medium, or when dramatic effect is desired to make a point.

When using graphics to communicate to a general audience, consider all the possibilities the visual medium has to offer you. Recall from Chapter 5, though, that one of the dangers of pictorial representation is that the visual aid may contain too much detail, especially if it is not drawn specifically for the purposes of your presentation. Visual aids, like text, should be focused and uncluttered when communicating with non-specialists.

EXERCISE 7.12   Burkholder and Rublee diagram the life cycle of the toxic dinoflagellate in Figure 2 of their 1994 Sea Grant Proposal (Chapter 10). Patrick Huyghe presents a brief verbal description of the life cycle on the last page of his *Discover* piece (1993). Assume you are working with Huyghe on a revision of this piece. Your task is to adapt Burkholder and Rublee's technical diagram for the general readership of *Discover*. Redraw this diagram, using both the original diagram and Huyghe's verbal description as your guide. You may add to or revise Huyghe's verbal description to accommodate your graphic if you wish.

7.9 Logic and organization in writing for public audiences

Be prepared for extensive editing by the editor of my article [handwritten margin note]

Unlike the research report and proposal, there is no standard structure for articles written for general audiences. Thus, you will want to consult the publication you are writing for to get a sense of style and formatting options. Major topics or points are often indicated by headings, as in Blaser's (1996) *Scientific American* article on *H. pylori,* contained in Chapter 9. When headings are used, they are topical rather than functional; that is, they provide clues to the content rather than the structure of the text. Functional headings (such as "Introduction," "Methods," "Results," "Discussion") are useful in genres governed by a standard structure, for they enable readers familiar with that structure to quickly locate the types of information they're most interested in. But such headings are of no use when no particular pattern of organization is expected by readers.

Topical headings help readers see at a glance what major topics or issues will be raised in each section of an article, and thus serve to introduce the unique pattern of organization used in a given essay. To do so, they must be clearly comprehensible to the general reader. Headings also provide an opportunity to catch readers' attention and pique their interest. Thus, authors writing for general audiences aim to create titles and headings that are not only informative but intriguing.

p. 206   *p. 184*   *p. 89* [handwritten margin notes]

EXERCISE 7.13   Compare the headings in Blaser's (1996) *Scientific American* piece with those of a standard research report such as Marshall and Warren (1984) or Graham et al. (1992). Then compare the title and headings of the *Scientific American* piece with the review article Blaser wrote for *Gastroenterology* (1987). (These readings are all contained in Chapter 9.) Write a one- or two-page analysis in which you use these sample articles to illustrate the differences between topical and functional headings, and the differences between specialized and general audiences.

p. 193 [handwritten margin note]

WILL BE LOOKING ALSO AT CONTENT OF ARTICLES [handwritten footer note]

Articles in general-reader publications may end with a list of suggested readings but usually do not include a formal reference list. These articles rarely report the results of individual experiments, focusing instead on the general outcomes of a body of research. Therefore they do not typically include citations of individual research reports. Such reports are occasionally included in a list of further readings (see Blaser [1996], Chapter 9), but it is not common practice to include sources written for research journal audiences when writing for newspapers or general-interest magazines. It is more helpful to those readers to list resources that are specifically designed for general readers and are readily available in local bookstores or public libraries.

Instead of reflecting on prior research reports to support their claims, authors often use interview data when writing for general audiences. Direct quotations are rarely used in other genres of scientific discourse (see Chapter 4), but they are legitimate and popular forms of evidence in general-reader publications such as newspapers and magazines. In fact, *quoting* can be understood to constitute another major strategy for adapting science to a general audience. Quotations from the "characters" involved can enliven a scientific story and thus are frequently used when the narrative strategy is employed. The Huyghe (1993) article in *Discover* includes numerous examples of this strategy. Another common use for quotations is to provide support or elaboration of a claim made in the essay, as when the author quotes a prominent authority's opinion of a new finding or its implications. In classical argumentation terms, quotations thus represent either *appeals to experience* (testimonials) or *appeals to authority*. Both types of testimony enhance the credibility and authenticity of the scientific story being told.

· ·

EXERCISE 7.14   Blaser (1996) does not include direct quotations in his *Scientific American* piece (Chapter 9), most likely because he is one of the researchers involved in this research and thus himself provides the authenticity that quotations would contribute. However, if you were to write on his topic for a general readership, what kinds of quotations would you want to obtain to supplement your article? In a paragraph or two, choose a target publication and audience for your piece, and explain whom you would like to quote, what kind of information you'd hope to obtain from them, and how you would use this material in your article.

· ·

ACTIVITIES AND ASSIGNMENTS

1. In this chapter we have discussed the use of visuals and the use of subtopics in Blaser's *Scientific American* essay (1996). Read this essay, included in Chapter 9. Identify other audience adaptation strategies being used, and describe how they work together.

2. Watch a local weather forecast on television. In a two- or three-page paper, identify and describe the audience strategies being used, and assess their effect on you as a member of the audience.

3. A. <u>Choose a paragraph from an expert text in your field</u> (e.g., from a textbook or journal). Select and analyze a general audience for which this topic might be appropriate. Adapt the paragraph for that audience using the audience adaptation strategies you have studied in this chapter. Be sure to specify the intended audience (by publication, age, education, background, interests, goals, etc.). Compare your adaptation with the original, and explain the similarities and differences.

B. Now select and analyze an appropriate administrative audience, and adapt the same piece for them (for example, suppose you are working in a lab and need to explain a piece of equipment or an experiment to the lab administrators so that they will continue funding your project). Note what changes you had to make in content and strategy from the piece written for a general audience. Compare this piece written for an administrative audience with the original text written for an expert audience; explain any similarities and differences between these texts.

4. Write a scientific essay for a popular magazine or newspaper that traditionally covers science (*Time, National Geographic, Scientific American, Discover,* the *New York Times,* etc.). Choose an essay from that magazine as a model. Analyze the audience for the magazine using the questions in Section 7. 2 as a guide, and examine your model for the audience adaptation strategies we have discussed. Use these strategies where necessary and appropriate in your essay. Create visuals designed for this journal and audience to include with your text. Be sure to include your model with your assignment.

5. Choose a particular public audience and occasion for an oral presentation on a topic in your field. For example, you might address a club, a junior high classroom, a congressional committee, the press. Give an oral presentation appropriate to this audience. The talk may be based on a previous topic you wrote about in this or other classes for an expert audience, or for Activity 4, but be sure to adapt it fully for an oral presentation to your new audience using the principles and strategies in this chapter. Remember to eliminate any terms, concepts, and details that don't fit your general audience's needs and interests; make sure other terms are appropriately explained. Design visuals for this audience. Before you present this talk in class, inform your listeners of the intended purpose, audience, and occasion for the presentation—who and where they are supposed to be as they role-play the target audience. Because they most likely represent a diverse public audience for your topic, they will be able to give you direct feedback about what they understood.

8

Considering Ethics
in Scientific Communication

The scientific research enterprise, like other human activities, is built on a foundation of trust. Scientists trust that the results reported by others are valid. Society trusts that the results of research reflect an honest attempt by scientists to describe the world accurately and without bias. The level of trust that has characterized science and its relationship with society has contributed to an unparalleled scientific productivity. But this trust will endure only if the scientific community devotes itself to exemplifying and transmitting the values associated with ethical scientific conduct.

(NAS 1995, Preface)

8.1 Scientific and social ethics

As members of society, scientists are influenced by the dominant values of the culture in which they live: truth, justice, freedom, progress, happiness. As societies in themselves, scientific disciplines also are governed by conventions, both written and unwritten, that come to constitute standards and principles of professional conduct and practice. That is, the conventions of your field constitute (or imply) a system of ethics. In exploring communication practices in your field, you have been learning about some of the conventions that shape your discipline. As you become a member of that discipline, it also is important that you understand the ethical nature of these conventions, and some of their implications for you as a scientist.

Scientists usually learn scientific ethics informally, by observing how scientists behave. But as the National Academy of Sciences remarks, "science has become so complex and so closely intertwined with society's needs that a more formal introduction to research ethics and the responsibilities that these commitments imply is also needed . . ."

(1995, Preface). While informal observation still remains the basis of learning ethics, across the country attempts are being made by scientists to integrate the teaching of ethics into science curricula (see "Conduct in science," Science 1995). As we conclude Part One of this book, we wish to reflect on the ethical dimension of the communication conventions we have discussed, as well as of the ethical implications of the socialization process you are now going through as a new member of the scientific community.

Since scientific disciplines are participants in the larger society in which they are situated, it only stands to reason that scientists would face some of the same ethical issues in communication faced by society at large. These might be organized under two categories: misconduct in publication (misrepresenting data or authorship) and bias in the gatekeeping institutions that oversee communication (prejudice, underrepresentation).

Occasionally we read in the newspapers about incidents of misconduct in scientific communication. For example, we read that Dr. Roger Poisson falsified and misrepresented data in a seminal report on breast cancer treatment on which major research and health recommendations had been based (Taubes 1995); about allegations that Dr. Robert Gallo "made a misstatement in a science publication" and erroneously claimed sole credit for the discovery of the HIV virus (Associated Press 1992); about charges by Sarvamangala Devi, an assistant researcher who discovered a vaccine that could be used in the treatment of AIDS patients, that she was being denied credit for her discovery by senior scientists at the National Institutes of Health because of her gender and ethnic background (Marshall 1995).

We also read about charges of bias in the gatekeeping institutions of science. We read about how the NIH stopped its funding for a controversial conference examining the relationship between genetics and crime after protest from black leaders (Associated Press 1992); about charges of elitism and bias against "low income students, women, minorities and the disabled" in science education (Cole 1990); and about continuing problems of gender representation in science (NAS 1989, 7)—even in the National Academy of Sciences itself (Angier 1992).

Given charges like these, science would appear to be a mere microcosm of society. According to Tom Hoshiko, professor of physiology and biophysics at Case Western University, "Most working scientists have experienced or know first-hand some cases of misconduct" (1991, 11). Because of alleged and reported cases of misconduct in science there is strong political pressure for congressional oversight of science—a pressure which scientists naturally resist, insisting that they can monitor themselves (Koshland 1989; Hoshiko 1991). Yet, misconduct in science is relatively rare. In 1989 Science editor Daniel Koshland Jr. estimated that while 3 million research articles had been published over the previous 10 years, only about a dozen major cases of fraud had been reported. There are perhaps several reasons for this.

Several government organizations—including the National Science Foundation, the National Institutes of Health's Office of Research Integrity, and the Public Health Service's Office of Integrity, among others—have articulated standards and policies for ethical conduct in science and the communication of science. As we noted in Chapter 6, funding agencies also have strict guidelines on such ethical concerns as the use of hazardous materials and the treatment of animal and human subjects. Funders expect proposed research to be designed and carried out according to the highest scientific and ethical codes.

In addition to this institutional oversight, scientists have had in place the system of checks and balances we discussed in Chapter 1, based on peer review and credit by publication, to guard against bias and fraud in science. While the efficacy and fairness of these internal checks have been questioned (Hoshiko 1991; Roy 1993), the checks have since the seventeenth century been designed to protect the rights of individual scientists, maintain the integrity of research, enhance the reputation of scientific institutions, and preserve the public trust in science (NAS 1995). Many journal publishers have been developing written policies concerning authorship and the allocation of credit (Hoshiko 1991). Guidelines for handling unpublished manuscripts and conflicts of interest are important for both journals and funding agencies. Journal and grant reviewers are expected to respect the confidentiality of the papers and proposals they review.

Finally, the values of scientific research itself—"honesty, skepticism, fairness, collegiality, openness" (NAS 1995, 21), or what Jacob Bronowski (1956) called the "Habit of Truth"—are supposed to maintain the ethical integrity of scientific research and communication. Because scientific communities are small societies, the behavior of scientists is strongly influenced by the opinion of their peers, with shame being a major deterrent of misconduct (Cohen 1995). As Hoshiko notes, "A scientist's most valuable possession is his or her reputation, which, once lost—whether deservedly or not—may be virtually impossible to regain" (1991, 11).

But by all accounts, the pressure is mounting: The very success and "affluence" of science (Hoshiko 1991, 11)—and the enormous pressures to compete for dwindling funds and publication and prizes and promotions and prestigious labs and equipment and research assistants and students and media attention—seem to be changing the ethical environment in which science is conducted (Cohen 1995). Scientists are human too. Scientists and the institutions they work for must continuously deal with ethical questions and conflicts related to research and communication.

EXERCISE 8.1   Obtain a copy of the research policies or guidelines for your department, lab, university, or professional association, or look at the special concerns addressed in a funding agency's RFP or the editorial guidelines for a journal in your field. What kinds of ethical issues are raised in these guidelines? Categorize these issues (e.g., humane treatment of subjects, misrepresenting data, authorship, sexual harassment). What do these categories tell you about the focus of ethical concern in your field?

8.2 The question of authorship

As the examples in Section 8.1 demonstrate, misconduct in scientific communication includes not only falsification (fudging, misrepresenting, or misreporting data and results) but also questions about authorship: plagiarism and the allocation of credit.

Plagiarism—the appropriation of other scientists' ideas, research, or statements without proper attribution—is a serious offense in science, since proper attribution is essential to the trust necessary among scientists for the free exchange of information. A scientist's reputation can be permanently damaged by acts of plagiarism. But as with other ethical questions, it is sometimes difficult to determine what constitutes plagiarism and what to do about it. Scientists must constantly make judgments about when formal acknowledgement of sources is required.

• •

EXERCISE 8.2   Read the following case study, "A Case of Plagiarism" (NAS 1995, 18), about a student who finds she has committed plagiarism; then consider the questions that the National Academy of Sciences asks. In addition to answering these questions, think about why NAS asks these particular questions; i.e., what ethical questions does NAS want you to focus on in the case? Are there some mitigating circumstances here, and if so, what are they?

> May is a second-year graduate student preparing the written portion of her qualifying exam. She incorporates whole sentences and paragraphs verbatim from several published papers. She does not use quotation marks, but the sources are suggested by statements like "(see . . . for more details)." The faculty on the qualifying exam committee note inconsistencies in the writing styles of different paragraphs of the text and check the sources, uncovering May's plagiarism.
>
> After discussion with the faculty, May's plagiarism (sic) is brought to the attention of the dean of the graduate school, whose responsibility it is to review such incidents. The graduate school regulations state that "plagiarism, that is, the failure in a dissertation, essay, or other written exercise to acknowledge ideas, research or language taken from others" is specifically prohibited. The dean expels May from the program with the stipulation that she can reapply for the next academic year.

1. Is plagiarism like this a common practice?
2. Are there circumstances that should have led to May's being forgiven for plagiarizing?
3. Should May be allowed to reapply to the program?

• •

Related to this concern about acknowledging the work of others are scientists' concerns about allocating credit appropriately among members of the research team itself. As illustrated in earlier chapters, scientists pay close attention to the "allocation of credit." In a standard research report, review, or grant proposal, credit is allocated in three places: the list of authors, the acknowledgments, and the citations. Credit for authorship is given to the primary and all secondary authors, including assistants, postdocs, and students (Cohen 1995, 1706)—but only to those who have played a direct or important role in research and/or writing (NAS 1995). Conventions about who is included as authors, and whether the primary author is listed first or last, differ across fields, journals, and research teams.

As we have already explored, the allocation of credit is an important social mechanism within science, for on it depends the whole system of personal incentive, professional collaboration, recognition, reward, and consensus building in science. The allocation of credit is thus an important ethical issue in scientific communication as well. Science is becoming a more collaborative enterprise. We noted in Chapter 3 that the average number of authors for articles in the *New England Journal of Medicine* is more than six, but some papers in the life sciences list more than 100 authors (Regalado 1995). In physics, the number of papers with over 50 authors rose dramatically between 1985 and 1994; over 150 papers had more than 100 authors, and about 30 papers had more than 500 authors (McDonald 1995)! In fact, the problem of the soaring number of authors has become so complex and so vexing that a number of scientists have decried this "author inflation" and called for reforms in the system (see Garfield 1995; McDonald 1995; Schwitters 1996). At the same time, science is becoming more competitive as research budgets shrink, the costs of doing science rise, and the need for recognition expands in order to obtain funding and attract other researchers and students (Cohen 1995).

These two trends—toward increased collaboration and increased competition—may well be in conflict. Sometimes the all-too-human drive for personal recognition and the increased pressure for funding work against the "ethic of collaboration" that is at the heart of science. Some observers of the Pons and Fleischmann case described in Chapter 1 ascribe their "fall from grace" to just such a conflict (e.g., Samios and Crease 1989). Despite the policies and conventions that have developed to guard against such incidents, scientists in all fields must continuously negotiate this tension between "the ethic of collaboration and the drive for glory" (Benditt 1995, 1705).

8.3 Scientific communication as moral responsibility

An ethical issue related to the allocation of credit is the sharing of information. If science is a social enterprise, the sharing of information becomes in effect a moral imperative for scientists. However, the drive for personal recognition that sometimes works against the ethic of collaboration may also work against the ethic of sharing. As Hoshiko notes, "Competition fosters self-protection and inhibits collegiality among investigators" (1991, 11). In addition, conflicts between the need for scientists to cooperate and the need to "protect" one's findings can be complicated by the requirements of corporate secrecy or national security, competition between labs or universities, and even personal animosities. We saw in Chapter 1 that some of these motives also may have come into play in the case of Pons and Fleischmann.

Many scientists believe that generosity in science is in the long run rewarded (Cohen 1995). Nevertheless, the social ethics of science are in a constant state of tension with the individualistic ethics of science. It is obvious that the peer review process works against competition and speed of publication—or, as in the case of Pons and Fleischmann, competition and the rush to publicize undercut peer review (Huizenga

1992). "The need for skeptical review of scientific results is one reason why free and open communication is so important in science," says the National Academy of Sciences (1995, 12). But the need for skeptical review and the need for free and open communication—the social mechanisms designed to maintain professional standards and foster the sharing of information in science—can and do create friction when they come into contact with an individual scientist's drive for priority and credit that is a major incentive in scientific research (NAS 1995).

The social responsibility of scientists to share their work with each other, then, sometimes becomes an ethical problem for which there is not always a clear-cut answer. Scientists are entitled to a "period of privacy," especially in the early stages of research (NAS 1995, 10). But during peer review or after the conference presentation or publication, scientists have a responsibility to share results openly, honestly, and accurately. Pride, competitiveness, the desire for originality, the fear of being scooped, the need for professional advancement (and the research funds that go with it), the pressure for secrecy from corporations or government, and, in a few cases, the profit motive—all potentially work against these responsibilities in science. Ways to better handle some of these ethical conflicts—through improvements in science education, further articulation of policy by funding agencies and professional societies, and more rigorous monitoring—are still being developed (see Hoshiko 1991; NAS 1995).

EXERCISE 8.3

In light of this discussion of competing ethics within science, review the case of Pons and Fleischmann in Figure 1.1. List the competing ethical factors in this case. Write a paragraph defending Pons and Fleischmann's actions, and also write a paragraph criticizing them. Different groups may be assigned parts of this exercise in preparation for the kind of ethical debate scientists, science educators, and policy makers are increasingly engaging in (Benditt 1995). Does this ethical examination of "conflicts of interest" in science give you a better understanding of the motives or increased sympathy for the plight of Pons and Fleischmann? Why or why not?

8.4 Scientific communication and public communication: An ethical conflict?

As we have discussed, a part of the mechanism by which scientists monitor themselves is peer review, which normally is prior to the release of a study to the public. We also discussed in Chapter 7 why scientists have an obligation to communicate with the public, and why the public has both a need and a right to know what is going on in science. Sometimes the need for peer review and the need to share information with the public come into conflict. This can become an ethical dilemma for the scientist as well.

While there are strong incentives to keep research secret until publication, some of the same factors may provide incentive to release results to the public before peer review and publication—both "positive factors" such as the desire to respond to urgent social need and "negative factors" such as competition, pride, and the desire for credit and fame.

In addition to violating the ethic of peer review embedded deep within the enterprise of science, there is also a stigma attached to (re)printing material in scientific journals that has already been "published" in the popular press. In fact, professional journals often will not publish work that has been released to the public first. Some scientists and editors are willing to reconsider the injunction against a press conference prior to scientific publication, or the embargo against publication in a scientific journal of a study that has been released to the public (Roy 1993), if the research is urgent or important enough to warrant its early release (Maddox 1989; CBE 1995). The problem is that it is usually only after scientific peer review and publication that urgency or importance can be determined—too late for the unfortunate scientists who are wrong, or the public that is misinformed and confused. As with most ethical issues, in specific cases the line in science between the obligation to monitor itself and the obligation to inform society is more often than not blurred.

• •

EXERCISE 8.4   The following case study is from the Report from the 1994 Annual Meeting of the Council of Biology Editors (CBE 1995, 14), at which editors of scientific publications were asked to respond to different scenarios concerning the "prepublication release of information." This two-part scenario was addressed to Richard Horton, editor of the *Lancet*. What are the ethical issues involved? What information do you need in order to make a decision? How do you think Horton will respond to each scenario, and why?

> Part 1. Dr. Gilbert believes he has found a new gene for Alzheimer's disease. He submits an abstract of his work to his professional society for presentation at its annual meeting. The abstract is not accepted, but he is invited to prepare a poster, which he does. The poster attracts a great deal of media attention at the meeting, and a story about it appears in the Sunday *New York Times,* complete with the figures and tables, as well as a summary of the study design and results. A similar story appears in *Newsweek.* A month later, the study report is submitted to *The Lancet* for publication. Horton believes the work is novel, interesting, and important. Is he disturbed by the prepublication release of information?

> Part 2. Dr. Sullivan believes she has found a new gene for bipolar disorder, and she presents her work orally at her professional society's annual meeting. The presentation is covered extensively in the popular media, although figures and tables are not available. Will this affect later publication in *The Lancet*? Suppose Sullivan elaborates on the study in a press conference after the presentation. Suppose she passes out copies of the paper she plans to submit to *The Lancet.* Suppose she publishes a slightly longer summary of the study in a book of proceedings of the meeting.

• •

It is not only the system of scientific checks and balances or the injunction against prior publication that prevents scientists from communicating with the public. Sometimes scientists are prevented from communicating with the public by the requirement of political or corporate secrecy. This sets up another ethical dilemma for scientists. The conflict between the need for secrecy in science and the public's right to know is not unlike the (unresolvable?) legal conflict over the need for national security versus the First Amendment right of free speech (for example, former government officials are prevented from publishing what they know about classified government activity—even illegal activity—until it has passed through government review). Sometimes, scientific and other information is kept from the public for legitimate security reasons. But sometimes information is kept from the public for reasons other than legitimate ones. The problem is, how do we, the public, know?

Thus, the requirement of secrecy can raise serious ethical and political questions about the power to control scientific information. On what grounds should information be kept from the public? What kind of information should be kept from the public, and under what circumstances? Perhaps most important, who in a democracy will make these decisions? The experts? The government? The Pentagon? The CIA? Executives of corporations? Average citizens? The press? Finally, how will these decisions be made? How much does the public need to know to make these decisions, and how can security simultaneously be maintained? Who does the scientist ultimately owe allegiance to? These are all vexing questions in our democratic society, and there are no clear answers.

Politicians have been known to "spin" or distort the data of a scientific report, or even to change it, for economic or political purposes. In at least some of these cases, the press has uncovered government misconduct and has reported it. For example, Philip Shabecoff (1989) revealed that the White House's Office of Management and Budget under President Bush changed the text of a scientific report on global warming, which was to be delivered to the Senate subcommittee on Science, Technology, and Space by the director of NASA's Goddard Institute for Space Studies. The wording and conclusion of the report were substantially altered to soften projections and thus justify the administration's tempered approach to international policy on the greenhouse effect. What outraged scientists and congressional members was that the OMB, a political office, not only misrepresented a scientist's conclusions, but actually tampered with his words.

Unfortunately, history also contains some extreme cases where those in power have used science for immoral ends. Katz (1992b, 1993) has examined what can happen when political or economic or corporate expediency is allowed to subvert the interests of science; he argues that when "the ethic of expediency" becomes the only goal of science or technology—becomes the value in science that overrides all other human values, as it did in Nazi Germany (or the former Soviet Union)—scientific communication can become an ethical weapon that can be used by the unscrupulous *against* society. As the National Academy of Sciences points out, the result is invariably a deterioration of science itself (see NAS 1995, 7).

Most of us probably never will have to face such extreme ethical circumstances. But in cases involving governmental or corporate misconduct, the scientist must be prepared to "blow the whistle." This presents another ethical dilemma in scientific communication. Scientists are often hesitant to act because of uncertainty about the appearance of wrongdoing, or from fear of reprisal. But according to the National Academy of Sciences, "someone who has witnessed misconduct [in science] has an unmistakable obligation to act" (1995, 18). However, *how* to act is often a pressing question. When do you "tell," and whom? Scientists may wish to seek advice from friends, colleagues, or supervisors before they act (see NAS 1995). At the time of this writing, the most spectacular example of whistle-blowing has been the allegations in the media by former top scientists of tobacco companies that their employers, contrary to their sworn statements, studied the addictive process of nicotine and deliberately manipulated its levels in cigarettes.

• •

EXERCISE 8.5   Find out whether there are specific policies, safeguards, and procedures for communicating ethical violations in your department, lab, university, or discipline. What kinds of ethical misconduct are covered by these policies? Note any specifics about the procedure, such as whom you are to communicate misconduct to, time periods for reporting, protection from reprisals, processes of adjudicating charges, and results of possible outcomes. In your opinion, are these policies, safeguards, and procedures adequate? Why or why not?

• •

• • • • • • • • • • • • • •

8.5 Scientific style and social responsibility

Throughout this book we have been exploring how scientific knowledge is constructed in and through the social process of communication. Because scientific knowledge is a social construction, scientists have a social and moral responsibility for the scientific knowledge they create. In the traditional view of science as a dispassionate description of truth, scientists did not have a responsibility for their discoveries: what they discovered was simply "there"; the job of the scientist was to dispassionately observe and report the "facts" (Brummett 1976). In the traditional view of science, it is up to others in society to decide how to use science, and how to deal with the consequences of the "facts" discovered by scientists.

However, if science is socially constructed as we argued in Chapter 1, then facts are not simply there but are construed by scientists based on the paradigms and values within which scientists work. In *Science and Human Values* (1956), Jacob Bronowski, chemist and man of letters, argued that scientific "facts" are concepts that organize appearances into laws. Thus for Bronowski, science and ethics are not different because both are "systems of concepts" that are tested in experience (40). "When judgment is recognized as a scientific tool, it is easier to see how science can be influenced by values," says the National Academy of Sciences; "values cannot—and should not—be separated from science. The desire to do good work is a human value. So is the conviction that standards of honesty and objectivity need to be maintained" (1995, 6).

In the contemporary view of science, scientists participate in the construction of "reality" through the process of communication (Brummett 1976). Whether it be an idea or a theory or an observation or a measurement, a method or a discovery or an invention or an experiment, scientists help shape the beliefs and concepts of a culture, and so bear responsibility for those beliefs and concepts. We have explored throughout this book how argumentation and persuasion are integral to the construction of scientific knowledge. Thus, scientists also have a moral imperative to create and use scientific arguments responsibly, by employing theories, methods, logic, and data acceptable in the field as well as to society at large.

In constructing responsible arguments, we need to carefully consider not only *what* we say but *how* we say it, for style has an ethical dimension as well. Take the traditional use of passive voice in science as a case in point. Passive voice has the effect of de-emphasizing the individual scientist or lab in favor of an *ethos* of neutrality and objectivity. Thus, the sentence "We collected the samples" becomes "The samples were collected" in passive construction. The effect of this style is that the agent, the human being conducting the experiment, is no longer apparent. This is not usually a problem. As we discussed in Chapter 3, for example, the AIP Style Manual (1990) recommends the use of passive voice in some methods descriptions where the agent is not important. Passive voice is used conventionally in many forums; thus using it appropriately is part of learning how to speak as a professional scientist. But the use of passive voice can become an ethical problem when it is used to hide responsibility, as sometimes happens when information is released to the public: "In the 1960's, the CIA tested LSD on civilians" can become "In the 1960's, LSD was tested on civilians."

In Chapter 7 we touched on the fact that miscommunication played decisive roles in the Challenger explosion, in the disarray that followed the Three Mile Island meltdown, and in failed attempts to site radioactive waste facilities around the country; in that chapter our focus was on scientists' ability to understand and appeal to audiences outside their discipline. Here our focus is on the role of style in these catastrophes; *how* something is said can become as much of an issue as what is said in scientific and technological accidents. A now-infamous example will serve as a case in point. As a horrified nation watched the space shuttle Challenger explode, NASA's media spokesperson announced: "A major malfunction has occurred." Many found the *ethos* of objectivity, of nonresponsibility, of calm scientific detachment—created by abstraction and jargon—inappropriate if not offensive because it ran counter to the way people actually felt; this style was continued by NASA officials in press conferences throughout the day (Havelock 1986), creating an impression of callousness that was difficult for NASA to overcome. Where public health, safety, and/or trust are at stake, and the accountability of science is put to the test, scientists must recognize and take pains to avoid even the appearance of evasiveness that can be created by style.

The use of scientific style, like the use of science, by others in society also can create an ethical problem for scientists. The scientific jargon sometimes used by government officials to support *political* positions can create an *ethos* of scientific objectivity while obfuscating meaning, as in the following example: "Understanding the role natural site features play as part of an integrated system in protecting public health and safety is an important part of the comprehensive site assessment activities" (North Carolina Low-Level Radioactive Waste Management Authority 1996, 1). The *ethos* of science is also used by advertisers to sell products ("9 out of 10 doctors recommend . . ."). The use of the *ethos* of objectivity created by such devices as passive voice, abstraction, jargon, and "context-free" statistics helps foster a false public image of scientists as omniscient and infallible. The National Academy of Sciences warns scientists not to set science up as a form of knowledge superior to other forms of knowledge (1995, 21). But the *ethos* of science created *by others* can raise some unforeseen social and political consequences for scientists as well.

As we discussed in Chapter 3, scientists understand (and indicate with the use of qualifiers) that scientific knowledge is always open-ended, uncertain, and subject to change. But because the public expects scientists to always be right, and scientific knowledge to be absolute, when scientists qualify their claims, or when scientists disagree, public audiences often become frustrated and confused and revert back to their own beliefs and previous convictions (Nelkin 1975). As the National Academy of Sciences points out:

> Many people harbor misconceptions about the nature and aims of science. They believe it to be a cold, impersonal search for truth devoid of human values. Scientists know these misconceptions are mistaken, but the misconceptions can be damaging. They can influence the way scientists are treated by others, discourage young people from pursuing interests in science, and, at worst, distort the science-based decisions that must be made in a technological society. Scientists must work to counter these feelings. (1989, 20)

• •

EXERCISE 8.6   Nelkin (1979) discusses a case in which the California State Legislature, deliberating about the safety of nuclear power, abandoned their attempt to seek the advice of experts after hearing 120 of them argue about the issue, deciding instead to rely on the opinion of the voters. How can scientists who disagree help avoid this response from the public and its official representatives? How should scientists disagree with each other? How should they explain their differences to the public?

• •

The traditional image of science as objective and absolute can be misleading and can engender a false faith in science. Though scientists hold more temperate views concerning the certainty of the results of their work, they must realize how their work is perceived and interpreted by others in society, and must take responsibility for correcting misperceptions. As the National Academy of Sciences states, "concern and involvement with the broader uses of scientific knowledge are essential if scientists are to retain the public's trust" (1995, 21). A part of the solution entails understanding how the public's image of science is the result of communication processes. In Chapter 7 we touched on how scientific information changes as it moves from the expert to the nonexpert forums because of the need for scientific information to adapt to the knowledge, values, goals, and concerns of public audiences. Fahnestock (1986) also has demonstrated that in this process of accommodation based on appeals to practical application and wonder, the level of certainty asserted rises. Thus, while an official of the Nuclear Regulatory Commission called the accidental release of nuclear radiation from the Three Mile Island Nuclear Power Plant into the atmosphere a "serious contamination problem on site," an official from the Pennsylvania Emergency Management Agency put a more positive—and certain—spin on things: "people have nothing to worry about. The radiation level is what people would get if they played golf in the sunshine" (Farrell and Goodnight 1981, 283). Whether they or journalists are fulfilling the social responsibility to inform the public of scientific developments, scientists must realize the image of science that is created in the process (NAS 1989, 21).

8.6 The ethics of style as socialization

There is another ethical dimension of style that you should be aware of, one that affects you personally as a new member of a scientific community. Because science is a social enterprise, governed both by explicit and tacit conventions, there is enormous pressure for new members of the field to abide by those conventions. As we mentioned in Chapter 2, in learning the conventions of your field, what you are engaged in is a process of socialization.

We all conform to some set of values; and as the National Academy of Sciences (1989, 1995) states, the values of science have produced a unprecedented period of scientific progress. But the need to conform to conventions sometimes comes into conflict with individuality, creativity, originality, and personal style. Conformity facilitates communication; by adhering to conventions, writers and speakers in a community fulfill the expectations of readers and listeners in the community. In Chapter 3 we saw how conformity to the IMRAD form facilitates communication: by adhering to structural conventions, researchers place information where it is expected and thus make information more accessible and the structure of the overall argument easier to follow. Conformity facilitates communication: by adhering to conventions, members of a scientific community make themselves more credible and better understood.

But by the same token, conformity by its very nature constrains individuality and communication. In adhering to these conventions, the engaged scientist does not necessarily say only what the community expects to hear, but does need to *acknowledge* what the community expects to hear as the way to introduce new ideas. Conventions exist for a reason, but we need to maintain an awareness of the rhetorical role of convention in science and the fact that conventions continually evolve. The use of active voice and first person pronouns in some fields—the recognition of the personal voice and *ethos* of the researcher(s) in a formal scientific text—is a profound shift in the social conventions of those fields. So too is the increasing use of email for the conduct of scientific research, which is beginning to alter the role of peer review in science, and perhaps eventually, of journal publication itself (NAS 1989; Brody 1996).

We should also point out the possible consequences of violating the conventions of your field. As we saw in the cases of Barry Marshall and of Pons and Fleischmann, failure to conform to the stylistic expectations of the field can result in difficulty being heard. In the long run, Marshall's case demonstrates that it *is* possible for individuals or nonconformists to be heard and ultimately to influence the community's "language"—but only if the person engages in free and open inquiry with the community, which Pons and Fleischmann apparently did not.

The ethical issue of style as socialization raises another question. In Chapter 7 and in this chapter we have talked about the need for scientists to communicate with the public. However, there is in some quarters in some fields professional pressure not to. Sometimes the efforts of scientists to communicate with the public damage their professional standing. Astronomer Carl Sagan was ridiculed and ostracized by some in his field after his television series and book, *Cosmos*. Some astronomers felt that his demeanor pandered to the public, debased the field of astronomy, and tarnished their image. Other scientists, such as Linus Pauling and B. F. Skinner, also have come in for criticism by their peers for statements made in public (Wilkes 1990).

Thus, again, the individual scientist is faced with another ethical dilemma, another moral choice: How to balance the need and desire to address the public with the need and desire to establish and maintain a successful career as a professional scientist? Should scientists have to sacrifice somewhat their professional *ethos,* as Carl Sagan did, to generate interest and enthusiasm for science in the public? Is there a way of both maintaining your professional *ethos* and communicating effectively with the public?

As we noted in Chapter 7, there are many examples of scientists from a variety of fields who have been able to write for a popular audience without damaging their professional *ethos.* Wilkes (1990) suggests that as long as scientists talk about their own field instead of presenting themselves as general authorities, they'll be on safe ground. However, the case of Carl Sagan, when contrasted with that of Jacob Bronowski, suggests that the answer may be more a matter of style than of subject matter. In his book and television program, Carl Sagan *was* talking about his own field! Jacob Bronowski, on the other hand, was a chemist who had a successful book and television series on Western civilization, *The Ascent of Man,* and remained well-regarded by his professional colleagues. The difference in the reception of their "popularizations" by their respective fields may be related in part to the fact that Sagan was criticized more for his "gee whiz" style—for his Mr. Wizard *ethos*—than for the scientific content of his presentations. Jacob Bronowski, on the other hand, was praised for his dignified style of presentation, although he spoke on subjects outside the field of chemistry.

While it's important to adapt your style to appeal to audience interests, moderation may be the key. It may be a matter of degree rather than of kind. But this is a complex issue. Both Albert Einstein and J. Robert Oppenheimer became outspoken advocates of world peace; but while Einstein was praised, Oppenheimer was destroyed by the politics of the cold war. Thus, context—the time and place and politics in which scientists live—as well as style undoubtedly plays a role in the professional reception of public pronouncements by peers.

The taboo against communicating with the public is lessening, thanks in no small measure to the National Academy of Sciences, which has pointed out the importance of the public to science itself (1989, 1995). Other organizations, such as the National Research Council (administered by NAS) and the National Academy of Engineering, help foster a climate of communication and cooperation across social and professional boundaries by gathering scientists and leaders from all parts of society to address scientific issues and give advice to government and the public.

moderation

Science is a social institution. As such, new scientists will want to learn and live by the conventions that scientists have established. Those conventions also constitute a system of ethics in your field, and behaving according to those ethics constitutes a process of socialization and increasing conformity. Socialization and conformity do not necessarily preclude individuality, as the Marshall case illustrates; but the process of joining a community has both moral obligations and moral implications. The scientist who recognizes the ethical dimension of these conventions will be in a better position not only to understand what is expected of her or him as a professional but also how to act in conflicting or difficult situations. Equally important, scientists who are aware of the conventions governing their fields and the assumptions underlying them are better able to modify those conventions and work for change in their community's professional practices—to modify conventions that have become outmoded and no longer reflect or serve the interests and values of scientists, science, or society.

ACTIVITIES AND ASSIGNMENTS

1. If you have access to the World Wide Web, log on to *Science*'s interactive web page, "*Science* On-Line" (http://www.aaas.org/science.html). In a section entitled "Beyond the Printed Page," *Science* presents a set of case studies illustrating particular ethical issues in science. Other sources of case studies include *On Being a Scientist* (1995; available from NAS in hard copy or via the National Academy Press web site: http://www.nap.edu/bookstore/) and *Science*'s special journal sections on "Conduct in Science" (the first was in Vol 268, June 23, 1995, 1705–18). Find a case study pertaining to an issue in scientific communication that you are interested in or that is related to your field. What ethical issues does the case raise? List the ethical conflicts involved. Write a paragraph or two on each side of the issue, or debate them in small groups or with your class. Make a list of the criteria being used to discuss the case. What scientific conventions are guiding the discussion? What do you learn from the case about being a professional scientist?

2. Choose one of the cases described in this book or a case you know from your own field. Read about it, and plan a presentation for your class. What issues were involved? How were they handled? Were they solved by the institution, the discipline, or the government? What was the outcome? Was it satisfactory?

3. Watch a commercial on television that seems to involve the testimony of scientists or health practitioners. Is an *ethos* of objectivity being created by advertisers for the purposes of selling products? If so, how? Do you feel it is appealing? Accurate? Ethical? On what do you base your answer? Can commercials be accurate and still appeal to their viewers' needs, interests, and values? How would you change this commercial to balance these conflicting needs?

9

Research on the "Ulcer Bug"

MARSHALL AND WARREN
AND COLLEAGUES

In 1983, a young Australian internist, Dr. Barry Marshall, reported that he and Dr. Robin Warren had identified a bacterium responsible for stomach ulcers—a revolutionary claim, given the prevailing assumption in the medical profession that ulcers were caused by stress and other psychological factors. Marshall and Warren's theory that ulcers were the result of bacterial infection in the stomach lining was greeted with both intense interest and skepticism by experts in the field. The theory was much debated in a number of scientific forums, including the letters sections of the *Lancet* and *Annals of Internal Medicine.* Seven of these research letters, including the original joint announcements by Warren and Marshall, are included in this chapter. These selections are followed by a full-length research report by Marshall and Warren (1984), which appeared in the *Lancet,* and a later report in *Annals of Internal Medicine* by another team, Graham et al. (1992), who studied the effects of Marshall and Warren's recommended treatment. The final entries in this chapter are a review of research on the topic written by Martin Blaser (1987) for *Gastroenterology,* and an essay Blaser (1996) wrote for a popular audience, published in *Scientific American.* The varying terminology in this set of texts reflects the fact that, although originally identified as a member of the *Campylobacter* genus, the organism was reclassified in 1989 as a new genus, *Helicobacter.*

UNIDENTIFIED CURVED BACILLI ON GASTRIC EPITHELIUM IN ACTIVE CHRONIC GASTRITIS

SIR,—Gastric microbiology has been sadly neglected. Half the patients coming to gastroscopy and biopsy show bacterial colonisation of their stomachs, a colonisation remarkable for the constancy of both the bacteria involved and the associated histological changes. During the past three years I have observed small curved and S-shaped bacilli in 135 gastric biopsy specimens. The bacteria were closely associated with the surface epithelium, both within and between the gastric pits. Distribution was continuous, patchy, or focal. They were difficult to see with haematoxylin and eosin stain, but stained well by the Warthin-Starry silver method (figure).

I have classified gastric biopsy findings according to the type of inflammation, regardless of other features, as "no inflammation", "chronic gastritis" (CG), or "active chronic gastritis" (ACG). CG shows more small round cells than normal while ACG is characterised by an increase in polymorphonuclear neutrophil leucocytes, besides the features of CG. It was unusual to find no inflammation. CG usually showed superficial oedema of the mucosa. The leucocytes in ACG were usually focal and superficial, in and near the surface epithelium. In many cases they only infiltrated the necks of occasional gastric glands. The superficial epithelium was often irregular, with reduced mucinogenesis and a cobblestone surface.

When there was no inflammation bacteria were rare. Bacteria were often found in CG, but were rarely numerous. The curved bacilli were almost always present in ACG, often in large numbers and often growing between the cells of the surface epithelium (figure). The constant morphology of these bacteria and their intimate relationship with the mucosal architecture contrasted with the heterogeneous bacteria often seen in the surface debris. There was normally a layer of mucous secretion on the surface of the mucosa. When this layer was intact, the debris was spread over it, while the curved bacilli were on the epithelium beneath, closely spread over the surface (figure).

The curved bacilli and the associated histological changes may be present in any part of the stomach, but they were seen most consistently in the gastric antrum. Inflammation, with no bacteria, occurred in mucosa near focal lesions such as carcinoma or peptic ulcer. In such cases, the leucocytes were spread through the full thickness of the nearby mucosa, in contrast to the superficial infiltration associated with the bacteria. Both the bacteria and the typical histological changes were commonly found in mucosa unaffected by the focal lesion.

The extraordinary features of these bacteria are that they are almost unknown to clinicians and pathologists alike, that they are closely associated with granulocyte infiltration, and that they are present in about half of our routine gastric biopsy specimens in numbers large enough to see on routine histology. The only other organism I have found actively growing in the stomach is *Candida*, sometimes seen in the floor of peptic ulcers. These bacteria were not mentioned in two major studies of gastrointestinal microbiology[1,2] possibly because of their unusual atmospheric requirements and slow growth in culture (described by Dr B. Marshall in the accompanying letter). They were mentioned in passing by Fung et al.[3]

How the bacteria survive is uncertain. There is a pH gradient from acid in the gastric lumen to near neutral in the mucosal vessels. The bacteria grow in close contact with the epithelium, presumably near the neutral end of this gradient, and are protected by the overlying mucus.

The identification and clinical significance of this bacterium remain uncertain. By light microscopy it resembles *Campylobacter jejuni* but cannot be classified by reference to *Bergey's Manual of*

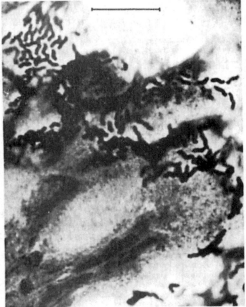

Curved bacilli on gastric epithelium.

Section is cut at acute angle to show bacteria on surface, forming network between epithelial cells. (Warthin-Starry silver stain; bar = 10 µm.)

Determinative Bacteriology. The stomach must not be viewed as a sterile organ with no permanent flora. Bacteria in numbers sufficient to see by light microscopy are closely associated with an active form of gastritis, a cause of considerable morbidity (dyspeptic disease). These organisms should be recognised and their significance investigated.

Department of Pathology,
Royal Perth Hospital,
Perth, Western Australia 6001   J. ROBIN WARREN

SIR,—The above description of S-shaped spiral bacteria in the gastric antrum, by my colleague Dr J. R. Warren, raises the following questions: why have they not been seen before; are they pathogens or merely commensals in a damaged mucosa; and are they campylobacters?

In 1938 Doenges[1] found "spirochaetes" in 43% of 242 stomachs at necropsy but drew no conclusions because autolysis had rendered most of the specimens unsuitable for pathological diagnosis. Freedburg and Barron[2] studied 35 partial gastrectomy specimens and found "spirochaetes" in 37%, after a long search. They concluded that the bacteria colonised the tissue near benign or malignant ulcers as non-pathogenic opportunists. When Palmer[3] examined 1140 gastric suction biopsy specimens he did not use silver stains, so, not surprisingly, he found "no structure which could reasonably be considered to be of a spirochaetal nature". He concluded that the gastric "spirochaetes" were oral contaminants which multiplied only in post mortem specimens or close to ulcers. Since that time, the spiral bacteria have rarely been mentioned, except as curiosities,[4] and the subject was not reopened with the

1. Gray JDA, Shiner M. Influence of gastric pH on gastric and jejunal flora. *Gut* 1967; **8**: 574–81.
2. Drasar BS, Shiner M, McLeod GM. Studies on the intestinal flora I: The bacterial flora of the gastrointestinal tract in healthy and achlorhydric persons. *Gastroenterology* 1969; **56**: 71–79.
3. Fung WP, Papadimitriou JM, Matz LR. Endoscopic, histological and ultrastructural correlations in chronic gastritis. *Am J Gastroenterol* 1979; **71**: 269–79.

1. Doenges JL. Spirochaetes in the gastric glands of *Macacus rhesus* and humans without definite history of related disease. *Proc Soc Exp Med Biol* 1938; **38**: 536–38.
2. Freedburg AS, Barron LE. The presence of spirochaetes in human gastric mucosa. *Am J Dig Dis* 1940; **7**: 443–45.
3. Palmer ED. Investigation of the gastric spirochaetes of the human? *Gastroenterology* 1954; **27**: 218–20.
4. Ito S. Anatomic structure of the gastric mucosa. In: Heidel US, Cody CF, eds. Handbook of physiology, section 6: Alimentary canal, vol II: Secretion. Washington, DC: American Physiological Society, 1967: 705–41.

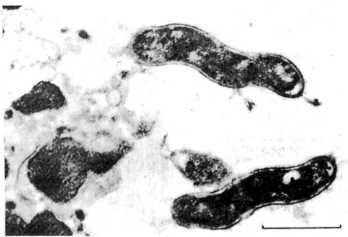

Fig 1—Thin-section micrograph showing spiral bacteria on surface of a mucous cell in gastric biopsy specimen. (Bar = 1 μm.)

advent of gastroscopic biopsy. Silver staining is not routine for mucosal biopsy specimens, and the bacteria have been overlooked.

In other mammals spiral gastric bacteria are well known and are thought to be commensals[5] (eg, Doenges[1] found them in all of forty-three monkeys). They usually have more than two spirals and inhabit the acid-secreting gastric fundus.[5] In cats they even occupy the canaliculi of the oxyntic cells, suggesting tolerance to acid.[6] The animal bacteria do not cause any inflammatory response, and no illness has ever been associated with them.

Investigation of gastric bacteria in man has been hampered by the false assumption that the bacteria were the same as those in animals and would therefore be acid-tolerant inhabitants of the fundus. Warren's bacteria are, however, shorter, with only one or two spirals and resemble campylobacters rather than spirochaetes. They live beneath the mucus of the gastric antrum well away from the

acid-secreting cells.

We have cultured the bacteria from antral biopsy specimens, using *Campylobacter* isolation techniques. They are microaerophilic and grow on moist chocolate agar at 37°C, showing up in 3–4 days as a faint transparent layer. They are about 0·5 μm in diameter and 2·5 μm in length, appearing as short spirals with one or two wavelengths (fig 1). The bacteria have smooth coats with up to five sheathed flagellae arising from one end (fig 2). In some cells, including dividing forms, flagellae may be seen at both ends and in negative stain preparations they have bulbous tips, presumably an artefact.[7]

These bacteria do not fit any known species either morphologically or biochemically. Similar sheathed flagellae have been described in vibrios[7] but micro-aerophilic vibrios have now

5. Lockard VG, Boler RK. Ultrastructure of a spiraled micro-organism in the gastric mucosa of dogs. *Am J Vet Res* 1970; **31:** 1453–62.
6. Vial JD, Orrego H. Electron microscope observations on the fine structure of parietal cells. *J Biophys Biochem Cytol* 1960; **7:** 367–72.

7. Glauert AM, Kerridge D, Horne RW. The fine structure and mode of attachment of the sheathed flagellum of *Vibrio metchnikovii*. *J Cell Biol* 1963; **18:** 327–36.
8. Shewan JM, Veron M. Genus I vibrio. In: Buchanan RE, Gibbons NE, eds. Bergey's manual of determinative microbiology, 8th ed. Baltimore: Williams & Wilkins, 1974: 341.

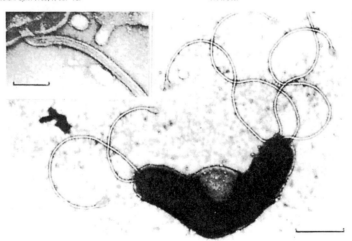

Fig 2—Negative stain micrograph of dividing bacterium from broth culture.

Multiple polar flagellae have terminal bulbs, (2% phosphotungstate, pH 6·8; bar = 1 μm.) Inset: detail showing sheathed flagellum and basal disc associated with plasma membrane. (3% ammonium molybdate, pH 6·5; bar = 100 nm.)

been transferred to the family Spirillaceae genus *Campylobacter*.[8] Campylobacters however, have "a single polar flagellum at one or both ends of the cell" and the campylobacter flagellum is unsheathed.[9] Warren's bacteria may be of the genus *Spirillum*.

The pathogenicity of these bacteria remains unproven but their association with polymorphonuclear infiltration in the human antrum is highly suspicious. If these bacteria are truly associated with antral gastritis, as described by Warren, they may have a part to play in other poorly understood, gastritis associated diseases (ie, peptic ulcer and gastric cancer).

I thank Miss Helen Royce for microbiological assistance, Dr J. A. Armstrong for electronmicroscopy, and Dr Warren for permission to use fig 1.

Department of Gastroenterology,
Royal Perth Hospital,
Perth, Western Australia 6001   BARRY MARSHALL

Campylobacter pylori Infection

To the Editor: We have difficulty with the conclusions reached by Perez-Perez and colleagues (1) on the use of *Campylobacter pylori* antibodies to determine the prevalence of *C. pylori* infection in healthy persons. The results of their IgG and IgA enzyme-linked immunosorbent assay (ELISA) tests have been used as another way to diagnose infection with *C. pylori*, but the current accepted "gold standard" for detecting this organism is the histologic examination or culture of endoscopically obtained gastric biopsy specimens. The ^{13}C-urea and ^{14}C-urea breath tests are other tests that can be used as a gold standard for detecting *C. pylori*; the advantage of these tests is that they probably do not depend on the sampling error that endoscopic biopsies may have (2-4).

Methodologic standards for determining the accuracy of diagnostic tests are well established (5, 6). The criterion that the results of the new test be compared with those of the old gold standard test was not fulfilled in all patients. Of the healthy controls, 166 never had endoscopy; in 29 patients who did have endoscopy, the antrum appeared macroscopically normal, and biopsy specimens were not obtained. *Campylobacter pylori* infection can be present in mucosa that appears endoscopically normal.

In Figure 2, the authors show the prevalence of *C. pylori* –specific serum antibodies based on the results of the ELISA tests. However, the data in this figure for the 166 healthy persons and the 29 persons who did not have endoscopy are extrapolations from the results in the patients shown to be positive for *C. pylori* who did have biopsy samples obtained. Whether most of these persons were indeed shown to be negative for *C. pylori* was never proved by any other test.

Although the ELISA tests the authors used may be an excellent noninvasive method for detecting *C. pylori* infection, we believe that the conclusions drawn by Perez-Perez and colleagues cannot be derived from their data.

S. J. O. Veldhuyzen van Zanten, MD, MPH
J. Goldie
R.H. Riddell, MD
Richard H. Hunt, MD
McMaster University Health Sciences Centre
Hamilton, Ontario L8N 3Z5

References
1. Perez-Perez GI, Dworkin BM, Chodos JE, Blaser MJ. *Campylobacter pylori* antibodies in humans. *Ann Intern Med.* 1988;109:11-7.
2. Graham DY, Klein PD, Evans DJ, et al. *Campylobacter pylori* detected noninvasively by the ^{14}C-urea breath test. *Lancet.* 1985;1:1174-7.
3. Bell GD, Weil J, Harrison G, et al. ^{14}C-urea breath analysis: a noninvasive test for *Campylobacter pylori* in the stomach. *Lancet.* 1987;1:1367-8.
4. Rauws EAJ, Tytgat GNJ, Langenberg W, van Royen E. Experience with ^{14}C-urea breath test in detecting *Campylobacter pylori*. In: Menge H, Gregor M, Tytgat GNJ, Marshall BJ. *Campylobacter pylori: Proceedings of the First International Symposium on Campylobacter pylori.* Berlin: Springer-Verlag; 1988:151-3.
5. Department of Clinical Epidemiology and Biostatistics, McMaster University Health Sciences Centre. How to read clinical journals: II. To learn about a diagnostic test. *Can Med Assoc J.* 1981;124:703-10.
6. Griner PF, Mayewski R, Mushlin AI, et al. Selection and interpretation of diagnostic tests and procedures: principles and applications. *Ann Intern Med.* 1981;94:557-92.

DUODENAL ULCER RELAPSE AFTER ERADICATION OF CAMPYLOBACTER PYLORI

SIR,—Dr Marshall and colleagues (Dec 24/31, p 1438) conclude that eradication of *Campylobacter pylori* led to a higher healing rate of duodenal ulcer and a longer remission. Both conclusions are unacceptable.

The healing rate of duodenal ulcer after 8 weeks of treatment with cimetidine 400 mg twice a day alone was 13/22 (59%) and, after colloidal bismuth subcitrate one tablet four times a day, 13/20 (65%). Such low 8-week healing rates have not been reported before—the expected figures should be about 90%. Marshall and colleagues did endoscopy at week 10 instead of week 8, and argue that the low healing rates were because of relapse during the 2 week interval. Their argument is not substantiated and is difficult to understand. It suggests that the 2-week relapse occurs equally quickly for bismuth as for cimetidine, and yet the 12-month relapse rates are different, as they have shown. The series of patients they studied appear unique.

The treatment with colloidal bismuth or cimetidine was not blinded to the patients or investigators. Hence, the title of the paper is misleading. Ulcer relapse was defined not only as the demonstration of an ulcer at scheduled endoscopy at weeks 14, 26, and 54 after treatment, but also as any recurrence of ulcer symptoms. Because bismuth but not cimetidine is associated with clearance of campylobacter, the design of this study does not safeguard against investigator or patient bias. The use of tinidazole and its placebo did not help because this drug would not of course differentiate cimetidine and bismuth, and because tinidazole was ineffective in the cimetidine group but effective in the bismuth group.

Eradication is a biased term. The distribution of campylobacter is patchy[1] and yet the two biopsy specimens taken for this purpose were regarded as adequate to justify the term "eradication".

Department of Medicine,
University of Hong Kong,
Queen Mary Hospital,
Hong Kong   S. K. LAM

1. Goodwin CS, Blincow E, Warren JR, et al. Evaluation of cultural techniques for isolating *Campylobacter pyloridis* from endoscopic biopsies of gastric mucosa. *J Clin Pathol* 1985; 138: 1127–31.

DUODENAL ULCER RELAPSE AFTER ERADICATION OF CAMPYLOBACTER PYLORI

SIR,—Professor Lam (Feb 18 p 384) criticises our study. We wish to reply.

Eradication of *Campylobacter pylori* in our patients was associated with a much longer remission of healed duodenal ulcers and the apparent rate of healing was also higher. The apparent ulcer-healing rate at follow-up endoscopy was low in our patient group in which *C pylori* was not eradicated (CP + ve patients), irrespective of the therapy. Our Table II shows healing in 44/72 (61%) CP + ve patients. Since 27/44 (61%) of these healed ulcers relapsed before 3 months, we cannot see why it is difficult to understand our suggestion that others probably relapsed before the first follow-up 2 weeks after therapy ended. Even if the argument is not substantiated, it is not unreasonable. The failure rates at 12 months were the same for both cimetidine (46/50, 92%) and colloidal bismuth subcitrate (CBS) (19/22, 86%). We agree that our patients are unique, as Lam says, since all patients are unique.

Every effort was taken to keep the trial blind. Follow-up was timed 2 weeks after therapy ceased (so that mouth-staining would not be seen by the endoscopist) and communication was avoided between the clinician and the patient before gastroscopy. More importantly the pathologist and microbiologist were independent; they had no knowledge of the therapy and neither the clinician nor the patient knew of the presence or absence of bacteria. Thus we do not agree that the title is misleading.

Lam mentions that our definition of ulcer relapse includes "any recurrence of ulcer symptoms". This is almost correct—our patients were treated as in normal practice. If they complained of symptoms requiring therapy, they were considered to have relapsed and usually (32/43, 74%) an ulcer was found on gastroscopy. We thought it unreasonable to call such patients "successfully treated" if they had a recurrence of symptoms but did not have an ulcer crater at endoscopy.

We agree that "bismuth not cimetidine is associated with clearance of campylobacter", but if Lam really believes we did not "safeguard against investigator or patient bias", we suggest he considers the following. There is no way to demonstrate CBS therapy on histology or microbiology, unless Lam thinks that the absence of gastritis and campylobacter, only seen with CBS therapy, causes observer bias. All bias can thus be eliminated by considering only the group of patients treated with CBS. This shows 22 CP + ve patients, 10 (45%) healed at follow-up and 3 (14%) still healed at 12 months; compare this with 24 CP − ve patients, 22 (92%) healed and 17 (71%) still healed at 12 months.

Tinidazole was not intended to "differentiate cimetidine and bismuth", nor for that matter was our study. If Lam is interested in tinidazole, he should examine our results for CBS with or without tinidazole, which show an improvement with tinidazole. We were investigating the effect of the eradication of *C pylori*, not the effect of cimetidine, CBS, or tinidazole per se—our results should be viewed from this aspect.

The distribution of campylobacter is patchy, as stated by Lam, but only at a microscopic level and, with improved methods, we rarely have contradictory results. In this series no CP + ve patient gave negative histological findings and culture without treatment and only 1 patient gave two successive negative biopsy specimens followed by a *C pylori* positive culture (probably a true re-infection). The series includes about 450 biopsies on 100 patients. Under these circumstances we consider "eradication" a correct and justified term.

Our work was designed to show the effect of the eradication of *C pylori* on duodenal ulcer relapse. It was never intended as a drug trial. Cimetidine is an ulcer-healing agent but will not eradicate *C pylori*. Our results suggest that eradication of *C pylori* is associated with a dramatic reduction in the relapse of healed ulcers.

Internal Medicine,
Health Sciences Center,
University of Virginia,
Charlottesville, Virginia 22908, USA

Royal Perth Hospital,
Perth, Western Australia

BARRY J. MARSHALL

J. ROBIN WARREN
C. STEWART GOODWIN

SIR,—We agree that serological tests with sonicates or crude extracts of *Campylobacter pylori* suffer from cross-reaction with other microorganisms, notably *C jejuni*. Our assay, however, did not show such cross-reaction when tested by indirect methods. We see no reason why direct tests should have been done.

Dr Graham and his colleagues do not seem to understand how our cut-off values were established. We compared two well-defined groups of patients with non-ulcer dyspepsia with and without campylobacter-associated gastritis, and found clear cut-offs indicating 100% positive and negative predictive values, respectively, leaving circumvention of the problem of cross-reactivity totally aside.

We are not aware of having stumbled into the trap of assuming that optical density is proportional to antibody titre. Extinction values do reflect the amount of antibody in serum. Measurements on test sera were divided by control values to calculate the P/N ratio on a linear scale. It might have been better if we had mentioned this P/N ratio explicitly in the legend of fig 1 in our paper.

Our discussion was indeed based on the premise that there is a safe and effective therapy for *C pylori* and that patients with *C pylori* infection should be treated. While we agree with Graham's criticism on the second part of this premise, we maintain that *C pylori* infection can be safely and effectively treated with colloidal bismuth subcitrate.

Despite his criticisms, Graham's line of thinking—as indicated by his quotations from the 1989 *Gastroenterology* paper with which his letter ends—almost entirely coincides with the views expressed in our paper. We tried to estimate, on the basis of morbidity figures in our hospital, the decrease in endoscopies that might result from adoption of the proposed strategy and we expressed doubts about the need to treat *C pylori* infection in the last sentence of our paper. So what conceptual errors did we make—or have Dr Graham's views been subject to dramatic change recently?

R. J. L. F. LOFFELD
E. STOBBERINGH
J. W. ARENDS

University Hospital Maastricht,
6210 BX Maastricht, Netherlands

Anti-*Helicobacter pylori* therapy: clearance, elimination, or eradication?

SIR,—The proceedings of the third workshop of the European *Helicobacter pylori* study group meeting held in Toledo, Spain, have been published as abstracts of the 16 oral communications and the 298 posters.[1] Not surprisingly no less than 61 abstracts relate to reporting of results of treatment of *H pylori* positive patients with various forms of antihelicobacter therapy. This includes various bismuth compounds, antibiotics, histamine antagonists, and omeprazole used either alone or in combination for periods ranging from a few days to several weeks.

Comparison of the results from various groups of workers is made even more difficult, or impossible, by the fact that the groups have either assessed their "success" rates while the patient is still on antihelicobacter treatment or is anything from 1 day to many months post therapy.

There is now general agreement[2,3] that "eradication" should only be applied if tests for *H pylori* are negative when repeated at least four weeks after treatment has been discontinued. In our experience with the ^{14}C-urea breath test,[4,5] and that of others with the ^{13}C-urea breath test,[6] within as little as 24–48 h of completing a course of, for instance, colloidal bismuth subcitrate, the organism that had been temporarily suppressed to undetectable levels rapidly multiplies to levels that again allow its detection. These rapid returns to positivity have been shown by restriction endonuclease DNA analysis to result from recrudescence and not re-infection.[7]

Several workers who report their results for success while on active treatment (or within 24 h of discontinuing it) use "clearance" or "elimination" of *H pylori* to describe their findings. I suggest both terms are inaccurate and misleading since what is in truth being measured is temporary suppression of the bacteria. Someone reading an article in which *H pylori* is described as cleared or eliminated could be forgiven for thinking both terms are synonomous with eradicated, which they certainly are not. If figures for clearance are to be reported then the figures for eradication should also be provided. How have the authors of the 61 abstracts from Toledo described their results?

11 abstracts provide only clearance data whereas 36 correctly report eradication figures. 4 abstracts show both clearance and eradication results, but in 1 these are described in terms of elimination of *H pylori* while on treatment and again elimination of the organism when re-assessed one month later. In 4 abstracts authors talk about eradication when they mean clearance and in 3 clearance is used when what they are describing is eradication. Finally 3 abstracts provide success rates at between 3 and 4 weeks after treatment and are probably describing eradication, but cannot be classified.

The importance of having a firm definition of clearance as well as of eradication is well illustrated in two papers from the same group[6,8], in which serial ^{13}C-urea breath tests were used to look at the effect of colloidal bismuth subcitrate on *H pylori* status. In the first[8] the clearance rate was shown as only 5/28 (18%) but the breath test was delayed until 3 days after treatment was stopped. In contrast, in a later publication this group reported their results[6] following 1, 2, or 4 weeks of bismuth treatment in which clearance of *H pylori* was described as being 10/15 (67%), 8/12 (75%), and 17/20 (85%), respectively, but here the post-treatment test was done within 24 h of stopping antihelicobacter treatment. Within a week of stopping treatment in all but 1 patient the urea breath test was again positive. If clearance had been assessed at this stage a figure of 1/47 (2·1%) would have been recorded.[6]

I therefore propose that we all concentrate on eradication results rather than clearance figures, which are of little clinical relevance. Results obtained more than 24 h and less than 4 weeks after anti-helicobacter treatment is stopped are by definition neither clearance nor eradication and should not be published.

Ipswich Hospital,
Ipswich IP4 5PD, UK

G. D. BELL

1. Gastroduodenal pathology and *Helicobacter pylori*. Third workshop of European *Helicobacter pylori* study group. *Rev Esp Enf Digest* 1990; 78 (suppl 1): 3–140.
2. Weil J, Bell GD, Jones PH, Gant P, Trowell JE, Harrison G. "Eradication" of *Campylobacter pylori*: are we being misled? *Lancet* 1988; ii: 1245.
3. Tytgat GNJ, Axon ATR, Dixon MF, Graham DY, Lee A, Marshall BJ. *Helicobacter pylori*: causal agent in peptic ulcer disease? World Congress of Gastroenterology Working Party report, 1990; 36–45.
4. Bell GD, Weil J, Harrison G, et al. ^{14}C-urea breath test analysis: a non-invasive test for *Campylobacter pylori* in the stomach. *Lancet* 1987; i: 1367–68.
5. Weil J, Bell GD, Harrison G, Trowell JE, Gant P, Jones PH. *Campylobacter pylori* survives high dose bismuth subcitrate (De-Nol) therapy. *Gut* 1988; 29: A1437.
6. Logan RPH, Gummett PA, Polson RJ, Baron JH, Misiewicz JJ. What length of treatment with tripotassium dicitrato-bismuthate (TDB) for *Helicobacter pylori*? *Gut* 1990; 31: A1178.
7. Langenberg W, Rauws EA, Widjojokusumo A, et al. Identification of *Campylobacter*

UNIDENTIFIED CURVED BACILLI IN THE STOMACH OF PATIENTS WITH GASTRITIS AND PEPTIC ULCERATION*

Barry J. Marshall   J. Robin Warren

Departments of Gastroenterology and Pathology, Royal Perth Hospital, Perth, Western Australia

Summary   Biopsy specimens were taken from intact areas of antral mucosa in 100 consecutive consenting patients presenting for gastroscopy. Spiral or curved bacilli were demonstrated in specimens from 58 patients. Bacilli cultured from 11 of these biopsies were gram-negative, flagellate, and microaerophilic and appeared to be a new species related to the genus *Campylobacter*. The bacteria were present in almost all patients with active chronic gastritis, duodenal ulcer, or gastric ulcer and thus may be an important factor in the aetiology of these diseases.

Introduction

GASTRIC spiral bacteria have been repeatedly observed, reported, and then forgotten for at least 45 years.[1-3] In 1940 Freedburg and Barron stated that "spirochaetes" could be found in up to 37% of gastrectomy specimens,[4] but examination of gastric suction biopsy material failed to confirm these findings.[5] The advent of fibreoptic biopsy techniques permitted biopsy of the antrum, and in 1975 Steer and Colin-Jones observed gram-negative bacilli in 80% of patients with gastric ulcer.[6] The curved bacilli they illustrated were said to be *Pseudomonas*, possibly a contaminant, and the bacteria were once more forgotten. The repeated demonstration of these bacteria in inflamed gastric antral mucosa[7] prompted us to do a pilot study in twenty patients. Typical curved bacilli were present in over half the biopsy specimens and the number of bacteria was closely related to the severity of the gastritis. The present study was designed to confirm the association between antral gastritis and the bacteria, to discover associated gastrointestinal diseases, to culture and identify the bacteria, and to find factors predisposing to infection.

---

*Based on paper read at Second International Workshop on Campylobacter Infections (Brussels, 1983).

Patients and Methods

Patients

All patients referred for gastroscopy on clinical grounds were eligible for the study which continued until there were 100 participants who gave informed consent and in whom biopsy was considered to be safe. The study was approved by our hospital's human rights committee.

Questionnaire

Where possible patients completed a clinical questionnaire designed to detect a source of infection or show any relationship with "known" causes of gastritis or *Campylobacter* infection, rather than give a detailed account of each patient's history. The emphasis was on animal contact, travel, diet, dental hygiene, and drugs, rather than symptoms.

Endoscopy

The gastroscopies were done by colleagues at the Royal Perth Hospital. Participants fasted for at least 4 h before endoscopy. An Olympus GIF-K fibreoptic gastroduodenoscope was used. Routine biopsies were done when indicated. For the study two extra specimens were taken from an area of intact antral mucosa, at a distance from any focal lesion such as an antral ulcer. When the mucosa appeared inflamed the specimens were taken from a red area, otherwise any part of the antrum was used. One biopsy was immediately fixed in phosphate-buffered formalin for histological examination, the other was placed in chilled anaerobic transport medium and taken to the microbiology laboratory within 1 h. In a few cases an extra specimen was taken for ultrastructural examination.

The gastroenterologist dictated his report soon after the endoscopy. We had not planned to analyse these reports so a standard terminology was not used and no special attention was paid to minor endoscopic lesions. Findings of doubtful clinical significance, such as mild endoscopic gastritis or duodenogastric bile reflux, may thus have been under-reported. (Hereafter the term "gastritis" refers to a histological grade of chronic gastritis unless stated otherwise.) Before we analysed the data, the endoscopy reports were coded for the major diagnoses.

Histopathology

Sections were stained with haematoxylin and eosin (H & E) and graded for gastritis (by J. R. W.) as 0 (normal), inflammatory cells rarely seen; 1 (normal), lymphoid cells present but within normal limits and with no other evidence of inflammation (see below); 2 (chronic), chronic gastritis; or 3 (active), active chronic gastritis.

Gradings were based solely on the type of inflammatory cells. Other types of mucosal change, such as gland atrophy or intestinal metaplasia, were noted separately, but were not used as evidence of inflammation. "Chronic gastritis" indicated inflammation with no increase in polymorphonuclear leucocytes (PMNs). There were either increased numbers of lymphoid cells or normal cell numbers with other evidence of inflammation such as oedema, congestion, or cell damage. The term "active" was used to indicate an increase in PMNs.[8] The gastritis was considered active if a few PMNs infiltrated one gland neck or pit, if occasional PMNs were scattered throughout the superficial epithelium, or if there was an obvious increase in PMNs in the lamina propria.

Later, sections stained with Warthin-Starry silver stain were examined for small curved bacilli on the surface epithelium. Numbers of bacteria were graded as 0, no characteristic bacteria; 1, occasional spiral bacteria found after searching; 2, scattered bacteria in most high-power fields or occasional groups of numerous bacteria; or 3, numerous bacteria in most high-power fields.

Microbiology

Tissue smears were Gram stained and examined for curved bacilli resembling *Campylobacter*. The remaining tissue was minced, plated on non-selective blood and chocolate agar, and cultured at 37°C under microaerophilic conditions as used for *Campylobacter* isolation.[9] At first plates were discarded after 2 days but when the first positive plate was noted after it had been left in the incubator for 6 days during the Easter holiday, cultures were done for 4 days.

Analysis of Results

Questionnaires, gastroscopy reports, and histopathology and microbiology results were coded independently in separate departments. Complete results for individual patients were not known until the statistician had received all the data. The findings were tested for positive correlation with the presence of either bacteria or gastritis, by the chi-squared method. Fisher's exact test of significance was used for all the 2 × 2 tables in this paper.

Results

In 12 weeks 184 patients were examined by the gastroenterology unit. Of the 84 patients excluded, 5 refused consent, 4 had contraindications to biopsy, and 75 patients, mostly unbooked cases, could not be invited to participate. These patients closely matched the study group for age, sex, and incidence of peptic ulcers (table I).

Questionnaires

99 patients completed the questionnaires. The only symptom which correlated with gastritis or bacteria was "burping" which was more common in patients with bacteria ($p = 0.03$) or gastritis ($p = 0.007$). This association remained when patients with peptic ulcer were excluded. None of the other questionnaire responses showed any relationship to the presence of gastric bacteria or gastritis.

Endoscopy

There was a very close correlation between both gastric ulcer and duodenal ulcer and the presence of the bacteria (table II). Most patients with peptic ulcer also had gastritis ($29/31$; $p = 0.0002$).

TABLE I—COMPARISON OF PARTICIPANTS WITH EXCLUDED PATIENTS

—	Study group (n = 100)	Exclusions (n = 84)
Mean age (range)	55 (20–88) yr	57 (18–88) yr
Males	63 (63%)	55 (65%)
Females	37 (37%)	29 (35%)
Gastric ulcer	22 (22%)	19 (23%)
Duodenal ulcer	13 (13%)	8 (10%)

TABLE II—ASSOCIATION OF BACTERIA WITH ENDOSCOPIC DIAGNOSES

Endoscopic appearance*	Total	With bacteria	p
Gastric ulcer	22	18 (77%)	0·0086
Duodenal ulcer	13	13 (100%)	0·00044
All ulcers	31	27 (87%)	0·00005
Oesophagus abnormal	34	14 (41%)	0·996
Gastritis†	42	23 (55%)	0·78
Duodenitis†	17	9 (53%)	0·77
Bile in stomach	12	7 (58%)	0·62
Normal	16	8 (50%)	0·84
Total	100	58 (58%)	

*More than one description applies to several patients (eg, 4 patients had both gastric and duodenal ulcers).
†Refers to endoscopic appearance, not histological inflammation.

TABLE III—HISTOLOGICAL GRADING OF GASTRITIS AND BACTERIA

Gastritis	Bacterial grade				
	Nil	1+	2+	3+	Total
Normal*	29	2	0	0	31
Chronic	12†	9	7	1	29
Active	2	5	15	18	40
Total	43	16	22	19	100

*Gastritis grades 0 and 1 normal.
†1 case showed bacteria on gram stained smear.

TABLE IV—RELATION BETWEEN GASTRITIS AND BACTERIA IN PATIENTS WITHOUT PEPTIC ULCER

Gastritis	Bacteria		
	No	Yes	Total
Normal	28	1	29
Chronic	8	12	20
Active	2	18	20
Total	38	31	69

Histopathology

Gastritis could usually be graded with confidence at low magnification. There was some difficulty with about 25 cases where the changes were mild or the specimens were small, superficial, or distorted. To ensure that gradings were reliable, single H & E sections from the last 40 cases were examined "blind" by another pathologist who agreed with the presence or absence of gastritis in 36 cases (90%), and gave an identical grading in 32.

Gradings for bacteria by silver staining were more straightforward. The bacteria stained well and were easily differentiated from contaminant bacteria or debris. Silver staining was the most sensitive method of detecting the spiral bacteria. Silver stained sections and Gram stained smears were both done in 96 cases and spiral bacteria were seen in 56 of them; 32 with both stains, 23 with silver alone, and 1 case with the Gram stain alone.

The correlation between gastritis and bacteria, defined by Gram and/or by silver staining, was remarkable (table III). Gastritis was present in 55/57 biopsy specimens with bacteria ($p = 2 \times 10^{-12}$). When the 31 patients with peptic ulcer were excluded, the correlation persisted, implying that the presence of bacteria was not secondary to an ulcer crater (table IV).

Microbiology

Specimens for culture were received from 96 patients and 11 were culture positive, all being seen with Gram and silver staining also. No spiral bacteria were grown from the first 34 cases, probably because the cultures were discarded too soon.

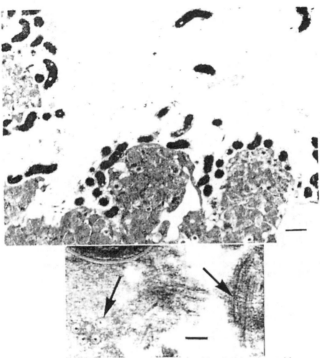

Electron micrograph from a mucosal biopsy with active chronic gastritis.

Upper: many profiles of sectioned pyloric campylobacter are located on the luminal aspect of mucus-secreting epithelial cells; plasma membranes are intact, but indented and almost devoid of microvilli (bar = 1 μm).

Lower: at higher magnification groups of transversely and longitudinally cut sheathed flagella are visible (arrows; bar = 100 nm).

The bacteria were S-shaped or curved gram-negative rods, 3 μm×0·5 μm, with up to 1½ wavelengths. In electron micrographs they had smooth coats and there were usually four sheathed flagella arising from one end of the cell. They grew best in a microaerophilic atmosphere at 37°C; a campylobacter gas generating kit was sufficient (Oxoid BR56). Moist chocolate or blood agar was the preferred medium. Growth was evident in 3 days as 1 mm diameter non-pigmented colonies. In artificial media the bacteria were usually larger and less curved than those seen on Gram stains of fresh tissue. They formed coccoid bodies in old cultures. The bacteria were oxidase +, catalase +, H_2S +, indole −, urease −, nitrate −, and did not ferment glucose. They were sensitive to tetracycline, erythromycin, kanamycin, gentamicin and penicillin, and resistant to nalidixic acid. DNA base analysis gave a guanine + cytosine content of 36 mol%, a value in the range for campylobacters.

Sources of Bias

The patient sample was from a defined population with gastric symptoms expected to have some gastroenterological abnormality. The biopsy tissue studied was from apparently intact mucosa—ie, not the sort of specimen a pathologist usually sees. We attempted to limit bias by making the study consecutive and blind, and were partly successful. The study was not strictly consecutive since 84 patients had to be excluded. However, gastroscopy reports and laboratory investigations were completed serially and usually independently ("blind") except that clinically relevant material was sent (to J. R. W.) with study biopsies, mainly from cases of gastric ulcer. However, an independent blind assessment of gastritis in 40 cases matched the study results well.

Discussion

The spiral bacteria of the human gastric antrum have never been cultured before, and their association with active chronic gastritis has not been described. They are a new species closely resembling campylobacters morphologically and in respect of atmospheric requirements and DNA base composition, but their flagellar morphology is not that of the genus *Campylobacter*.[9] Campylobacters have a single unsheathed flagellum at one or both ends of the cell whereas the new organism has four sheathed flagella at one end.[7,10] If it is premature to talk of *"Campylobacter pyloridis"*[11] perhaps the name "pyloric campylobacter" will do to define the site where these organisms are commonly found and to indicate the similarity to known *Campylobacter* spp.

There was no well-defined clinical syndrome associated with pyloric campylobacter. Only "burping" was significantly associated. Others have described this symptom in patients with non-ulcer dyspepsia and PMN infiltration of the antrum is also common in such patients.[12,13] We expected abdominal pain to correlate with pyloric campylobacter or gastritis, but it did not. Perhaps, since most patients undergoing gastroscopy have pain (75% in our study) the question "Do you have abdominal pain—yes or no?" was too general.

Much of the questionnaire was designed to select likely sources or causes of pyloric campylobacter infection. For example, bacteria might have colonised patients who already had gastritis and were taking antacids, milk, or cimetidine, thus impairing their "gastric acid barrier" and predisposing them to infection.[14] Animal contact and carious teeth were also considered as sources of infection. Campylobacters are commensals of domestic and farm animals (C $coli$, C $jejuni$), and they also inhabit the human mouth (C $sputorum$ ss $sputorum$).[15] We found no evidence that any of these factors predisposed to the infection.

The absence of a relation between "known causes" of gastritis and the presence of histological gastritis has been noted by others. For example, analgesic abusers often have no gastritis, even when a gastric ulcer is present;[16] alcohol consumption is not clearly related to gastritis;[17] the quantity of bile in the stomach (duodenogastric reflux) is not obviously related to the state of gastric mucosa;[18] autoimmune disease is an unlikely cause, since gastric autoantibodies are uncommon except in pernicious anaemia, where the main histological changes are in the body of the stomach, not the antrum.[19] Gastric ulcer seems an unlikely primary cause of antral gastritis because the gastritis remains after successful treatment of the ulcer with cimetidine or carbenoxolone, and gastritis is just as common in patients with duodenal ulcer as with gastric ulcer.[6,20–23] Thus, the aetiology of chronic gastritis remains uncertain.

We have found a close association between pyloric campylobacter and antral gastritis. When PMN infiltrated the mucosa the bacteria were almost always present (38/40). In the absence of inflammation they were rare (2/31), suggesting that they are not commensals. The bacteria were not cultured unless the patient had histological evidence of both gastritis and pyloric campylobacter. We know of no other disease state where, in the absence of complicating factors such as ulceration (table IV), bacteria and PMNs are so intimately related without the bacteria being pathogenic.

How does pyloric campylobacter survive? The bacteria were usually in close contact with the mucosa, often in grooves between cells, within acinus-like infoldings of the epithelium or within the mucosal pits (figure). The surface mucus coating was superficial to the bacteria and any foreign material or organisms from the oral flora were present above the mucus, rarely mixed with it, and not beneath it: the mucus appeared to form a stable layer over the spiral bacteria. The antrum secretes mainly mucus, and the deeper levels of the surface mucus coating are slightly alkaline.[24] Thus pyloric campylobacter grows in a near-neutral environment, in close contact with the mucosa and protected from the bactericidal gastric juice. The absence of these bacteria from past reports of gastric microbiology may be because only gastric juice was cultured.[25,26] Even salmonellae cannot survive the low intragastric pH for more than a few minutes.[14] Where gastric biopsy material has been cultured,[6,27,28] microaerophilic techniques were not used and pyloric campylobacter did not grow.

Peptic ulcer was the only endoscopic finding associated with histological gastritis and pyloric campylobacter. This was surprising since the bacteria were not prominent on gastric ulcer borders and in duodenal ulcer no correlation would be expected. Perhaps the mucus coating is deficient or unstable near ulcer borders, thus allowing damage to the bacteria as well as the mucosa. Within a few millimetres of an ulcer, both pyloric campylobacter and gastritis were usually present. Other studies have shown continuing gastritis after ulcer healing with cimetidine and we have observed the persistence of pyloric campylobacter colonisation in such

patients. The failure of the H_2 receptor antagonists to prevent ulcer relapse is attributed to an underlying ulcer diathesis which is unaffected by therapy. A bacterial aetiology, with continuing gastritis, could be the explanation. The diathesis may be a myth. Of ulcer-healing agents the only one thought to improve relapse rates is tripotassium dicitrato-bismuthate.[29] This compound is bactericidal to pyloric campylobacter and in patients treated with it the gastritis improved and the bacteria disappeared.[30]

The aetiology of peptic ulceration is unknown but until now a bacterial cause has not really been considered. We have found colonisation of the gastric antrum with pyloric campylobacter in over half of a series of cases at routine endoscopy. The bacteria were present almost exclusively in patients with chronic antral gastritis and were also common in those with peptic ulceration of the stomach or duodenum. Although cause-and-effect cannot be proved in a study of this kind, we believe that pyloric campylobacter is aetiologically related to chronic antral gastritis and, probably, to peptic ulceration also.

We thank Dr T. E. Waters, Dr C. R. Sanderson, and the gastroenterology unit staff for the biopsies, Miss Helen Royce and Dr D. I. Annear for the microbiological studies, Mr Peter Rogers and Dr L. Sly for supplying the G & C data, Dr J. A. Armstrong for the electron microscopy, Dr R. Glancy for reviewing slides, Miss Joan Bot for the silver stains, Mrs Rose Rendell of Raine Medical Statistics Unit UWA, and Ms Maureen Humphries, secretary, and, for travel support, Fremantle Hospital.

Correspondence should be addressed to: B. M., Department of Microbiology, Fremantle Hospital, PO Box 480, Fremantle 6160, Western Australia.

REFERENCES

1. Doenges JL. Spirochaetes in gastric glands of macacus rhesus and humans without definite history of related disease. Proc Soc Exp Biol Med 1938; 38: 536–38.
2. Ito S. Anatomic structure of the gastric mucosa. In: Heidel US, Cody CF, eds. Handbook of physiology, section 6: Alimentary canal, vol II: secretion. Washington, DC: American Physiological Society, 1967: 705–41.
3. Fung WP, Papadimitriou JM, Matz LR. Endoscopic, histological and ultrastructural correlations in chronic gastritis. Am J Gastroenterol 1979; 71: 269–79.
4. Freedburg AS, Barron LE. The presence of spirochaetes in human gastric mucosa. Am J Dig Dis 1940; 7: 443–45.
5. Palmer ED. Investigation of the gastric spirochaetes of the human. Gastroenterology 1954; 27: 218–20.
6. Steer HW, Colin-Jones DG. Mucosal changes in gastric ulceration and their response to carbenoxolone sodium. Gut 1975; 16: 590–97.
7. Warren JR, Marshall B. Unidentified curved bacilli on gastric epithelium in active chronic gastritis. Lancet 1983; i: 1273–75.
8. Whitehead R, Truelove SC, Gear MWL. The histological diagnosis of chronic gastritis in fibreoptic gastroscope biopsy specimens. J Clin Pathol 1972; 25: 1–11.
9. Kaplan RL. Campylobacter. In: Lenette E, Balows A, Hausler WJ, Truant JP, eds. Manual of clinical microbiology, 3rd ed. Washington, DC: American Society for Microbiology, 1980: 235–41.
10. Pead PJ. Electron microscopy of Campylobacter jejuni. J Med Microbiol 1979; 12: 383–85.
11. Skirrow MB. Taxonomy and biotyping: Morphological aspects. In: Pearson AD, Skirrow MB, Rowe B, Davies JR, Jones DM, eds. Campylobacter II: Proceedings of the Second International Workshop on Campylobacter Infections. London: Public Health Laboratory Service, 1983: 36.
12. Crean GP, Card WI, Beattie AD, Holden RJ, James WB, Knill-Jones RP, Lucas RW, Spiegelhalter D. Ulcer-like dyspepsia. Scand J Gastroenterol 1982; 17 (suppl 79): 9–15.
13. Greenlaw R, Sheahan DG, Deluca V, Miller D, Myerson E, Myerson P. Gastroduodenitis: a broader concept of peptic ulcer disease. Dig Dis Sci 1980; 25: 660–72.
14. Giannela RA, Broitman SA, Zamcheck N. Gastric acid barrier to ingested microorganisms in man: Studies in vivo and in vitro. Gut 1972; 13: 251–56.
15. Blaser MJ, Reller LB. Campylobacter enteritis. N Engl J Med 1981; 305: 1444–52.
16. MacDonald WC. Correlation of mucosal histology and aspirin intake in chronic gastric ulcer. Gastroenterology 1973; 65: 381–89.
17. Wolff G. Does alcohol cause chronic gastritis? Scand J Gastroenterol 1970; 5: 289–91.
18. Goldner FH, Boyce HW. Relationship of bile in the stomach to gastritis. Gastrointest Endosc 1976; 22: 197–99.
19. Whitehead R. Mucosal biopsy of the gastrointestinal tract. In: Benrington JL, ed. Major problems in pathology: Vol III, 2nd ed. Philadelphia: WB Saunders, 1979: 15.
20. Gilmore HM, Forrest JAH, Pettes MR, Logan RFA, Heading RC. Effect of short and long term cimetidine on histological duodenitis and gastritis. Gut 1978; 19: 981.
21. McIntrye RLE, Piris J, Truelove SC. Effect of cimetidine on chronic gastritis in gastric ulcer patients. Aust NZ J Med 1982; 12: 106.
22. Schrager J, Spink R, Mitra S. The antrum in patients with duodenal and gastric ulcers. Gut 1967; 8: 497–508.

B. J. MARSHALL AND J. R. WARREN: REFERENCES—*continued*

23. Magnus HA. Gastritis. In: Jones FA, ed. Modern trends in gastroenterology. London: Butterworth, 1952: 323–51.
24. Allen A, Garner G. Mucus and bicarbonate secretion in the stomach and their possible role in mucosal protection. *Gut* 1980; **21**: 249–62.
25. Draser BS, Shiner M, McLeod GM. Studies on the intestinal flora I: The bacterial flora of the gastrointestinal tract in healthy and achlorhydric persons. *Gastroenterology* 1969; **56**: 71–79.
26. Enander LK, Nilsson F, Ryden AC, Schwan A. The aerobic and anaerobic flora of the gastric remnant more than 15 years after Billroth II resection. *Scand J Gastroenterol* 1982; **17**: 715–20.
27. Mackay IR, Hislop IG. Chronic gastritis and gastric ulcer. *Gut* 1966; **7**: 228–33.
28. Rollason TP, Stone J, Rhodes JM. Spiral organisms in endoscopic biopsies of the human stomach. *J Clin Pathol* 1984; **37**: 23–26.
29. Martin DF, May SJ, Tweedle DE, Hollanders D, Ravenscroft MM, Miller JP. Difference in relapse rates of duodenal ulcer after healing with cimetidine or tripotassium di-citrato bismuthate. *Lancet* 1980; i: 7–10.
30. Marshall B, Hislop I, Glancy R, Armstrong J. Histological improvement of active chronic gastritis in patients treated with De-Nol. *Aust NZ J Med* (in press) (abstr).

Effect of Treatment of *Helicobacter pylori* Infection on the Long-term Recurrence of Gastric or Duodenal Ulcer

A Randomized, Controlled Study

David Y. Graham, MD; Ginger M. Lew, PA-C; Peter D. Klein, PhD; Dolores G. Evans, PhD; Doyle J. Evans, Jr., PhD; Zahid A. Saeed, MD; and Hoda M. Malaty, MD

■ *Objective:* To determine the effect of treating *Helicobacter pylori* infection on the recurrence of gastric and duodenal ulcer disease.

■ *Design:* Follow-up of up to 2 years in patients with healed ulcers who had participated in randomized, controlled trials.

■ *Setting:* A Veterans Affairs hospital.

■ *Participants:* A total of 109 patients infected with *H. pylori* who had a recently healed duodenal (83 patients) or gastric ulcer (26 patients) as confirmed by endoscopy.

■ *Intervention:* Patients received ranitidine, 300 mg, or ranitidine plus triple therapy. Triple therapy consisted of tetracycline, 2 g; metronidazole, 750 mg; and bismuth subsalicylate, 5 or 8 tablets (151 mg bismuth per tablet) and was administered for the first 2 weeks of treatment; ranitidine therapy was continued until the ulcer had healed or 16 weeks had elapsed. After ulcer healing, no maintenance antiulcer therapy was given.

■ *Measurements:* Endoscopy to assess ulcer recurrence was done at 3-month intervals or when a patient developed symptoms, for a maximum of 2 years.

■ *Results:* The probability of recurrence for patients who received triple therapy plus ranitidine was significantly lower than that for patients who received ranitidine alone: for patients with duodenal ulcer, 12% (95% CI, 1% to 24%) compared with 95% (CI, 84% to 100%); for patients with gastric ulcer, 13% (CI, 4% to 31%) compared with 74% (44% to 100%). Fifty percent of patients who received ranitidine alone for healing of duodenal or gastric ulcer had a relapse within 12 weeks of healing. Ulcer recurrence in the triple therapy group was related to the failure to eradicate *H. pylori* and to the use of nonsteroidal anti-inflammatory drugs.

■ *Conclusions:* Eradication of *H. pylori* infection markedly changes the natural history of peptic ulcer in patients with duodenal or gastric ulcer. Most peptic ulcers associated with *H. pylori* infection are curable.

Annals of Internal Medicine. 1992;**116**:705-708.

From Baylor College of Medicine, the Veterans Affairs Medical Center, and the U.S. Department of Agriculture/Agricultural Research Center Children's Nutrition Research Center. For current author addresses, see end of text.

Peptic ulcer disease is a chronic disease characterized by frequent recurrences. The continuation of antiulcer therapy after ulcer healing results in a reduced rate of ulcer recurrence but does not affect the natural history of the disease, because the expected pattern of rapid recurrence resumes when maintenance therapy is discontinued (1). Recent studies have suggested that the eradication of *Helicobacter pylori* infection affects the natural history of duodenal ulcer disease such that the rate of recurrence decreases markedly (2-6). However, the interpretation of these results has been complicated by the fact that several of the larger studies did not use control groups or any form of blinding (3, 5, 6). In addition, studies of the effect of *H. pylori* eradication in patients with gastric ulcer have not been done. We report the results of a randomized, controlled trial in which we evaluated the effect of therapy designed to eradicate *H. pylori* on the pattern of ulcer recurrence in patients with duodenal or gastric ulcer.

Methods

Our study took place between September 1988 and October 1990 at a single Veterans Affairs hospital. All patients whose *H. pylori*-associated active duodenal or gastric ulcer had healed during randomized trials comparing ranitidine and ranitidine plus "triple therapy" were invited to participate in this follow-up study. During the initial studies, patients were randomly assigned to either ranitidine alone (300 mg once daily in the evening) or to ranitidine plus triple therapy. Triple therapy consisted of bismuth subsalicylate and two antimicrobial agents: tetracycline hydrochloride, 500 mg four times a day and metronidazole, 250 mg thrice daily. Bismuth subsalicylate tablets containing 151 mg bismuth per tablet (Pepto-Bismol, Proctor & Gamble, Cincinnati, Ohio) were administered for the first 2 weeks of therapy, and patients received 5 or 8 tablets.

Two groups of patients were entered into our follow-up study. Patients with healed duodenal ulcers came from a randomized study of 146 patients, 105 of whom have been previously described (7); of these 146 patients, 112 experienced ulcer healing, 24 were lost to follow-up during the 16 weeks of therapy, and 10 had no ulcer healing after 16 weeks of treatment. Of the 112 patients in whom documented healing occurred, 83 (74%) agreed to enter the follow-up study. Patients with healed gastric ulcers came from a randomized trial of 41 patients; of these 41 patients, 31 had ulcer healing, 9 were lost to follow-up during the 16 weeks of therapy, and 1 patient had no ulcer healing after 16 weeks of treatment. Of the 31 patients in whom documented healing occurred, 26 (84%) agreed to enter the follow-up study. In sum, 109 patients with healed

Table 1. Demographic and Clinical Characteristics of Patients

Variable	Patients with Duodenal Ulcer		Patients with Gastric Ulcer	
	Ranitidine Alone	Triple Therapy plus Ranitidine	Ranitidine Alone	Triple Therapy plus Ranitidine
Patients, n	36	47	11	15
Median age (range), y	61 (31-85)	58 (29-79)	66 (43-76)	60 (27-67)
Male gender, %	97	100	100	100
Race, n(%)				
White	25 (69)	30 (64)	7 (64)	8 (53)
Black	11 (30)	17 (36)	3 (27)	7 (47)
Other	0	0	1 (9)	0
Recent NSAID* use, n(%)	6 (17)	11 (23)	2 (18)	5 (33)
Daily aspirin (1 tablet), n(%)	3 (8)	4 (8.5)	1 (9)	1 (6)
Smoker†, n(%)	17 (47)	34 (72)	6 (54)	11 (73)
Alcohol use				
1 or more drinks/wk, n(%)	7 (19)	15 (32)	2 (18)	4 (27)
H. pylori infection, %‡	100	100	100	100

* NSAID = nonsteroidal anti-inflammatory drugs.

† $P = 0.03$ for the difference between the ranitidine and triple therapy groups among patients with duodenal ulcer.

‡ Infection at entry into the ulcer healing study was confirmed by at least two of the following: urea breath test, histologic evaluation, culture, and serologic testing.

peptic ulcers (83 with duodenal ulcers and 26 with gastric ulcers) were included in our long-term follow-up study.

Patients were followed for up to 2 years. During this time, patients received no antiulcer medications (including antacids). Twenty-four patients regularly receiving nonsteroidal anti-inflammatory drugs were allowed to continue them if they wished. Patient follow-up visits were scheduled for 1 month

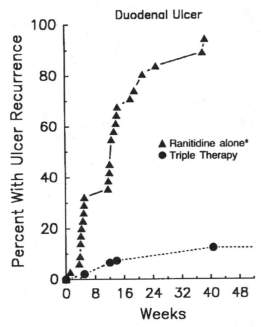

Figure 1. Lifetable recurrence of duodenal ulcers for the year after successful healing with ranitidine alone or triple therapy plus ranitidine. No maintenance therapy was given; the recurrence rate of ulcers in patients healed with ranitidine alone was significantly greater ($P < 0.01$) than in those who received triple therapy plus ranitidine. * The only patient in the ranitidine alone group who had not developed recurrent ulcer by October 1990 was then followed for a total of 16 months without ulcer recurrence (*see* text).

after therapy, 3 months after therapy, and every 3 months for up to 2 years. Patients were also instructed to return for endoscopy if symptoms recurred. The endoscopist was blinded to the treatment status of the patients.

Patients originally assigned to receive ranitidine alone who experienced ulcer recurrence were crossed over to receive triple therapy plus ranitidine; after ulcer healing occurred with this latter therapy, patients were offered follow-up using the protocol described above. These patients were termed "crossover follow-up" patients.

Ulcers were identified by endoscopy using Fujinon videoendoscopes (Fujinon, Inc., Wayne, New Jersey). A video still (ProMavica, Sony Corporation of America, Sony Park Ridge, New Jersey) of each ulcer was made so that the site and characteristics of the ulcer could be reviewed before subsequent endoscopic procedures. One video disk was assigned to each patient. An ulcer was defined as a circumscribed break in the duodenal mucosa that measured at least 5 mm in diameter, had apparent depth, and was covered by an exudate.

All patients were assessed for *H. pylori* infection by the ^{13}C-urea breath test (8, 9); by a sensitive and specific enzyme-linked immunosorbent assay (ELISA) for IgG antibody against the high-molecular-weight, cell-associated proteins of *H. pylori* (10); by culture; and by histologic evaluation of antral mucosal biopsy specimens. Eradication was defined by no evidence of *H. pylori* infection (by urea breath test, culture, or histologic evaluation) 1 or more months after discontinuing triple therapy. Patients were tested every 3 months and when symptomatic.

The protocol was approved by the Institutional Review Board at the Veteran Affairs Medical Center and Baylor College of Medicine. Written informed consent was obtained before patient entry.

Statistical Analysis

Ulcer recurrence was calculated by the lifetable method (Lifetest procedure, SAS/STAT software release 6.04, SAS Institute, Inc., Cary, North Carolina). Categorical data were evaluated by chi-square test with the Yates correction or by the Fisher exact test. All P values ≤ 0.05 (two-tailed) were considered to be significant. Ninety-five percent confidence intervals are given when appropriate.

Results

We followed 83 patients with duodenal ulcer and 26 patients with gastric ulcer (median age, 62 years). The sample was 98% men. The two groups of patients (assigned to ranitidine alone or ranitidine plus triple therapy) had similar demographic and clinical characteristics (Table 1). The percentage of smokers and alcohol users was higher in the group receiving triple therapy plus ranitidine than in the group receiving ranitidine alone. Only one significant difference was found between the treatment groups: Among patients with duodenal ulcer, the group receiving triple therapy had more smokers than the group receiving ranitidine alone (*P* = 0.03). All patients had active *H. pylori* infection before the start of ulcer therapy. All 47 patients treated with ranitidine alone were still infected at the end of therapy. In contrast, *H. pylori* was eradicated in 55 of 62 patients (89%) receiving triple therapy.

The lifetable probability of ulcer recurrence 1 year after ulcer healing (Figures 1 and 2) was significantly lower for patients who received triple therapy plus ranitidine (12% [CI, 1% to 24%] for patients with duodenal ulcer and 13% [CI, 4% to 31%] for patients with gastric ulcer) compared with those who received ranitidine alone (95% [CI, 84% to 100%] for patients with duodenal ulcer and 74% [CI, 44% to 100%] for patients with gastric ulcer) (*P* = 0.001). The median duration of follow-up for patients who had received triple therapy plus ranitidine was 38 weeks (range, 4 to 108 weeks) for patients with duodenal ulcer and 52 weeks (range, 12 to 95 weeks) for patients with gastric ulcer.

Fifty percent of patients with either duodenal or gastric ulcer who experienced healing with ranitidine alone had a recurrence within 12 weeks (*see* Figures 1 and 2). Seventy-five percent of recurrences were symptomatic. At the end of the study period, only three patients in the ranitidine alone group (one with duodenal ulcer and two with gastric ulcer) had not had ulcer recurrence. Follow-up on these three patients after the study was completed showed the following: One patient with duodenal ulcer was lost to follow-up after 16 months, and two patients with gastric ulcer were last seen at the 15-month follow-up visit (one was lost to follow-up and the other died of an unrelated illness).

Infection with *H. pylori* was a strong predictor of ulcer recurrence. All 47 patients whose ulcers healed while receiving ranitidine alone still had *H. pylori* infection at the end of therapy, and, by lifetable analysis, 95% of them developed recurrent ulcers by the end of 1 year. None of the patients in whom *H. pylori* was eradicated became reinfected during the study period. *Helicobacter pylori* infection was not eradicated in seven patients who received triple therapy. Of these seven patients, four experienced ulcer recurrence and three were lost to follow-up (two patients after 6 months and one patient after 1 year).

Three patients with duodenal ulcer in whom *H. pylori* infection was eradicated after triple therapy still developed recurrent duodenal ulcers (two patients after 3 months and one patient after 9 months). All three patients were using nonsteroidal anti-inflammatory drugs (ibuprofen, piroxicam, or salsalate). In addition, two patients with recurrent gastric ulcer who were also receiving such drugs had persistent *H. pylori* infection.

Ten patients with duodenal ulcer who experienced ulcer recurrence after healing with ranitidine alone were crossed over to receive ranitidine plus triple therapy after the completion of the randomized trial. Triple ther-

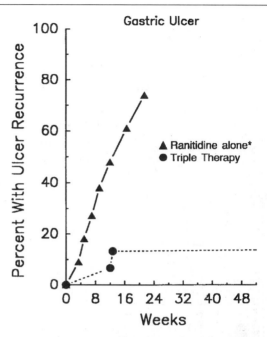

Figure 2. Lifetable recurrence of gastric ulcers for the year after successful healing with ranitidine alone or triple therapy plus ranitidine. No maintenance therapy was given; the recurrence rate of ulcers in patients healed with ranitidine alone was significantly greater (*P* < 0.01) than in those who received triple therapy plus ranitidine. * The two patients in the ranitidine alone group who had not developed recurrent ulcers by October 1990 were then followed for a total of 15 months without ulcer recurrence (*see* text).

apy resulted in the eradication of *H. pylori* infection in these patients. After ulcer healing, the patients were followed for a median of 44 weeks (range, 23 to 116 weeks), and none experienced ulcer recurrence. Four other patients with duodenal ulcer refractory to ranitidine alone were crossed over to receive ranitidine plus triple therapy. *Helicobacter pylori* infection was eradicated in all four patients. After ulcer healing, the patients were followed for a median of 40 weeks, and none experienced ulcer recurrence.

Discussion

Recent studies have shown that the eradication of *H. pylori* infection is associated with healing of gastritis (11) and a marked reduction in the rate of recurrence of duodenal ulcers (2-6). The protocols of these studies have varied, but the results have been the same; the eradication of *H. pylori* infection changes the natural history of duodenal ulcer disease, and factors that contribute to rapid ulcer recurrence, such as smoking, seem to no longer to pose a risk (6). Our study confirms previous findings in patients with duodenal ulcers and extends the findings to patients with gastric ulcers. Patients in whom *H. pylori* infection was eradicated remained asymptomatic and ulcer free.

Smoking, alcohol use, and male gender have all been

described as risk factors for ulcer recurrence (1, 12). In our study, smoking and alcohol use were more frequent in the group that received triple therapy plus ranitidine, a factor that could have biased our results. However, our study confirms the observation that smoking is not a risk factor for ulcer recurrence after the eradication of *H. pylori* infection (6). In our patients, the only factors associated with ulcer recurrence were *H. pylori* infection and the continued use of nonsteroidal anti-inflammatory drugs.

Our study was single-blind, with the endoscopist blinded to initial therapy. Although some may argue that the lack of double-blinding introduced an important bias into our study, no objective data support such a contention, and we believe such a scenario extremely unlikely, especially considering the equipment now available for studying the gastroduodenal mucosa.

Three previous reports have used the word "cure" in the title or discussion (4-6). These studies, taken together, provide compelling evidence for the hypothesis that peptic ulcer, either duodenal or gastric, is the end result of a bacterial infection. We believe that, eventually, anti-*H. pylori* agents will be part of the therapy for *H. pylori*-associated ulcer disease. The universal introduction of such therapy may be delayed because several safe and effective therapies are currently available for healing peptic ulcers and because ulcer relapse can be greatly reduced by maintenance therapy with histamine-2-receptor antagonists (1, 12). In addition, there are still concerns that the benefits of therapy (that is, reduced recurrence) may not yet outweigh such side effects as antibiotic-associated diarrhea or the development of widespread antibiotic resistance in *H. pylori* and other bacteria (13). We have observed that triple therapy is often not effective in patients who have previously received metronidazole (unpublished data), and compliance with the complicated treatment protocols remains a major problem (14). Simpler protocols and improved therapies are needed. The eradication of the infection may also not yield a true cure because the patient vulnerable to additional *H. pylori* encounters may acquire a new infection and experience recurrence of the original disease. We recommend that patients with resistant ulcers (defined as failure to heal in 12 weeks), those with ulcer-associated complications, and those with symptoms severe enough to be candidates for surgery receive triple therapy for *H. pylori* infection.

Grant Support: In part by the Department of Veterans Affairs, by grant DK 39919 from the National Institute of Diabetes and Digestive and Kidney Diseases, by the U.S. Department of Agriculture/Agricultural Research Service Children's Nutrition Research Center, and by Hilda Schwartz.

Requests for Reprints: David Y. Graham, MD, Veterans Affairs Medical Center (111D), 2002 Holcombe Boulevard, Houston, TX 77030.

Current Author Addresses: Drs. Graham, Evans, Evans, Jr., Saeed, Malaty, and Ms. Lew: Veterans Affairs Medical Center (111D), 2002 Holcombe Boulevard, Houston, TX 77030.
Dr. Klein: Children's Nutrition Research Center, 1100 Bates Street, Houston, TX 77030.

References

1. **Sontag SJ.** Current status of maintenance therapy in peptic ulcer disease. Am J Gastroenterol. 1988;83:607-17.
2. **Coghlan JG, Gilligan D, Humphries H, McKenna D, Dooley C, Sweeney E, et al.** *Campylobacter pylori* and recurrence of duodenal ulcers—a 12-month follow-up study. Lancet. 1987;2:1109-11.
3. **Lambert JR, Borromeo M, Korman MG, Hansky J, Eaves ER.** Effect of colloidal bismuth (De-Nol) on healing and relapse of duodenal ulcers-role of *Campylobacter pyloridis* [Abstract]. Gastroenterology. 1987;92:1489.
4. **Marshall BJ, Goodwin CS, Warren JR, Murray R, Blincow ED, Blackbourn SJ, et al.** Prospective double-blind trial of duodenal ulcer relapse after eradication of *Campylobacter pylori*. Lancet. 1988;2:1437-42.
5. **Rauws EA, Tytgat GN.** Cure of duodenal ulcer associated with eradication of *Helicobacter pylori*. Lancet. 1990;335:1233-5.
6. **George LL, Borody TJ, Andrews P, Devine M, Moore-Jones D, Walton M, et al.** Cure of duodenal ulcer after eradication of *Helicobacter pylori*. Med J Aust. 1990;153:145-9.
7. **Graham DY, Lew GM, Evans DG, Evans DJ Jr, Klein PD.** Effect of triple therapy (antibiotics plus bismuth) on duodenal ulcer healing with ranitidine. A randomized controlled trial. Ann Intern Med. 1991;115:266-9.
8. **Graham DY, Klein PD, Evans DJ Jr., Evans DG, Alpert LC, Opekun AR, et al.** *Campylobacter pylori* detected noninvasively by the 13C-urea breath test. Lancet. 1987;1:1174-7.
9. **Klein PD, Graham DY.** *Campylobacter pylori* detection by the ^{13}C-urea breath test. In: *Campylobacter pylori* and Gastroduodenal Disease. Rathbone BJ, Heatley V, eds. Blackwell Scientific Publications, Oxford, 1989, pp. 94-106.
10. **Evans DJ Jr, Evans DG, Graham DY, Klein PD.** A sensitive and specific serologic test for detection of *Campylobacter pylori* infection. Gastroenterology. 1989;96:1004-8.
11. **Rauws EA, Langenberg W, Houthoff HJ, Zanen HC, Tytgat GN.** *Campylobacter pyloridis*-associated chronic active antral gastritis: a prospective study of its prevalence and the effects of antibacterial and antiulcer treatment. Gastroenterology. 1988;94:33-40.
12. **Van Deventer GM, Elashoff JD, Reedy TJ, Schneidman D, Walsh JH.** A randomized study of maintenance therapy with ranitidine to prevent the recurrence of duodenal ulcer. N Engl J Med. 1989;320:1113-9.
13. **Graham DY, Börsch GM.** The who's and when's of therapy for *Helicobacter pylori* [Editorial]. Am J Gastroenterol. 1990;85:1552-5.
14. **Graham DY, Lew GM, Malaty HM, Evans DG, Evans DJ Jr, Klein PD, et al.** Factors influencing the eradication of *Helicobacter pylori* with triple therapy. Gastroenterology. 1992;102:493-6.

Gastric *Campylobacter*-like Organisms, Gastritis, and Peptic Ulcer Disease

MARTIN J. BLASER
Medical Service, Veterans Administration Medical Center, and Division of Infectious Diseases, Department of Medicine, University of Colorado School of Medicine, Denver, Colorado

Although the presence of gastric bacteria has been long established, the recognition and isolation of Campylobacter pylori and similar organisms has opened a new era in the understanding of inflammatory gastroduodenal conditions. Visualization or isolation of gastric Campylobacter-like organisms (GCLOs) is significantly associated with histologic evidence of gastritis, especially of the antrum. Correlation with peptic ulceration also exists but probably is due to concurrent antral gastritis. Outbreaks of hypochlorhydria with concomitant gastritis have been attributed to GCLO infection, and a human volunteer became ill after ingesting C. pylori. Despite rapid microbiologic characterization of the organisms and the epidemiology, pathology, and serology of infection, the pathogenetic significance of GCLOs remains unknown. Whether GCLOs cause, colonize, or worsen gastritis must be considered an unanswered question at present. The efficacy of antimicrobial treatment of GCLO infection on the natural history of gastritis is not presently resolved. Nevertheless, GCLOs are at the least an important marker of inflammatory gastroduodenal disease, and attempts to ascertain their clinical significance are clearly warranted.

Peptic ulcer disease and other inflammatory gastroduodenal conditions are among the most common maladies of humans throughout the world (1). In most cases, their etiologies cannot be discerned and, despite extensive investigation, the pathophysiology of these processes remains obscure. Although the presence of gastric bacteria has long been known, their significance has been uncertain. The development of fiberoptic endoscopy, permitting collection of fresh clinical specimens, has ushered in a new era for the management and investigation of gastroduodenal inflammatory conditions. Gastric bacteria now are being observed with regularity (2–4), and recently, Marshall and Warren (5,6) were able to isolate a spiral bacterium that had never been cultivated before. This organism, which they called *Campylobacter pyloridis*, has since been isolated by many other investigators (7–9). The field has moved quickly, and a review of its current status is appropriate.

Historical Developments

After Bottcher's observation of bacteria in the human stomach in 1874, similar spiral organisms were identified in the stomachs of animals (10,11) and in patients with gastric carcinoma (12), but their visualization in patients with nonmalignant conditions was variable (13–17). By 1939, Doenges had found several types of spiral organisms in 43% of 242 stained human stomach autopsy specimens (18). The "encouraging results" of Gorham (cited in 16), who gave bismuth intramuscularly to treat chronic peptic ulcers, were considered to be due to its antibacterial action.

More than 30 yr later, using electron microscopy, Steer visualized curved bacteria on the surface of the gastric epithelium (2) in biopsy specimens obtained from patients with gastric ulceration but not from normal subjects (19). By light microscopy, Rollason and colleagues (4) observed spiral organisms in 42.6% of stained specimens of fresh gastric tissue from 310 consecutive endoscopic biopsies. Organisms were present on the surface of the gastric

Received September 12, 1986. Accepted February 2, 1987.

Address requests for reprints to: Martin J. Blaser, M.D., Infectious Disease Section (111L), Veterans Administration Medical Center, 1055 Clermont Street, Denver, Colorado 80220.

Dr. Blaser is a Clinical Investigator of the Veterans Administration.

The author thanks Dr. William R. Brown and Dr. Dennis Ahnen for review of the manuscript and Dr. Wen-lan L. Wang and Dr. Bruce Dunn for providing the clinical specimens.

0016-5085/87/$3.50

Abbreviation used in this paper: GCLO, gastric *Campylobacter*-like organism.

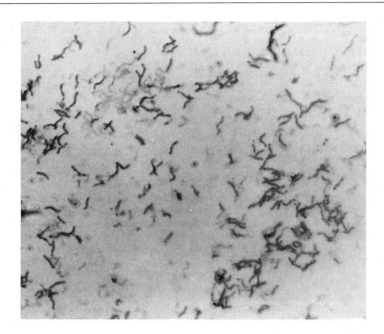

Figure 1. Gram stain of *C. pylori* in pure culture after 96-h incubation on chocolate agar (original magnification, ×1000).

mucosa, with similar frequency in the cardia, body, and antrum, were most readily seen in the necks and bases of the gastric glands, but did not invade the epithelial cells or lamina propria. Organisms were faintly gram-negative but were intensely stained using the Warthin–Starry silver impregnation method. By electron microscopy, they were a homogeneous population of curved rods, up to 6 μm in length, but they failed to grow under anaerobic culture.

In 1980, Dr. J. R. Warren in Perth, Australia observed similar curved and S-shaped bacilli, most often in the gastric antrum and associated with active gastritis (3); by light microscopy they resembled *Campylobacter jejuni*. As such, using techniques developed for culture of campylobacters, Marshall first isolated microaerophilic curved bacteria from gastric antral biopsy specimens. Subsequently, Marshall and Warren (6) studied biopsy specimens from 100 patients who had been referred for gastroscopy. Spiral bacteria were visualized by silver staining in 87% of 31 patients who had gastric or duodenal ulcers, most of whom also had gastritis. Bacteria were seen in 81% of 69 patients with acute or chronic gastritis, but in only 2 (6%) of 31 patients without gastritis. Of greatest significance is that 11 specimens from 96 patients again yielded a curved microaerophilic organism, which originally was called *C. pyloridis* (6,20) but is now known as *C. pylori*.

Microbiologic Characteristics of *Campylobacter pylori* and Related Organisms

Campylobacter pylori are small nonsporulating gram-negative bacteria with flagellae at one end. They are curved rods, 3.5 μm long and 0.5–1 μm wide, with a spiral periodicity (Figure 1); Gram stain may show U-shaped or circular cells. These spiral organisms differ from spirochetes in having the rigid cell walls and flagella characteristic of most gram-negative bacteria. The guanine plus cytosine content of *C. pylori* (35.8–37.1 mol %) is within the range for *Campylobacter* species but less than for *Spirillum* or *Vibrio* (20). By electron microscopy, however, the structure of *C. pylori* is different from that of other campylobacters and more closely resembles *Spirillum* (21). Several other morphologic and biochemical characteristics (7,21–26) are markedly different from those of other *Campylobacter* species (Table 1) and other enterobacteria; most notably, *C. pylori* is a strong producer of the enzyme urease. *Campylobacter pylori* are fastidious and do not grow aerobically or anaerobically and when incubated below 30°C. Growth is poor in most liquid media; either a blood or hemin source appears essential (24,27–29). Best growth is on chocolate or blood agar plates and takes 2–5 days. Because the taxonomic status of *C. pylori* is not settled at present (30) and because of phenotypic variability, it is

Table 1. Characteristics Distinguishing Gastric Campylobacter-like Organisms From Intestinal Campylobacters

	C. pylori	GCLO-2	Ferret GCLO	C. jejuni	C. laridis	C. fetus
Optimal growth temperature (°C)	37	37	ND	42	42	37
Hippurate hydrolysis	–	+	–	+	–	–
Susceptibility to cephalothin	S	S	R	R	R	R
Susceptibility to naladixic acid	R	R	S	S	R	R
Urease	+	–	+	–	–	–
C-19 cyclopropane on GLC	+	+	ND	+	–	–
Nitrate reductase	–	–	+	+	+	+

C., *Campylobacter*; GLC, gas-liquid chromatography; GCLO, gastric *Campylobacter*-like organism; ND, not determined; R, resistant; S, sensitive.

preferable at this time to call this family of organisms gastric *Campylobacter*-like organisms (GCLOs) unless *C. pylori* is specifically isolated and identified.

Kasper and Dickgiesser (31) isolated *C. pylori* from 39% of 328 patients who had antral biopsies, but they isolated a different organism, which they called GCLO-2, from 6 (2%) other patients. Gastric *Campylobacter*-like organism-2 colonies are 1–5 mm in size, are similar in appearance to colonies of *C. jejuni*, and grow heavily within 48 h but have a number of biochemical differences (Table 1). Although deoxyribonucleic acid hybridization studies have not been performed, available data suggest that GCLO-2 represents a new species, closely related to but distinctive from *C. jejuni* (22,32). In addition to GCLO-2, cigar-shaped organisms have been seen on the surface of the mucus in biopsy specimens obtained from patients with peptic ulcers, but these organisms could not be cultivated; *C. pylori* were grown from all but one (33), which raises the possibility that these are aberrant forms of *C. pylori*. Strongly urease-positive GCLOs are apparently present in the stomachs of many (34), if not all (35), laboratory-raised ferrets, but these organisms also are distinct from *C. pylori* (Table 1). Other urease-positive thermophilic campylobacters have been isolated from river and sea water and shellfish, but not from the human specimens (36). Whether any of these organisms are related to *C. pylori* or are associated with gastritis remains to be determined.

The antigenic nature of *C. pylori* has not been completely defined. Most strains appear to share a major surface antigen with *C. jejuni* (33). Group antigens are present, but some antigenic heterogeneity clearly exists (37,38,39). By sodium dodecyl sulfate-polyacrylamide-gel electrophoresis, organisms appear homogeneous (26), with major bands in the regions of 14–21, 29–33, 50–55, 60–68, and 95–105 kilodaltons, and are dissimilar from other *Campylobacter* species (39), but relative proportions of proteins among strains vary (40). By Western blotting, essentially all these bands were antigenic to

naturally infected humans and immunized rabbits. A major surface-exposed 60-kilodalton antigen was detected by radioimmunoprecipitation and immunoblotting (40).

Pathological Associations With Gastric *Campylobacter*-like Organism Infection

Association of Gastric Campylobacter-like Organisms With Gastritis

Investigators on four continents have now identified GCLOs in gastric biopsy specimens and have shown an association between the presence of GCLOs and gastritis diagnosed by histology in adults (Table 2). Although methods employed in these studies to document the presence of GCLOs have varied, as have the definitions of gastritis used, it is notable that in all but one study the GCLO detection rate was significantly greater in patients with gastritis than in those without. The exception occurred in a small study in Australia in which nearly equal rates of GCLO detection were found in the two groups (44). Whether idiopathic antral gastritis in children is specifically associated with the presence of GCLOs is not yet settled (53–55). In attempts to answer the important question of whether GCLOs are present in healthy persons, three endoscopic studies of asymptomatic volunteers have been reported. Despite absence of symptoms or risk factors and the young age of the volunteers [mean ages 27–30 yr), 20.4% were found to have histologic gastritis; all of these subjects had GCLO present (Table 2). In contrast, no gastritis was found in 79.6% of the subjects and GCLOs were not detected in any of these cases. In total, the volunteer studies indicate that gastritis may be present in asymptomatic young adults, that this gastritis is associated with the presence of GCLOs, but that in the majority of subjects neither gastritis nor GCLOs are present.

Table 2. Association of Gastritis With Presence of Gastric Campylobacter-like Organisms in Adults

	Method(s) for detecting GCLO	Patients with gastritis		Patients without gastritis	
		Total No.	With GCLO (%)	Total No.	With GCLO (%)
Symptomatic patients					
Eleven studies (6–9,24,33,41–45)[a]	Culture and histology	523	76.9	191	9.4
Five studies (46–50)	Histology alone	169	72.2	35	5.7
One study (51)	Culture alone	86	70.9	133	11.3
Total (17 studies)	Any of the above	778	75.2	359	9.7
Asymptomatic volunteers					
Three studies (7,45,52)	Culture and histology	11	100	43	0

GLCO, gastric *Campylobacter*-like organism. [a] Parenthetical values represent reference numbers.

Location of Gastric Campylobacter-like Organisms

Gastric *Campylobacter*-like organisms have been found on the luminal aspect of surface mucus-secreting cells and within the gastric pits, but in most studies not invading tissue (9,48) (Figure 2). Organisms always are beneath or within the mucus layer (56), but GCLOs are not seen in biopsy specimens in which mucus totally covers the specimen (33). Most organisms beneath the mucus layer, adjacent to the epithelial cell surface, appear to lie within 2 µm of an intercellular junction (56). Gastric *Campylobacter*-like organism colonization has been associated with epithelial cell flattening and decreased intracellular mucin (42,43). By electron microscopy, organisms appear to be in close contact or partially fused with epithelial cell membranes; some bacteria appear to be covered by epithelial microvilli or engulfed within endocytotic vacuoles (48,57), similar to enteropathogenic strains of *Escherichia coli*. In an electron microscopic study of 1 patient, some neutrophils located between intact gastric epithelial cells had spiral bacteria within their phagocytic vacuoles (58). In no cases have GCLOs been seen invading beyond the epithelium. Clearly, GCLOs are predominantly luminal, but whether or not the reports of tissue involvement can be substantiated will have great bearing on our assessment of their pathogenetic role and clinical relevance.

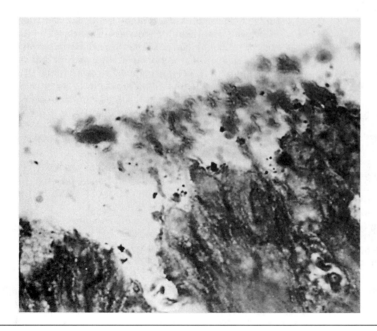

Figure 2. Gram stain (counterstained with carbol fuchsin) of gastric antral biopsy specimen from a patient with acute gastritis (magnification, ×1000). Numerous spiral bacteria can be seen adjacent to but not invading the mucosal tissue.

Association With Gastric Versus Duodenal Mucosa

Gastric *Campylobacter*-like organisms are largely associated with abnormal gastric mucosa, but their distribution is irregular (27). Several investigators (9,47,59) have noted that GCLOs are present in the duodenum only in association with gastric metaplasia and not with normal duodenal mucosa. Conversely, GCLOs do not overlie intestinal metaplasia in the antrum but overlie the surrounding inflamed gastric mucosa (24,60). Patients with intestinal metaplasia have fewer GCLOs than patients with the same degree of gastritis but without intestinal metaplasia (24,43). These associations could be due to the presence of receptors for GCLOs on gastric cells and their absence from small intestinal cells, but no direct evidence for this hypothesis has been advanced.

Associations by Type and Severity of Histology

Rathbone and colleagues (38) examined more than 150 antral and fundal biopsy samples and found that when *C. pylori* were present there always was abnormal mucosa somewhere in the stomach; *C. pylori* was present on normal fundal mucosa only when antral gastritis also was seen. In 103 Peruvian patients with gastritis, the cardia was as frequently and as densely colonized by GCLOs as was the antrum (43). In the study of healthy adult volunteers by Peterson et al. (50), GCLOs were commonly found adjacent to endoscopically and histologically normal fundal mucosa, but in the antrum, the presence of GCLOs was significantly associated with the histologic lesion of acute gastritis. These observations suggest that GCLOs are normal residents in healthy fundic mucosa and colonize the antrum when acute gastritis develops. Alternatively, if GCLOs are indeed pathogens, when introduced into the stomach they may only be capable of inducing acute gastritis in the antrum.

In autoimmune pernicious anemia, the principal lesion is atrophic gastritis of the fundus (type A) with loss of normal epithelial cells, whereas antral gastritis (type B) is typically considered "peptic" in origin (61). Gastric *Campylobacter*-like organisms were observed only in small numbers and only in the fundus in 3 of 14 patients with pernicious anemia, but in large numbers and in the antrum in 13 of 14 age- and sex-matched controls with peptic ulceration (62). The rarity of GCLOs in pernicious anemia argues against the possibility that GCLOs secondarily colonize inflamed gastric tissue; however, 11 of 14 patients had intestinal metaplasia, which might also explain the low detection rate (24,60). Gastric *Campylobacter*-like organisms are present significantly more often in patients with chronic superficial gastritis than in patients with gastritis due to alkaline bile reflux from the duodenum (63), especially when it occurs after major resections for duodenal ulcers (64).

Attempts to correlate the severity or chronicity of gastritis with the number of GCLOs present have yielded variable results. Among 51 patients who underwent endoscopy, GCLOs were isolated from 68% of biopsy specimens classified as actively inflamed and in 62% classified as "quiescent" (33); similar results were reported by Jones et al. (9). In contrast, Marshall and Warren (6) found GCLOs in 95% of patients with active gastritis but only 58% of patients with chronic or quiescent gastritis, and McNulty and Watson (8) found a strong correlation between the number of bacteria present and the severity of gastritis. In other studies, the presence of GCLOs was best associated with severe (65) or active gastritis [defined as having neutrophils present in the specimen (66)] and less well with chronic [no neutrophils present (66)] or moderately severe (65) gastritis. Acute, purulent, antral gastritis associated with the presence of GCLOs also has been reported (67).

Relation of Gastric *Campylobacter*-like Organisms to Peptic Ulcer Disease

Antral gastritis is nearly always present in patients with benign gastric ulcers (61,68), and usually persists after ulcer healing (69), but the pathogenic relationship is controversial (61,68,69). Duodenal ulceration also is highly associated with antral gastritis and with gastric metaplasia in the duodenal mucosa (68). By scanning electron microscopy of biopsy specimens from patients with duodenal ulcers, spiral (and comma) shaped bacteria are present in the antrum and in areas of gastric metaplasia in the duodenum (70). Gastric *Campylobacter*-like organisms have now been identified in the gastric antra of a substantial proportion of patients with either gastric or duodenal ulcers (Table 3), and in the latter condition the frequency of GCLO infections is significantly greater than that in patients with gastritis alone; however, this association of GCLOs with duodenal ulceration may be due to coexisting severe antral gastritis (65). In a study of 108 unselected endoscopy patients, isolation of GCLOs was associated with a past history of ulcer disease or active ulceration, and with histologic appearance of gastritis (26). In one study (33), 20 of 21 persons with duodenal ulcers had an abnormal antral biopsy specimen, and GCLOs were found in 17; GCLOs were absent from the only normal antral specimen. At the

Table 3. Association of Gastric Campylobacter-like Organism in the Antrum and Presence of Peptic Ulceration[a]

Condition	No. of patients	Percentage with GCLO[b]
Duodenal ulcer	119	77.0 ± 10[c]
Gastric ulcer	111	65.5 ± 5
Gastritis without ulceration	238	60.5 ± 9[c]

GCLO, gastric *Campylobacter*-like organism. [a] From References 6–8, 24, 41, 42, 49, and 71. [b] Values expressed as mean ± SEM. [c] $p = 0.003$, two-tailed paired *t*-test.

least, the presence of GCLOs may be a marker of those persons who are at high risk for duodenal ulcer.

Epidemic Hypochlorhydria

Sixteen healthy persons and 1 person with the Zollinger–Ellison syndrome undergoing gastric secretion studies in Dallas, Texas became rapidly and profoundly hypochlorhydric (72). Nine subjects had had a mild illness with abdominal pain, nausea, and vomiting a few days before hypochlorhydria was detected. Severe fundal and, to a lesser extent, antral gastritis was present, but parietal cells appeared normal and parietal cell antibodies were not found. No duodenitis was found. Serum gastrin levels were normal or high but pepsinogen levels were elevated. Acid secretion returned to baseline levels after a mean of 126 days in parallel with an improved histologic appearance of the mucosa. Serologic, cultural, and microscopic studies were unable to define an etiology for this outbreak, which was believed to be due to transmission of an infectious agent from a contaminated pH electrode. Apparently, Dr. Marshall has since observed GCLOs on biopsy specimens from these subjects (73).

Similarly, in a British study of gastric secretion, 4 of 6 previously healthy subjects developed hypochlorhydria after a transient illness with nausea, vomiting, and abdominal pain (74). Biopsy specimens showed active gastritis with polymorphonuclear leukocyte infiltration, which was most severe in the antrum. Decreased basal and peak acid output was observed during an 8-mo follow-up, with evidence of progressive recovery of function. Follow-up biopsy at 8 mo showed clearing of inflammation in 1 patient, but chronic gastritis with plasma cells and lymphocytes in the other 3. No conventional microbial agents were found, but, again, GCLOs were observed on later staining of tissues (cited by Dr. Marshall). Although these two outbreaks strongly point to a transmissable agent as the cause of gastritis leading to hypochlorhydria, and are compatible with

GCLOs being the transmitted agents, they do not rule out the possibility that another agent caused this illness and GCLOs were secondary colonizers.

Serology of Gastric *Campylobacter*-like Organism Infection

Serologic studies of *C. pylori* infection have been performed using a variety of antigens and antibody-detection techniques (9,38,65,75–85; Perez GP and Blaser MJ, unpublished observations). Although the results of all these studies are not in full agreement, the major findings are as follows: (a) patients with gastritis or peptic ulcer disease or dyspeptic symptoms have elevated *C. pylori*-specific serum antibody levels compared with healthy controls; (b) specific serum immunoglobulin A and immunoglobulin G levels appear to correlate best with infection; (c) specific antibodies may be found in gastric secretions; (d) gastritis severity and level of antibody response do not correlate; (e) both in patients with gastritis and in control populations, the percentage of persons with antibodies present rises with age; and (f) once present, specific antibodies persist for years.

Epidemiology of Gastric *Campylobacter*-Like Organism Infections

The epidemiology of GCLO infections is gradually becoming clearer. Studies on four continents suggest that GCLOs are ubiquitous (6–8,24,42,43), usually in association with gastritis. Affected populations have ranged from well-to-do persons in developed countries to indigent persons in developing countries (43). During the outbreak of hypochlorhydria in Dallas, among 37 subjects exposed to the contaminated pH electrode, those who developed hypochlorhydria were significantly younger than those who remained normal (72). These data suggest an age-related immunity to the etiologic agent. The prevalence of GCLO infection, as documented by histologic (66) and serologic studies (77,80–82), rises with age, as does gastritis (86). The source and transmission of GCLOs are unknown. Spiral organisms are common in the mouth and the lower intestinal tract in all mammals, and thus are potential reservoirs for GCLOs. However, neither culture nor histologic studies have yet identified these organisms outside the stomach and duodenum (24). Attempts to isolate GCLOs from bile, saliva, gingival mucosa, colonic contents, feces, duodenal or jejunal fluid, and urethral and vaginal swabs have been negative (27,42,87). Gastric *Campylobacter*-like organisms were not isolated from the nasopharynges or saliva of patients with known *C. pyloridis* infection or gastritis (42,57).

Among 9 GCLO-positive patients, both gastritis and organisms still were present after a mean follow-up of 17 wk; however, 22 GCLO-negative persons remained negative after a mean follow-up of 11 wk (88). Restriction endonuclease analysis of *C. pylori* deoxyribonucleic acid revealed that isolates from 16 patients showed different patterns, indicating that the sources of infection are heterogeneous (89). Multiple isolates from the same patient, however, had identical patterns regardless of colony type, interval between collecting specimens (up to 2 yr), and antimicrobial treatment. After antimicrobial treatment and apparent elimination of GCLOs, recurrence of organisms thus represented recrudescence of the original infection rather than new infection. These and the serologic data suggest that exposure to an infective dose of GCLOs does not occur frequently, and that when organisms are present, they persist for at least several months and probably for years.

Pathophysiology of Gastric *Campylobacter*-Like Organism Infection

The pathophysiology of GCLO infection is not well understood, but preliminary studies have shed light on the organism's niche and relationship to tissue damage. That large numbers of organisms are present adjacent to inflamed tissues suggests that the bacteria are not just passively traversing the stomach from the oropharynx (90); large numbers suggest in situ multiplication. In comparison to *Escherichia coli*, *C. pylori* are able to remain motile in a highly viscous environment, suggesting that these organisms are adapted to the mucus layer of the gastrointestinal tract (56). Gastric *Campylobacter*-like organisms, which are exquisitely sensitive to gastric acidity (24,42), live within mucus and between the mucus layer and the gastric mucosal surface in an environment that has a nearly neutral pH. Colonization of the mucosa appears to be independent of gastric pH (43), possibly because the mucus layer protects GCLOs from gastric acid. Because generation of ammonia may buffer gastric acid, the high urease activity of these organisms also may be adaptive. The presence of these organisms under gastric mucus and their marked sensitivity to acid suggest that these are not directly implanted onto the mucosa after ingestion of contaminated food. An alternative hypothesis is that these organisms colonize esophageal or small intestinal epithelium and migrate to the stomach underneath the protective mucus layer.

In contrast to *C. jejuni*, *C. pylori* does not invade HeLa cells in tissue culture (70). By immunofluorescent staining of clinical specimens, *C. pylori* always is confined to the luminal surface of the gastric mucus-secreting cells (70), but many of the adjacent epithelial cells have lost their microvilli (91). Gastric inflammation without bacteria present occurs in mucosa near focal lesions, such as carcinoma or peptic ulcer. In those cases, leukocytes are seen in the full thickness of the mucosa, whereas only superficial infiltration is seen in association with the bacteria (3). The number of GCLOs and of intraepithelial polymorphonuclear cells present in a biopsy specimen appear highly associated (70), and neutrophils present in the gastric lumen may ingest GCLOs (91). Colonization of affected areas is patchy, with heavily colonized areas adjacent to those with no colonization. Depletion of mucin in the epithelial glands is found in colonized areas but not in noncolonized areas (43). Whether this association is causal is not known, but mucus depletion could expose the gastric epithelium to gastric acid, pepsin, and bile salts.

An intraperitoneal injection of *C. pylori* into 2 rats caused no ill effects (92). No other animal studies have been reported, but a human volunteer ingested 100 million colony forming units of a *C. pylori* strain originally isolated from a patient with nonulcer dyspepsia (92). After 7 days he noted epigastric fullness, vomited once, and had "putrid breath and morning hunger" until taking an antimicrobial agent (tinidazole) to which the organism was susceptible; no further symptoms were noted. Gastric biopsy before the ingestion appeared entirely normal but follow-up biopsy on day 10 showed large numbers of GCLOs, depletion of mucus in epithelial cells, and apparent infiltration of the lamina propria with neutrophils. Repeat biopsy 4 days later, before treatment, showed normalization of findings. No serologic response to infection was detected.

Diagnosis of Gastric *Campylobacter*-like Organism Infection

Diagnosis of GCLO infection can now be made by isolation of the organisms, visualization of organisms in gastric or duodenal specimens, serologic testing, or biochemical assays based on their metabolic activities.

Isolation of GCLOs primarily depends on culture of mucosal tissue, owing to their location under the gastric mucus, rather than from gastric juice where they are present only in very low numbers (7). Yields from culture are highest when specimens are transported in glucose (57,93), are ground (27), or are incubated within 2 h of endoscopy (44). Optimal incubation is at 37°C in an atmosphere containing 5% oxygen, 7% carbon dioxide, 8% hydrogen, and 80% nitrogen with high humidity (27). On subculture, candle jars may be used (33). When gastric antral biopsy specimens were incubated on *Campylobacter*-selective and -nonselective media, GCLOs were isolated (24) in concentrations ranging from about 10,000 to 10 million per gram. Although GCLOs were present in pure culture in 65% of specimens yielding any growth, the other organisms (streptococci, *Enterobacteriaceae*, diphtheroids, and occasional anaerobic bacteria), present in 35% of specimens, overgrew the more fastidious and slow-growing GCLOs in the nonselective medium. Thus, use of a selective medium is justified for attempts at primary isolation (27) even though some *C. pylori* strains may be susceptible to the antimicrobial agents present. One such medium contains brain-heart infusion agar, 7% horse blood, 6 mg/L vancomycin, 20 mg/L naladixic acid, and 2 mg/L amphotericin B (27). Colonial morphology and urease production can be used to make a rapid presumptive identification of *C. pylori* (56). Microbiologic characteristics of the GCLOs for use in identifying the organisms to a species level include vibrio-like morphology on Gram stain, assessment of oxidase, catalase, and urease tests, and antimicrobial susceptibility (Table 1).

Histologic identification of GCLOs in gastric mucosal specimens is highly correlated with culture-positivity (7–9,28,29,33,42,66). Best visualization of GCLOs is in the prepyloric region of the stomach (70), and can be seen on the surface of the gastric epithelium by means of H&E, Giemsa, acridine-orange, and Gram or silver stains. Gastric *Campylobacter*-like organisms stain intensely with silver, such as in the Warthin–Starry technique, but these methods are difficult to standardize. In contrast, the H&E and Giemsa stains are easy to perform but examination of the specimens are slow and tedious. The acridine-orange stain, which is easy to perform and can be done on formalin-fixed paraffin-embedded specimens (94), is sensitive and specific (66). Gram staining of cytology specimens obtained by brushing the gastric mucosa also is simple and may be more effective at showing GCLOs than silver staining of biopsy specimens (43). Examination of minced tissue specimens by phase contrast microscopy is another rapid, simple, sensitive, and inexpensive test (29). The development of a monoclonal antibody specific for *C. pylori* should aid in the use of rapid immunofluorescent detection of these organisms in tissue (95).

Campylobacter pylori are rapidly and strongly urease-positive (7,24,96,97), and there is an excellent correlation between isolation of *C. pylori* and the presence of a positive urease reaction when gastric biopsy specimens are placed in Christenson's medium (96). Whereas *C. pylori* are positive, intestinal campylobacters are negative. A commercially available test in Australia (CLO-test; Delta West Ltd., Canning Vale, Western Australia) detects preformed urease in gastric or duodenal biopsy specimens in 15 min to 3 h (98). Alternatively, a biopsy specimen may be crushed and immediately placed in a urea broth test: most are positive within 1 h (99). The high urease activity is the basis for detection in a noninvasive assay; after a liquid test meal, given to delay gastric emptying, [^{13}C]urea was administered orally to healthy control subjects and persons with GCLO identified by culture (100). In patients with GCLO infection, urea-derived $^{13}CO_2$ appeared in the breath within 20 min and accumulated for more than 100 min, whereas normal subjects showed no release of labeled CO_2. Similarly, *C. pylori*-infected patients have low urea and high ammonia levels in gastric juice compared with uninfected controls (101).

Treatment of Gastric *Campylobacter*-like Organism Infections

Because of the limited effectiveness of acid reduction in the long-term treatment of peptic conditions (102), attention has turned to antimicrobial treatment of GCLOs as an alternative approach. If indeed GCLOs are causative, antimicrobial treatment should be effective; moreover, its efficacy would provide further evidence that these organisms have an etiologic role. *Campylobacter pylori* are susceptible to erythromycin, tetracycline, penicillin, ampicillin, cefoxitin, ciprofloxacin, gentamicin, cephalothin, clindamycin, and kanamycin, but resistant to naladixic acid, sulfonamides, and trimethoprim; some strains are susceptible to metronidazole (6,9,103,104). Spiramycin, cimetidine, ranitidine, and sucralfate have no apparent effect on GCLOs at clinically achievable levels (88,105), but the *C. pylori* isolation rate from patients with peptic ulcer and dyspepsia was significantly lower in those treated with cimetidine than those not treated (105).

Two bismuth salts, tripotassium dicitratobismuthate and bismuth sodium tartrate (Pepto-Bismol, Procter and Gamble, Cincinnati, Ohio), inhibited *C. pylori* at concentrations (between 4 and 32 μg/ml) that may be reached in the gastric lumen (104). Although bismuth compounds have been used for the relief of gastric disorders for >200 yr, the basis of their activity is poorly understood. Bismuth salts have a wide variety of actions that include an antacid effect, coating of the gastric mucosa, decreasing gastric and intestinal motility, increasing mucus secretion, absorbing fluids, and inhibiting the growth of a variety of microorganisms (106). Bismuth preparations are effective in the treatment of traveler's diarrhea (107), and upper gastrointestinal disturbances including nausea, heartburn, or pain. A variety of blinded and placebo-controlled trials have shown the efficacy of bismuth salts for treatment of everything from acute indigestion to duodenal or gastric ulcers (108–112). All of these studies were performed before the identification of GCLOs and thus did not include microbiologic studies.

Lambert and colleagues (113) treated 44 patients who had C. pylori present in the gastric antrum with tripotassium dicitratobismuthate for either 4 or 8 wk, or with cimetidine or placebo. Antral biopsy specimens obtained before the start of and 1 wk after the conclusion of the trial were assessed blindly for the presence of inflammatory cells, atrophy, metaplasia, and dysplasia. The pretrial gastritis scores for all four groups were similar, and the presence of GCLOs was strongly correlated with inflammation. Bacteria were eliminated from 69% and 73% of patients treated with tripotassium dicitratobismuthate for 4 and 8 wk, respectively, but from no patients in the other two groups. The gastritis score improved significantly in patients from whom the bacteria were eliminated, but no histologic improvement was seen in the other patients. Although supporting the role of GCLOs in the pathogenesis of chronic gastritis, the limited data presented do not indicate whether the trial was randomized or blinded. Furthermore, the effect of tripotassium dicitratobismuthate on gastritis in patients in whom GCLOs could not be detected was not examined. This is an important control group in helping to determine whether elimination of C. pylori causes or follows healing of inflammation.

In another preliminary study, treatment with either bismuth subcitrate or amoxicillin was associated with the elimination of GCLOs and histologic improvement of gastritis in the majority of subjects (88). However, relapse was common in the weeks after therapy. In another study, Pepto-Bismol treatment was associated with both clearance of organisms and histologic improvement in gastritis, whereas neither occurred after erythromycin or placebo treatment (114). In a third study, electron microscopy performed 40–100 min after tripotassium dicitratobismuthate treatment showed that GCLOs had detached from the gastric epithelium and that many of the organisms had lysed (115). In a fourth study (116), neither cimetidine nor erythromycin treatment changed the number of GCLOs in the gastric mucosa, whereas tripotassium dicitratobismuthate was effective. Gastric Campylobacter-like organisms remained present 6 mo later, however, and serum antibody levels were unaffected by treatment. The use of amoxycillin and bismuth for combination therapy also has been explored (115).

Furazolidone is a furan derivative with broad-spectrum antibacterial activity including activity toward many Campylobacter species (117) and C. pylori specifically (118). Direct inhibition of gastric secretion also has been reported. Zheng and colleagues (119) treated 70 patients who had endoscopically confirmed peptic ulcers with either furazolidone or placebo for 2 wk. Ulcer healing and reduced upper gastrointestinal pain were significantly more common in the furazolidone-treated group. From an earlier uncontrolled series, 74% responded favorably to furazolidone, and relapse rates over the next 4 yr were considerably lower than in an untreated group. Similarly, metronidazole, to which many C. pylori strains are susceptible, also has been reported as having antiulcer activity (120,121).

Conclusions

The best information on the natural history of primary GCLO infection comes from the voluntary ingestion of C. pylori by a single subject. The results suggested that he suffered an acute but self-limited gastritis, and that C. pylori was causative. Indirect support for this concept comes from the two epidemics of hypochlorhydria in volunteers who apparently ingested contaminated gastric juices. It is clear that these subjects suffered from an acute gastritis, and apparently GCLOs were visualized on smears of injured tissue. Their gastritis, however, was more fundal than antral, and although C. pylori were present, it is not certain that they were causal.

There is unquestionably a striking association between the presence of GCLOs and chronic antral gastritis. In favor of a pathogenic role for the GCLO is (a) the serologic response of infected hosts to C. pylori, (b) that ingestion of GCLOs by polymorphonuclear leukocytes has been observed, and (c) that C. pylori is apparently able to cause an acute gastritis. None of these observations, however, proves causality of chronic gastritis, and finding similar organisms in healthy ferrets (34,35) suggests that they may not be pathogenic. That the human volunteer was apparently able to clear his infection, and that the presence of these organisms appears age-related argues that these may be secondary colonizers in persons with reduced gastric acidity or other deficits. Overgrowth of conventional bacterial organisms in the stomachs of patients with reduced acidity is a well-described phenomenon (122,123). Although these acid-sensitive organisms appear to reside in a niche largely protected from low pH, it may be that hypochlorhydria of any etiology removes a major inhibitory factor on their growth. Hypochlorhydria probably is not the only factor that permits multiplication of these organisms in vivo; the presence of tissue damaged by other causes may be another. That cimetidine-treated patients with peptic disease had a lower isolation rate of *C. pylori* than untreated patients (105) also suggests that the organisms may be colonizers of damaged mucosa. Clearly, GCLOs have a tropism for gastric mucosa, as confirmed in a complementary fashion by the observations concerning gastric metaplasia into the duodenum and intestinal metaplasia into the stomach. At present, however, there is insufficient evidence to conclude that the GCLOs cause chronic gastritis or peptic ulcer disease. Nevertheless, GCLOs are certainly a marker for these processes, and ascertainment of their presence is at least potentially valuable in that regard.

For the question of causality to be fairly tested, better definition and characterization of the term "gastritis" is needed. It is not clear from the literature how many different histologic types have been recognized, what their natural histories are, and with which types GCLOs are associated. The presence of GCLOs in patients who have gastritis due to known causes, such as ingestion of nonsteroidal antiinflammatory agents, would suggest that colonization is secondary to injury; their absence over time would suggest that injury does not necessarily predispose to colonization. Results of such studies have not yet been reported. Further volunteer studies will probably best help determine the role of GCLOs in acute gastritis and provide some information on the natural history of these infections. Before such studies are done, it will be necessary to ensure that the organisms can be eradicated from their gastric niche. Finally, double-blinded placebo-controlled studies using treatment modalities that only have antibacterial effects, that are effective against most GCLOs, and that are not inactivated in the gastric environment can help solve the problem of causality for chronic gastritis. Bismuth compounds, because of their varied effects, are probably not satisfactory for such trials.

The rediscovery of spiral gastric organisms and the cultivation of *Campylobacter pylori* has opened a new era in gastric microbiology that has great clinical relevance. The next few years may provide answers to the perplexing problem of chronic idiopathic gastritis and possibly provide insights into the pathogenesis of peptic ulcer disease. Studies to clarify the role of GCLOs in the pathogenesis of these conditions should be a high priority. At the present, however, clinicians might wait for more definitive clinical studies before attempting to obtain cultures from affected patients or initiating specific antimicrobial treatment.

References

1. Langman MJS. The epidemiology of chronic digestive diseases. London: Edward Arnold, 1979.
2. Steer HW. Ultrastructure of cell migration through the gastric epithelium and its relationship to bacteria. J Clin Pathol 1975;28:639–46.
3. Warren JR, Marshall B. Unidentified cured bacilli on gastric epithelium in active chronic gastritis. Lancet 1983;i:1273.
4. Rollason TP, Stone J, Rhode JM. Spiral organisms in endoscopic biopsies of the human stomach. J Clin Pathol 1984; 37:23–6.
5. Marshall BJ. Unidentified curved bacilli on gastric epithelium in active chronic gastritis. Lancet 1983;ii:1273–5.
6. Marshall BJ, Warren JR. Unidentified curved bacilli in the stomach of patients with gastritis and peptic ulceration. Lancet 1984;i:1311–4.
7. Langenberg M-L, Tytgat GNJ, Schipper MEI, Rietra PJGM, Zanen HC. Campylobacter-like organisms in the stomach of patients and healthy individuals. Lancet 1984;i:1348.
8. McNulty CMA, Watson DM. Spiral bacteria of the gastric antrum. Lancet 1984;i:1068–9.
9. Jones DM, Lessells AM, Eldridge J. Campylobacter-like organisms on the gastric mucosa: culture, histological, and serological studies. J Clin Pathol 1984;37:1002–6.
10. Bizzozero G. Ueber die schlauchformigen drusen des magendarmkanals und die beziehungen ihres epithels zu dem oberflachenepithel der schleimhaut. Arch f mikr Anast 1893;42:82.
11. Salomon H. Ueber das spirillam des Saugertiermagens und sein Verhalten zu den Belegzellen. Zentrablatt fur Bacteriologie 1896;19:433.
12. Krienitz W. Ueber das Auftreten von mageninhalt bei carcinoma ventriculi. Dtsch Med Wochenschr 1906;22:872.
13. Rosenow EC, Sandford AH. The bacteriology of ulcer of the stomach and duodenum in man. J Infect Dis 1915;17:210–6.
14. Celler HL, Thalheimer W. Bacteriological and experimental studies on gastric ulcer. J Exp Med 1916;23:791–800.
15. Appelmans R, Vassiliadis P. Etude sur la flore microbienne des ulcers gastro-duodeneaux et des cancers gastriques. Rev Belge Sci Med 1932;4:198–203.
16. Freedburg AS, Barron LE. The presence of spirochetes in the human gastric mucosa. Am J Dig Dis 1940;7:443–5.
17. Seeley GP, Colp R. The bacteriology of peptic ulcers and gastric malignancies: possible bearing on complications following gastric surgery. Surgery 1941;10:369–80.
18. Doenges JL. Spirochaetes in the gastric glands of *Macacus rhesus* and of man without related disease. Arch Pathol 1939;27:469.
19. Steer HW, Colin-Jones DG. Mucosal changes in gastric ulceration and their response to carbenoxolone sodium. Gut 1975;16:590–7.
20. Marshall BJ, Royce H, Annear DI, et al. Original isolation of *Campylobacter pyloridis* from human gastric mucosa. Microbios Lett 1984;25:83–8.
21. Jones DM, Curry A, Fox AJ. An ultrastructural study of the gastric Campylobacter-like organism "Campylobacter pyloridis". J Gen Microbiol 1985;131:2335–41.
22. Goodwin S, Blincow E, Armstrong J, McCulloch R, Collins D. Campylobacter pyloridis is unique: GCLO-2 is an ordinary campylobacter. Lancet 1985;ii:39.
23. Goodwin CS, McCulloch RK, Armstrong JA, Wee SH. Unusual cellular fatty acids and distinctive ultrastructure in a new spiral bacterium (*Campylobacter pyloridis*) from the human gastric mucosa. J Med Microbiol 1985;19:257–67.
24. Buck GE, Gourley WK, Lee WK, Subramanyan K, Latimer

JM, DiNuzzo AR. Relation of *Campylobacter pyloridis* to gastritis and peptic ulcer. J Infect Dis 1986;153:664–9.

25. Hudson MJ, Wait R. Cellular fatty acids of campylobacter species with particular reference to *Campylobacter pyloridis*. Pearson AD, Skirrow MB, Lior H, Rowe B, eds. *Campylobacter III*. Proceedings of the third international workshop on campylobacter infections. London: PHLS, 1985:198–9.

26. Pearson AD, Bamforth J, Booth L, et al. Polyacrylamide gel electrophoresis of spiral bacteria from the gastric antrum. Lancet 1984;i:1349–50.

27. Goodwin CS, Blincow E, Warren JR, Waters TE, Sanderson CR, Easton L. Evaluation of cultural techniques for isolating *Campylobacter pyloridis* from endoscopic biopsies of gastric mucosa. J Clin Pathol 1985;138:1127–31.

28. Kasper G, Dickgiesser N. Isolation of campylobacter-like bacteria from gastric epithelium. Infection 1984;12:179–80.

29. Pinkard KJ, Harrison B, Capstick JA, Medley G, Lambert JR. Detection of *Campylobacter pyloridis* in gastric mucosa by phase contrast microscopy. J Clin Pathol 1986;39:112–3.

30. Curry A, Jones DM, Eldridge J, Fox AJ. Ultrastructure of *Campylobacter pyloridis*—not a campylobacter? Pearson AD, Skirrow MB, Lior H, Rowe B, eds. Campylobacter III. Proceedings of the Third International Workshop on Campylobacter infections. London: PHLS, 1985:195.

31. Kasper G, Dickgiesser N. Isolation from gastric epithelium of *Campylobacter*-like bacteria that are distinct from "*Campylobacter pyloridis*". Lancet 1985;i:111–2.

32. Kasper G, Owen RJ. Characteristics of a new group of campylobacter-like organisms (CLOs) from gastric epithelium. Pearson AD, Skirrow MB, Lior H, Rowe B, eds. *Campylobacter III*. Proceedings of the Third International Workshop on *Campylobacter* infections. London: PHLS, 1985:203.

33. Price AB, Levi J, Dolby, et al. *Campylobacter pyloridis* in peptic ulcer disease: microbiology, pathology, and scanning electron microscopy. Gut 1985;26:1183–8.

34. Fox JG, Edrise BM, Cabot EB, Beaucage C, Murphy JC, Prostak KS. *Campylobacter*-like organisms isolated from gastric mucosa of ferrets. Am J Vet Res 1986;47:236–9.

35. Rathbone BJ, West AP, Wyatt JI, Johnson AW, Tompkins DS, Heatley RV. *Campylobacter pyloridis*, urease and gastric ulcers. Lancet 1986;ii:400–1.

36. Bolton FJ, Holt AV, Hutchinson DN. Urease-positive thermophilic campylobacters. Lancet 1985;i:1217.

37. Lior H, Pearson AD, Woodward DL, Hawtin P. Biochemical and serological characteristics of *Campylobacter pyloridis*. Pearson AD, Skirrow MB, Lior H, Rowe B, eds. Campylobacter III. Proceedings of the Third International Workshop on Campylobacter infections. London: PHLS, 1985:196–7.

38. Rathbone BJ, Wyatt JI, Worsley BW, Trejdosiewicz LK, Heatley RV, Losowsky MS. Immune response to *Campylobacter pyloridis*. Lancet 1985;i:1217.

39. Perez-Perez GI, Blaser MJ. Conservation and diversity of *Campylobacter pyloridis* major antigens. Infect Immun 1987;55:1256–63.

40. Newell DG. The outer membrane proteins and surface antigens of *Campylobacter pyloridis*. Pearson AD, Skirrow MB, Lior H, Rowe B, eds. Campylobacter III. Proceedings of the Third International Workshop on Campylobacter infections. London: PHLS, 1985:199–200.

41. Burnett RA, Forrest JAH, Girdwood RWA, Fricker CR. *Campylobacter*-like organisms in the stomach of patients and healthy individuals. Lancet 1984;i:1349.

42. Marshall BJ, McGechie DB, Rogers PA, Glancy RJ. Pyloric campylobacter infection and gastroduodenal disease. Med J Aust 1985;142:439–44.

43. Gilman RJ, Leon-Barua R, Koch J, et al. Rapid identification of pyloridisc campylobacter in Peruvians with gastritis. Dig Dis Sci 1986;31:1089–94.

44. Arnot RS. Gastritis and campylobacter infection. Med J Aust 1985;142:100–1.

45. Pettross CW, Cohen H, Appleman MD, Valenzuela JE, Chandrasoma P. *Campylobacter pyloridis*: relationship to peptic disease, gastric inflammation and other conditions (abstr). Gastroenterology 1986;90:1585.

46. Lopez-Brea M, Jimenez ML, Blanco M, Pajares JM. Isolation of *Campylobacter pyloridis* from patients with and without gastroduodenal pathology. Pearson AD, Skirrow MB, Lior H, Rowe B, eds. Campylobacter III. Proceedings of the Third International Workshop on Campylobacter infections. London: PHLS, 1985:193–194.

47. Thomas JM, Poynter D, Gooding C, et al. Gastric spiral bacteria. Lancet 1984;ii:100.

48. Tricottet V, Bruneval P, Vire O, Camilleri JP. *Campylobacter*-like organisms and surface epithelium abnormalities in active, chronic gastritis in humans: an ultrastructural study. Ultrastruct Pathol 1986;10:113–22.

49. Kalenic S, Faliseva V, Scukanec-Spoljar M, Vodopija I. *Campylobacter pyloridis* in the gastric mucosa of patients with gastritis and peptic ulcer. Pearson AD, Skirrow MB, Lior H, Rowe B, eds. Campylobacter III. Proceedings of the Third International Workshop on Campylobacter infections. London: PHLS, 1985:193.

50. Peterson WL, Lee EL, Feldman M. Gastric *Campylobacter*-like organisms in healthy humans: correlation with endoscopic appearance and mucosal histology (abstr). Gastroenterology 1986;90:1585.

51. Pearson AD, Ireland A, Holdstock G, et al. Clinical and pathological correlates of *Campylobacter pyloridis* isolated from gastric biopsy specimens. Pearson AD, Skirrow MB, Lior H, Rowe B, eds. Campylobacter III. Proceedings of the Third International Workshop on Campylobacter infections. London: PHLS, 1985:181–2.

52. Barthel JS, Westblum TU, Havey AD, Gonzalez FJ, Everett ED. Pyloric *Campylobacter*-like organisms (PCLOs) in asymptomatic volunteers (abstr). Gastroenterology 1986;90:1338.

53. Drumm B, O'Brien A, Cutz E, Sherman P. *Campylobacter pyloridis* are associated with primary antral gastritis in the pediatric population (abstr). Gastroenterology 1986;90:1399.

54. Cadranel S, Goosens H, DeBoeck M, Malengreau A, Rodesch P, Butzler JP. *Campylobacter pyloridis* in children. Lancet 1986;i:735–6.

55. Hill R, Pearman J, Worthy P, Caruso V, Goodwin S, Blincow E. *Campylobacter pyloridis* and gastritis in children. Lancet 1986;i:387.

56. Hazell SL, Lee A, Brady L, Hennessey W. *Campylobacter pyloridis* and gastritis: association with intracellular spaces and adaptation to an environment of mucus as important factors in colonization of the gastric epithelium. J Infect Dis 1986;153:658–63.

57. Goodwin CS, Armstrong JA, Marshall BJ. *Campylobacter pyloridis*, gastritis, and peptic ulceration. J Clin Pathol 1986;39:353–65.

58. Shoushia S, Bull TR, Parkins RA. Gastric spiral bacteria. Lancet 1984;ii:101.

59. Phillips AD, Hine KR, Holmes GKT, Woodings DF. Gastric spiral bacteria. Lancet 1984;ii:101.

60. Thomas JM. *Campylobacter*-like organisms in gastritis. Lancet 1984;ii:1217.

61. Strickland RG, Mackay IR. A reappraisal of the nature and significance of chronic atrophic gastritis. Am J Dig Dis 1973;18:426–40.

62. O'Connor HJ, Axon ATR, Dixon MF. *Campylobacter*-like

organisms unusual in type A (pernicious anaemia) gastritis. Lancet 1984;ii:1091.

63. O'Connor JH, Wyatt JI, Dixon MF, Axon ATR. *Campylobacter*-like organisms and reflux gastritis. J Clin Pathol 1986; 39:531–4.

64. O'Connor HJ, Dixon MF, Wyatt JI, et al. Effect of duodenal ulcer surgery and enterogastric reflux on *Campylobacter pyloridis*. Lancet 1986;ii:1178–81.

65. von Wulffen H, Heesemann J, Butzow GH, Loning T, Laufs R. Detection of *Campylobacter pyloridis* in patients with antrum gastritis and peptic ulcers by culture, complement fixation test, and immunoblot. J Clin Microbiol 1986;24: 716–20.

66. Rawles JW, Paull G, Yardley JH, et al. Gastric *Campylobacter*-like organisms (CLO) in a U.S. hospital population (abstr). Gastroenterology 1986;90:1599.

67. Salmeron M, Desplaces N, Lavergne A, Houdart R. *Campylobacter*-like organisms and acute purulent gastritis. Lancet 1986;ii:459.

68. Greenlaw R, Sheahan DG, DeLuca V, Miller D, Myerson D, Myerson P. Gastroduodenitis. A broader concept of peptic ulcer disease. Dig Dis Sci 1980;25:660–72.

69. Gear MWL, Truelove SC, Whitehead R. Gastric ulcer and gastritis. Gut 1971;12:639–42.

70. Steer H, Newell DG. Mucosa-related bacteria in benign peptic ulceration. Pearson AD, Skirrow MB, Lior H, Rowe B, eds. Campylobacter III. Proceedings of the Third International Workshop on Campylobacter infections. London: PHLS, 1985:173–4.

71. Bohnen J, Krajden S, Kempston J, Andeson J, Karmali M. *Campylobacter pyloridis* in Toronto. Pearson AD, Skirrow MB, Lior H, Rowe B, eds. Campylobacter III. Proceedings of the Third International Workshop on Campylobacter infections. London: PHLS, 1985:175–7.

72. Ramsey EJ, Carey KV, Peterson WL, et al. Epidemic gastritis with hypochlorhydria. Gastroenterology 1979;76:1449–57.

73. Marshall BJ. *Campylobacter pyloridis* and gastritis. J Infect Dis 1986;153:650–7.

74. Gledhill T, Leicester RJ, Addis B, et al. Epidemic hypochlorhydria. Br Med J 1985;290:1383–6.

75. Morris A, Nicholson G, Lloyd G, Haines D, Rogers A, Taylor D. Seroepidemiology of *Campylobacter pyloridis*. NZ Med J 1986;99:657–9.

76. Rathbone BJ, Trejdosiewicz LK, Heatley RV, Losowsky MS, Wyatt JI, Worsley BW. Epidemic hypochlorhydria. Br Med J 1985;291:52–3.

77. Kaldor J, Tee W, McCarthy P, Watson J, Dwyer B. Immune response to *Campylobacter pyloridis* in patients with peptic ulceration. Lancet 1985;i:921.

78. Eldridge J, Lessells AM, Jones DM. Antibody to spiral organisms on gastric mucosa. Lancet 1984;i:1237.

79. McNulty CAM, Crump B, Gearty J, et al. The distribution of and serological response to *Campylobacter pyloridis* in the stomach and duodenum. Pearson AD, Skirrow MB, Lior H, Rowe B, eds. Campylobacter III. Proceedings of the Third International Workshop on Campylobacter infections. London: PHLS, 1985:174–5.

80. Marshall BJ, McGechie DB, Francis GJ, Utley PJ. Pyloric campylobacter serology. Lancet 1984;ii:281.

81. Hawtin P, Pearson AD, McBride H, Gibson J, Booth L. Specific IgG and IgA responses to *Campylobacter pyloridis* in man. Pearson AD, Skirrow MB, Lior H, Rowe B, eds. Campylobacter III. Proceedings of the Third International Workshop on Campylobacter infections. London: PHLS, 1985:186–7.

82. Hutchinson DN, Bolton FJ, Hinchliffe PM, Holt AV. Distribution in various clinical groups of antibody to *Campylobacter pyloridis* detected by ELISA, complement fixation and microagglutination tests. Pearson AD, Skirrow MB, Lior H, Rowe B, eds. Campylobacter III. Proceedings of the Third International Workshop on Campylobacter infections. London: PHLS, 1985:185.

83. Marshall BJ, Whisson M, Francis G, McGechie D. Correlation between symptoms of dyspepsia and *Campylobacter pyloridis* serology in Western Australian blood donors. Pearson AD, Skirrow MB, Lior H, Rowe B, eds. Campylobacter III. Proceedings of the Third International Workshop on Campylobacter infections. London: PHLS, 1985:188–9.

84. Morris A, Lloyd G, Nocholson G. *Campylobacter pyloridis* serology among gasteoendoscopy clinic staff. NZ Med J 1986;99:820–1.

85. Ishii E, Inoue H, Tsuyuguchi T, et al. *Campylobacter*-like organisms in cases of stomach diseases in Japan. Pearson AD, Skirrow MB, Lior H, Rowe B, eds. Campylobacter III. Proceedings of the Third International Workshop on Campylobacter infections. London: PHLS, 1985:179–180.

86. Siurala M, Isokoski M, Varis K, Kekki M. Prevalence of gastritis in a rural population. Bioptic study of subjects selected at random. Scand J Gastroenterology 1968;3: 211–23.

87. Eldridge J, Jones DM, Sethi P. The occurrence of antibody to the *Campylobacter pyloridis* in various groups of individuals. Pearson AD, Skirrow MB, Lior H, Rowe B, eds. Campylobacter III. Proceedings of the Third International Workshop on Campylobacter infections. London: PHLS, 1985: 183–4.

88. Langenberg M-L, Rauws EAJ, Schipper MEI, et al. The pathogenic role of *Campylobacter pyloridis* studied by attempts to eliminate these organisms. Pearson AD, Skirrow MB, Lior H, Rowe B, eds. Campylobacter III. Proceedings of the Third International Workshop on Campylobacter infections. London: PHLS, 1985:162–3.

89. Langenberg W, Rauws EA, Widjojokusumo A, Tytgat GNJ, Zanen HC. Identification of *Campylobacter pyloridis* isolates by restriction endonuclease DNA analysis. J Clin Microbiol 1986;24:414–7.

90. Fricker CR. Adherence of bacteria associated with active chronic gastritis to plastics used in the manufacture of fibreoptic endoscopes. Lancet 1984;i:800.

91. Lee WK, Gourley WK, Buck GE, Subraanyam K. A study of *Campylobacter pyloridis* in patients with gastritis and gastric ulcer. Pearson AD, Skirrow MB, Lior H, Rowe B, eds. Campylobacter III. Proceedings of the Third International Workshop on Campylobacter infections. London: PHLS, 1985:178–9.

92. Marshall BJ, Armstrong JA, McGechie DB, Clancy RJ. Attempt to fulfil Koch's postulates for pyloridisc campylobacter. Med J Aust 1985;142:436–9.

93. Morris A, Gillies M, Patton K. Detection of *Campylobacter pyloridis* infection. NZ Med J 1986;99:336.

94. Walters LL, Budin RE, Paull G. Acridine-orange to identify *Campylobacter pyloridis* in formalin fixed paraffin-embedded gastric biopsies. Lancet 1986;i 42.

95. Engstrand L, Pahlson C, Gustavsson S, Schwan A. Monoclonal antibodies for rapid identification of *Campylobacter pyloridis*. Lancet 1986;ii:1403–4.

96. McNulty CAM, Wise R. Rapid diagnosis of *Campylobacter*-associated gastritis. Lancet 1985;i:1443–4.

97. Owen RJ, Martin SR, Borman P. Rapid urea hydrolysis by gastric campylobacters. Lancet 1985;i:111.

98. Morris A, McIntyre D, Rose T, Nicholson G. Rapid diagnosis of *Campylobacter pyloridis* infection. Lancet 1986;i:149.

99. McNulty CAM, Wise R. Rapid diagnosis of *Campylobacter pyloridis* gastritis. Lancet 1986;i:387.

100. Graham DY, Klein PD, Evans DG, et al. Rapid noninvasive diagnosis of gastric *Campylobacter* by a ^{13}C urea breath test (abstr). Gastroenterology 1986;90:1435.

101. Marshall BJ, Langton SR. Urea hydrolysis in patients with *Campylobacter pyloridis* infection. Lancet 1986;i:965–6.

102. Nyren O, Adami H-O, Bates S, et al. Absence of therapeutic benefit from antacids or cimetidine in non-ulcer dyspepsia. N Engl J Med 1986;314:339–43.

103. Kasper G, Dickgiesser N. Antibiotic sensitivity of "*Campylobacter pyloridis*". Eur J Clin Microbiol 1984;3:444.

104. McNulty CAM, Dent J, Wise R. Susceptibility of clinical isolates of *Campylobacter pyloridis* to 11 antimicrobial agents. Pearson AD, Skirrow MB, Lior H, Rowe B, eds. Campylobacter III. Proceedings of the Third International Workshop on Campylobacter infections. London: PHLS, 1985:169–70.

105. Goodwin CS, Blake P, Blincow E. The minimum inhibitory and bactericidal concentrations of antibiotics and anti-ulcer agents against *Campylobacter pyloridis*. J Antimicrob Chemother 1986;17:309–14.

106. Wilson TR. The pharmacology of tri-potassium di-citrato bismuthate (TDB). Postgrad Med J 1975;51 (Suppl 5):18–21.

107. Ericsson CD, DuPont HL, Johnson PC. Nonantibiotic therapy for travelers' diarrhea. Rev Infect Dis 1986;8:S202–6.

108. Hailey FJ, Newsom JH. Evaluation of bismuth subsalicylate in relieving symptoms of indigestion. Arch Intern Med 1984;144:269–72.

109. Weiss G, Serfontein WJ. The efficacy of a bismuth-protein-complex compound in the treatment of gastric and duodenal ulcers. S Afr Med J 1971;45:467–70.

110. Moshal MG. A double-blind gastroscopic study of a bismuth-peptide complex in gastric ulceration. S Afr Med J 1974;48:1610–1.

111. Salmon PR, Brown P, Williams R, Read AE. Evaluation of colloidal bismuth (De-Nol) in the treatment of duodenal ulcer employing endoscopic selection and follow-up. Gut 1974;15:189–93.

112. Martin DF, Hollanders D, May SJ, Ravenscroft MM, Tweedle DEF, Miller JP. Differences in relapse rates of duodenal ulcer after healing with cimetidine or tripotassium dicitrato bismuthate. Lancet 1981;i:7–10.

113. Lambert JR, Dunn KL, Turner H, Korman MG. Effect on histological gastritis following eradication of *Campylobacter pyloridis* (abstr). Gastroenterology 1986;90:1509.

114. McNulty CAM, Gearty JC, Crump B, et al. *Campylobacter pyloridis* and associated gastritis: investigator blind, placebo controlled trial of bismuth salicylate and erythromycin ethylsuccinate. Br Med J 1986;293:645–9.

115. Marshall BJ, Armstrong JA, McGechie DB, Francis GJ. The antibacterial action of bismuth: early results of antibacterial regimens in the treatment of duodenal ulcer. Pearson AD, Skirrow MB, Lior H, Rowe B, eds. Campylobacter III. Proceedings of the Third International Workshop on Campylobacter infections. London: PHLS, 1985:165–6.

116. Jones DM, Eldridge J, Whorwell PJ, Miller JP. The effects of various anti-ulcer regimens and antibiotics on the presence of *Campylobacter pyloridis* and its antibody. Pearson AD, Skirrow MB, Lior H, Rowe B, eds. Campylobacter III. Proceedings of the Third International Workshop on Campylobacter infections. London: PHLS, 1985:161.

117. Wang W-LL, Reller LB, Blaser MJ. Comparison of antimicrobial susceptibility patterns of *Campylobacter jejuni* and *Campylobacter coli*. Antimicrob Agents Chemother 1984; 26:351–3.

118. Howden A, Boswell P, Tovey F. In-vitro sensitivity of *Campylobacter pyloridis* to furazolidone. Lancet 1986; ii:1035.

119. Zheng Z-T, Wang Z-Y, Chu YX, et al. Double-blind short term trial of furazolidone in peptic ulcer. Lancet 1985; i:1048–9.

120. Quintero Diaz M, Sotto Escobar A. Metronidazole versus cimetidine in the treatment of gastroduodenal ulcer. Lancet 1986;i:907.

121. Shirokova KI, Filomonow RM, Poliakova LV. Metronidazole in the treatment of peptic ulcer. Klin Med (Mosk) 1981; 59:48–50.

122. Drasar BS, Shiner M, McLoed GM. The bacterial flora of the gastrointestinal tract in healthy and achlorhydric persons. Gastroenterology 1969;56:71–9.

123. Reed PL, Haines K, Smith PLR, House FR, Walters CL. Gastric juice nitrosamines in health and gastroduodenal disease. Lancet 1981;ii:550–2.

The Bacteria behind Ulcers

One half to one third of the world's population harbors Helicobacter pylori, *"slow" bacteria that infect the stomach and can cause ulcers and cancer there*

by Martin J. Blaser

mystery
notice
"story"
mode

In 1979 J. Robin Warren, a pathologist at the Royal Perth Hospital in Australia, made a puzzling observation. As he examined tissue specimens from patients who had undergone stomach biopsies, he noticed that several samples had large numbers of curved and spiral-shaped bacteria. Ordinarily, stomach acid would destroy such organisms before they could settle in the stomach. But those Warren saw lay underneath the organ's thick mucus layer—a lining that coats the stomach's tissues and protects them from acid. Warren also noted that the bacteria were present only in tissue samples that were inflamed. Wondering whether the microbes might somehow be related to the irritation, he looked to the literature for clues and learned that German pathologists had witnessed similar organisms a century earlier. Because they could not grow the bacteria in culture, though, their findings had been ignored and then forgotten.

Warren, aided by an enthusiastic young trainee named Barry J. Marshall, also had difficulty growing the unknown bacteria in culture. He began his efforts in 1981. By April 1982 the two men had

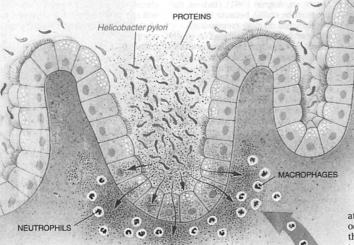

attempted to culture samples from 30-odd patients—all without success. Then the Easter holidays arrived. The hospital laboratory staff accidentally held some of the culture plates for five days instead of the usual two. On the fifth day, colonies emerged. The workers christened them *Campylobacter pyloridis* because they resembled pathogenic bacteria of the *Campylobacter* genus found in the intestinal tract. Early in 1983 Warren and Marshall published their first report, and within months scientists around the world had isolated

ULCER-CAUSING BACTERIA (*Helicobacter pylori*) live in the mucus layer (*pale yellow*) lining the stomach. There they are partially protected from the stomach's acid (*pink*). The organisms secrete proteins that interact with the stomach's epithelial cells and attract macrophages and neutrophils, cells that cause inflammation (*left*). The bacteria further produce urease, an enzyme that helps to break down urea into ammonia and carbon dioxide; ammonia can neutralize stomach acid (*center*). H. pylori also secrete toxins that contribute to the formation of stomach ulcers (*right*). The microbes typically collect in the regions shown in the diagram at the far right.

Educated science audience

notice clarity of explanations

the bacteria. They found that it did not, in fact, fit into the *Campylobacter* genus, and so a new genus, *Helicobacter*, was created. These researchers also confirmed Warren's initial finding, namely that *Helicobacter pylori* infection is strongly associated with persistent stomach inflammation, termed chronic superficial gastritis.

The link led to a new question: Did acquire chronic superficial gastritis. Left untreated, both the infection and the inflammation last for decades, even a lifetime. Moreover, this condition can lead to ulcers in the stomach and in the duodenum, the stretch of small intestine leading away from the stomach. *H. pylori* may be responsible for several forms of stomach cancer as well.

More than 40 years ago doctors rec-

essary for ulcers to form, it is not sufficient to explain their occurrence—most patients with ulcers have normal amounts of stomach acid, and some people who have high acid levels never acquire ulcers.

Nevertheless, the stress-acid theory of ulcers gained further credibility in the 1970s, when safe and effective agents to reduce gastric acid were introduced.

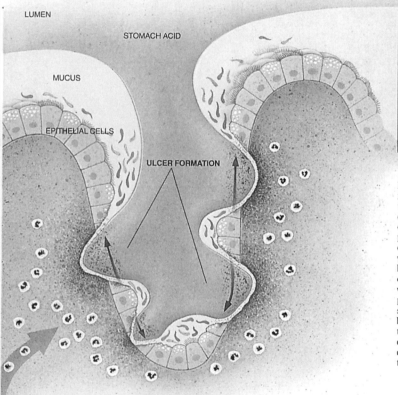

LUMEN

STOMACH ACID

MUCUS

EPITHELIAL CELLS

ULCER FORMATION

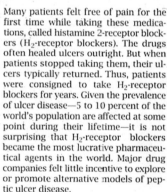

Many patients felt free of pain for the first time while taking these medications, called histamine 2-receptor blockers (H_2-receptor blockers). The drugs often healed ulcers outright. But when patients stopped taking them, their ulcers typically returned. Thus, patients were consigned to take H_2-receptor blockers for years. Given the prevalence of ulcer disease—5 to 10 percent of the world's population are affected at some point during their lifetime—it is not surprising that H_2-receptor blockers became the most lucrative pharmaceutical agents in the world. Major drug companies felt little incentive to explore or promote alternative models of peptic ulcer disease.

economic tie-in & public interest

The Bacteria and Ulcer Disease

In fact, ulcers can result from medications called nonsteroidal anti-inflammatory agents, which include aspirin and are often used to treat chronic arthritis. But all the evidence now indicates that *H. pylori* cause almost all cases of ulcer disease that are not medication related. Indeed, nearly all patients having such ulcers are infected by *H. pylori,* versus some 30 percent of age-matched control subjects in the U.S., for example. Nearly all individuals with ulcers in the duodenum have *H. pylori* present there. Studies show that *H. pylori* infection and chronic gastritis increase from three to 12 times the risk of a peptic ulcer developing within 10 to 20 years of infection with the bacte-

the inflamed tissue somehow invite *H. pylori* to colonize there, or did the organisms actually cause the inflammation? Research proved the second hypothesis correct. In one of the studies, two male volunteers—Marshall included—actually ingested the organisms [*see box on next page*]. Both had healthy stomachs to start and subsequently developed gastritis. Similarly, when animals ingested *H. pylori*, gastritis ensued. In other investigations, antibiotics suppressed the infection and alleviated the irritation. If the organisms were eradicated, the inflammation went away, but if the infection recurred, so did the gastritis. We now know that virtually all people infected by *H. pylori*

ognized that most people with peptic ulcer disease also had chronic superficial gastritis. For a variety of reasons, though, when the link between *H. pylori* infection and gastritis was established, the medical profession did not guess that the bacteria might prompt peptic ulcer disease as well. Generations of medical students had learned instead that stress made the stomach produce more acid, which in turn brought on ulcers. The theory stemmed from work carried out by the German scientist K. Schwartz. In 1910, after noting that duodenal ulcers arose only in those individuals who had acid in their stomachs, he coined the phrase "No acid, no ulcer." Although gastric acidity is nec-

Don't Try This at Home

MICHAEL BAILEY University of Virginia

Barry J. Marshall (*left*) of the Royal Perth Hospital in Australia made headlines after he announced in 1985 that he had ingested *Helicobacter pylori*. Marshall hoped to demonstrate that the bacteria could cause peptic ulcer disease. Marshall did in fact develop a severe case of gastritis, but the painful inflammation vanished without treatment.

Two years later Arthur J. Morris and Gordon I. Nicholson of the University of Auckland in New Zealand reported the case of another volunteer who wasn't so lucky. This man, a healthy 29-year-old, showed signs of infection for only 10 days, but his condition lasted much longer. On the 67th day of infection, the volunteer started treatment with bismuth subsalicylate (Pepto-Bismol). Five weeks later a biopsy indicated that the medication had worked. But a second biopsy taken nine months after the first showed that both the infection and the gastritis had recurred. Only when the subject received two different antibiotics as well as bismuth subcitrate was his infection finally cured three years later.   —M.J.B.

ria. Most important, antimicrobial medications can cure *H. pylori* infection and gastritis, thus markedly lowering the chances that a patient's ulcers will return. But few people can overcome *H. pylori* infection without specific antibiotic treatment.

notice definition When someone is exposed to *H. pylori*, his or her immune system reacts by making antibodies, molecules that can bind to and incapacitate some of the invaders. These antibodies cannot eliminate the microbes, but a blood test readily reveals the presence of antibodies, and so it is simple to detect infection. Surveys consistently show that one third to one half of the world's population carry *H. pylori*. In the U.S. and western Europe, children rarely become infected, but more than half of all 60-year-olds have the bacteria. In contrast,

60 to 70 percent of the children in developing countries show positive test results by age 10, and the infection rate remains high for adults. *H. pylori* infection is also common among institutionalized children.

Although it is as yet unclear how the organisms pass from one person to another, poor sanitation and crowding clearly facilitate the process. As living conditions have improved in many parts of the world during the past century, the rate of *H. pylori* infection has decreased, and the average age at which the infection is acquired has risen. Gastric cancer has also become progressively less common during the past 80 years. At the start of the 20th century, it was the leading cause of death from cancer in the U.S. and many other developed countries. Now it is far down on

the list. The causes for its decline are not well understood, but we have reason to believe that the drop in *H. pylori* infection rates deserves some credit.

A Connection to Cancer

In the 1970s Pelayo Correa, now at Louisiana State University Medical Center, proposed that gastric cancer resulted from a series of changes in the stomach taking place over a long period. In Correa's model, a normal stomach would initially succumb to chronic superficial gastritis for unknown reasons. We now know that *H. pylori* are to blame. In the second step—lasting for perhaps several decades—this gastritis would cause more serious harm in the form of a lesion, called atrophic gastritis. This lesion might then lead to fur-

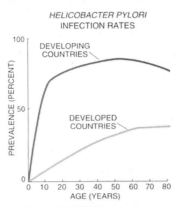

HELICOBACTER PYLORI INFECTION RATES

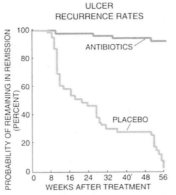

ULCER RECURRENCE RATES

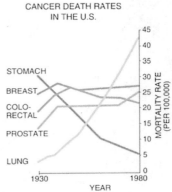

CANCER DEATH RATES IN THE U.S.

RATES OF INFECTION with *H. pylori* vary throughout the world. In developed countries, the infection is rare among children, but its prevalence rises with age. In developing countries, far more people are infected in all age groups (*left*). Supporting the fact that such infections cause ulcer disease, Enno Hentschel and his colleagues at Hanusch Hospital in Vienna found that antimicrobial therapy dramatically decreased the chance that a duodenal ulcer would recur (*center*). As infection rates have declined during the past century in the U.S., so, too, have the number of deaths from stomach cancer (*right*)—suggesting that *H. pylori* infection can, under some circumstances, cause that disease as well.

ther changes, among them intestinal metaplasia and dysplasia, conditions that typically precede cancer. The big mystery since finding *H. pylori* has been: Could the bacteria account for the second transition—from superficial gastritis to atrophic gastritis and possibly cancer—in Correa's model?

The first real evidence linking *H. pylori* and gastric cancer came in 1991 from three separate studies. All three had similar designs and reached the same conclusions, but I will outline the one in which I participated, working with Abraham Nomura of Kuakini Medical Center in Honolulu. I must first give some background. In 1942, a year after the bombing of Pearl Harbor, the selective service system registered young Japanese-American men in Hawaii for military service. In the mid-1960s medical investigators in Hawaii examined a large group of these men—those born between 1900 and 1919—to gain information on the epidemiology of heart disease, cancer and other ailments. By the late 1960s they had assembled a cohort of about 8,000 men, administered questionnaires and obtained and frozen blood samples. They then tracked and monitored these men for particular diseases they might develop.

For many reasons, by the time we began our study we had sufficient information on only 5,924 men from this original group. Among them, however, 137 men, or more than 2 percent, had acquired gastric cancer between 1968 and 1989. We then focused on 109 of these patients, each of whom was matched with a healthy member of the cohort. Next, we examined the blood samples frozen in the 1960s for antibodies to *H. pylori*. One strength of this study was that the samples had been taken from these men, on average, 13 years before they were diagnosed with cancer. With the results in hand, we asked the critical question: Was evidence of preexisting *H. pylori* infection associated with gastric cancer? The answer was a strong yes. Those men who had a prior infection had been six times more likely to acquire cancer during the 21-year follow-up period than had men showing no signs of infection. If we confined our analysis to cancers affecting the lower part of the stomach—an area where *H. pylori* often collect—the risk became 12 times as great.

The other two studies, led by Julie Parsonnet of Stanford University and by David Forman of the Imperial Cancer Research Fund in London, produced like findings but revealed slightly lower risks. Over the past five years, further epidemiological and pathological investigations have confirmed the associa-

tion of *H. pylori* infection and gastric cancer. In June 1994 the International Agency for Research in Cancer, an arm of the World Health Organization, declared that *H. pylori* is a class-1 carcinogen—the most dangerous rank given to cancer-causing agents. An uncommon cancer of the stomach, called gastric lymphoma, also appears to be largely caused by *H. pylori*. Recent evidence suggests that antimicrobial treatment to cure *H. pylori* infection may bring about regression in a subset of tumors of this kind, which is an exciting development in both clinical medicine and cancer biology.

How Persistence Takes Place

Certainly most bacteria cannot survive in an acidic environment, but *H. pylori* are not the only exception. Since that bacteria's discovery, scientists have isolated 11 other organisms from

dioxide. Fostering the production of ammonia may be one way helicobacters neutralize the acid in their local environment, further securing their survival.

An interesting puzzle involves what *H. pylori* eat. There are two obvious guesses: the mucus in which it lives and the food its human host ingests. But Denise Kirschner of Texas A&M University and I constructed a mathematical model showing that *H. pylori* would not be able to persist for years relying on those nutrient sources. In our model, the mathematics of persistence in the stomach requires some regulated interaction between the host cells and the bacteria. Inflammation provides one such interaction, and so I have proposed that *H. pylori* might trigger inflammation for the purpose of acquiring nutrients. An apparent paradox in *H. pylori* biology is that although the organisms do not invade the gastric tissue, they can cause irritation there. Rather, as we

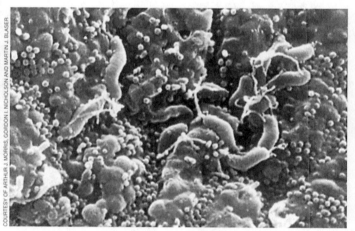

CURVED ORGANISMS, magnified 8,000 times, are *H. pylori* in the stomach of the second human volunteer to ingest the bacteria. The man had consumed the microbes 463 days earlier and developed chronic superficial gastritis as a result.

the stomachs of other primates, dogs, cats, rodents, ferrets and even cheetahs. These bacteria, for now considered to be members of the *Helicobacter* family, seem to have a common ancestor. All are spiral-shaped and highly motile (they swim well)—properties enabling them to resist the muscle contractions that regularly empty the stomach. They grow best at oxygen levels of 5 percent, matching the level found in the stomach's mucus layer (ambient air is 21 percent oxygen). In addition, these microbes all manufacture large amounts of an enzyme called urease, which cleaves urea into ammonia and carbon

and others have found, the microbes release chemicals that the stomach tissue absorbs. These compounds attract phagocytic cells, such as leukocytes and macrophages, that induce gastritis.

The host is not entirely passive while *H. pylori* bombard it with noxious substances. Humans mount an immune response, primarily by making antibodies to the microbe. This response apparently does not function well, though, because the infection and the antibodies almost inevitably coexist for decades. In essence, faced with a pathogen that cannot be easily destroyed, humans had two evolutionary options: we could have evolved to fight *H. pylori* infection to its death, possibly involving the abrogation of normal gastric function, or we could have become tolerant and tried to ignore the organisms. I believe the choice was made long ago in favor of tolerance. The response to other persistent pathogens—such as the microbes responsible for malaria and leprosy—may follow the same paradigm, in which it is adaptive for the host to dampen its immune reaction.

Fortunately, it is not in *H. pylori*'s best interest to take advantage of this passivity, growing to overwhelming numbers and ultimately killing its host. First, doing so would limit the infection's opportunity to spread. Second, even in a steady state, *H. pylori* reaches vast numbers (from 10^7 to 10^{10} cells) in the stomach. And third, further growth might exhaust the mechanisms keeping the immune system in check, leading to severe inflammation, atrophic gastritis and, eventually, a loss of gastric acidity. When low acidity occurs, bacteria from the intestines, such as *Escherichia coli*, are free to move upstream and colonize the stomach. Although *H. pylori* can easily live longer than *E. coli* in an acid environment, *E. coli* crowds *H. pylori* out of more neutral surroundings. So to avoid any competition with intestinal bacteria, *H. pylori* must not cause too much inflammation, thereby upsetting the acid levels in the stomach.

Are *H. pylori* symbionts that have only recently evolved into disease-causing organisms? Or are they pathogens on the long and as yet incomplete road toward symbiosis? We do not yet know, but we can learn from the biology of *Mycobacterium tuberculosis,* the agent responsible for tuberculosis. It, too, infects about one third of the world's population. But as in *H. pylori* infection, only 10 percent of all infected people become sick at some point in their life; the other 90 percent experience no symptoms whatsoever. The possible explanations fall into several main categories. Differences among microbial strains or among hosts could explain why some infected people acquire certain diseases and others do not. Environmental cofactors, such as whether someone eats well or smokes, could influence the course of infection. And the age at which someone acquires an infection might alter the risks. Each of these categories affects the outcome of *H. pylori* infection, but I will describe in the next section the microbial differences.

Not All Bacteria Are Created Equal

Given its abundance throughout the world, it is not surprising that *H. pylori* are highly diverse at the genetic level. The sundry strains share many structural, biochemical and physiological characteristics, but they are not all equally virulent. Differences among them are associated with variations in two genes. One encodes a large protein that 60 percent of all strains produce. Our group at Vanderbilt University, comprising Murali Tummuru, Timothy L. Cover and myself, and a group at the company Biocine in Italy, led by Antonello Covacci and Rino Rappuoli, identified and cloned the gene nearly simultaneously in 1993 and by agreement called it *cagA*. Among patients suffering from chronic superficial gastritis alone, about 50 to 60 percent are infected by *H. pylori* strains having the *cagA* gene. In contrast, nearly all individuals with duodenal ulcers bear cagA strains. Recently we reexamined the results of the Hawaiian study and found that infection by a cagA strain was associated with a doubled risk of gastric cancer. Research done by Jean E. Crabtree of Leeds University in England and by the Vanderbilt group has shown that persons infected by cagA strains experience more severe inflammation and tissue injury than do those infected by strains lacking the *cagA* gene.

The other *H. pylori* gene that seems to influence disease encodes for a toxin. In 1988 Robert D. Leunk, working for Procter & Gamble—the makers of bismuth subsalicylate (Pepto-Bismol)—reported that a broth containing *H. pylori* could induce the formation of vacuoles, or small holes, in tissue cultures. In my group, Cover had clearly shown that a toxin caused this damage and that it was being made not only by *H. pylori* grown in the laboratory but also by those residing in human hosts. In 1991 we purified the toxin and confirmed Leunk's finding that only 50 to 60 percent of *H. pylori* strains produced it. Our paper was published in May 1992 and included a brief sequence of some of the amino acids that encode for the mature toxin. Based on that scanty information, within the next year four groups—two in the U.S., including our own, one in Italy and one in Germany—were able to clone the gene, which we all agreed to name *vacA*. The race to publish was on. Each of our four papers appeared in separate journals within a three-month period.

Lest this sounds like duplicated labor, I should point out that each team had in fact solved a different aspect of the problem. We learned, for example, that virtually all *H. pylori* strains possess *vacA*, whether or not they produce the toxin when grown in culture. We also discovered that there is an extraordinary amount of strain-to-strain variability in *vacA* itself. In addition, broth from toxin-producing strains inoculated directly into the stomach of mice brought

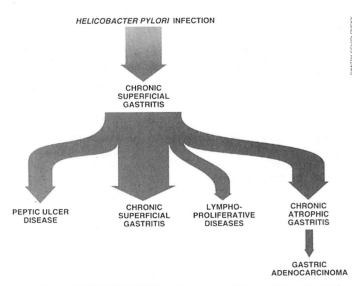

HELICOBACTER PYLORI INFECTION

CHRONIC SUPERFICIAL GASTRITIS

PEPTIC ULCER DISEASE

CHRONIC SUPERFICIAL GASTRITIS

LYMPHO-PROLIFERATIVE DISEASES

CHRONIC ATROPHIC GASTRITIS

GASTRIC ADENOCARCINOMA

H. PYLORI INFECTION PROGRESSES to chronic superficial gastritis within months. Left untreated, the condition persists for life in most people. A small fraction, however, may develop peptic ulcer disease, lymphoproliferative diseases or severe chronic atrophic gastritis, leading to adenocarcinoma of the stomach.

DIMITRY SCHIDLOVSKY

Which Treatment Strategy Should You Choose?

	OLD MODEL	NEW MODEL
CAUSE	**Excess stomach acid** eats through tissues and causes inflammation	*Helicobacter pylori* bacteria secrete toxins and cause inflammation in the stomach, bringing about damage
TREATMENTS	**Bland diet**, including dairy products every hour, small meals, no citrus or spicy foods and no alcohol or caffeine **H_2-receptor blockers** lessen blood levels of histamine, which increases the production of stomach acid **Surgery** to remove ulcers that do not respond to medication or that bleed uncontrollably. In the 1970s, it was the most common operation a surgical resident learned. Now it is increasingly rare	**Antibiotic regimen.** In February 1994 an NIH panel endorsed a two-week course of antibiotics for treating ulcer disease: amoxicillin or tetracycline, metronidazole (Flagyl) and bismuth subsalicylate (Pepto-Bismol). In December 1995 an FDA advisory committee recommended approval of two new four-week treatments, involving clarithromycin (Biaxin) with either omeprazole (Prilosec) or ranitidine bismuth citrate (Tritec). One-week therapies are also highly effective
SUCCESS	Patients who stop taking H_2-receptor blockers face a **50 percent chance** that their ulcers will recur within six months and a 95 percent chance that they will reappear within two years	**No recurrence** after the underlying bacterial infection is eliminated
COST	H_2-receptor blockers cost from $60 to $100 per month, adding up to **thousands of dollars** over decades of care. Surgery can cost as much as $18,000	**Less than $200** for a standard one-week therapy

ROBERTO OSTI (drawings); LISA BURNETT

about substantial injury. Strains that produce the toxin are some 30 to 40 percent overrepresented in ulcer patients compared with those having gastritis alone. And toxigenic strains usually but not always contain *cagA*, which is located far away from *vacA* on the chromosome.

Slow-Acting Bacteria and Disease

Over the past 15 years, researchers and physicians have learned a good deal about *H. pylori*. This knowledge has revolutionized our understanding of gastritis, formerly thought to represent the aging stomach, and of peptic ulcer disease and gastric cancer. It has made possible new treatments and screening methods. In addition, a new field of study has emerged—the microbiology and immunology of the human stomach—that will undoubtedly reveal more about persistent infections within mucosal surfaces.

But let us extrapolate from these findings. Consider that slow-acting bacteria, *H. pylori*, cause a chronic inflammatory process, peptic ulcer disease, that was heretofore considered metabolic. And also keep in mind that this infection greatly enhances the risk of neoplasms developing, such as adenocarcinomas and lymphomas. It seems reasonable, then, to suggest that persistent microbes may be involved in the etiology of other chronic inflammatory diseases of unknown origin, such as ulcerative colitis, Crohn's disease, sarcoidosis, Wegener's granulomatosis, systemic lupus erythematosus and psoriasis, as well as various neoplasms, including carcinomas of the colon, pancreas and prostate. I believe *H. pylori* are very likely the first in a class of slow-acting bacteria that may well account for a number of perplexing diseases that we are facing today.

The Author

MARTIN J. BLASER has been the Addison B. Scoville Professor of Medicine and director of the Division of Infectious Diseases at Vanderbilt University and at the Nashville Veterans Affairs Medical Center since 1989. He has worked at the Rockefeller University, the University of Colorado, the Denver Veterans Administration Medical Center, the Centers for Disease Control and Prevention, and St. Paul's Hospital in Addis Ababa, Ethiopia. He received a B.A. with honors in economics from the University of Pennsylvania in 1969 and an M.D. from New York University in 1973. He holds several patents and is a member of numerous professional societies and editorial boards. He has written more than 300 articles and edited several books.

Further Reading

UNIDENTIFIED CURVED BACILLI IN THE STOMACH OF PATIENTS WITH GASTRITIS AND PEPTIC ULCERATION. Barry J. Marshall and J. Robin Warren in *Lancet*, No. 8390, pages 1311–1315; June 16, 1984.

HELICOBACTER PYLORI INFECTION AND GASTRIC CARCINOMA AMONG JAPANESE AMERICANS IN HAWAII. A. Nomura, G. N. Stemmermann, P.-H. Chyou, I. Kato, G. I. Perez-Perez and M. J. Blaser in *New England Journal of Medicine*, Vol. 325, No. 16, pages 1132–1136; October 17, 1991.

HUMAN GASTRIC CARCINOGENESIS: A MULTISTEP AND MULTIFACTORIAL PROCESS. Pelayo Correa in *Cancer Research*, Vol. 52, No. 24, pages 6735–6740; December 15, 1992.

EFFECT OF RANITIDINE AND AMOXICILLIN PLUS METRONIDAZOLE ON THE ERADICATION OF HELICOBACTER PYLORI AND THE RECURRENCE OF DUODENAL ULCER. Enno Hentschel et al. in *New England Journal of Medicine*, Vol. 328, No. 5, pages 308–312; February 4, 1993.

REGRESSION OF PRIMARY LOW-GRADE B-CELL GASTRIC LYMPHOMA OF MUCOSA-ASSOCIATED LYMPHOID TISSUE TYPE AFTER ERADICATION OF HELICOBACTER PYLORI. A. C. Wotherspoon et al. in *Lancet*, Vol. 342, No. 8871, pages 575–577; September 4, 1993.

PARASITISM BY THE "SLOW" BACTERIUM HELICOBACTER PYLORI LEADS TO ALTERED GASTRIC HOMEOSTASIS AND NEOPLASIA. Martin J. Blaser and Julie Parsonnet in *Journal of Clinical Investigation*, Vol. 94, No. 1, pages 4–8; July 1994.

10

Research on Predatory Algae

BURKHOLDER AND COLLEAGUES

The United States has been stunned in recent years by reports of massive unexplained fish kills in estuaries along the southeastern coast. In 1988, a team headed by Dr. JoAnn M. Burkholder, an aquatic botanist at North Carolina State University, discovered a new genus of "predatory" algae present at the site of several of these disasters: a dinoflagellate that kills fish by releasing a lethal neurotoxin into the water. This startling discovery was first reported in a letter to *Nature* (1992), the first document in this chapter. This letter is cited in the second document, a full-length report in the *Journal of Plankton Research* in 1995 in which Burkholder and her colleagues reported a later study exploring the possibility of controlling the algae with a plankton predator. The third entry is an RFP from the National Sea Grant College Program, followed by a 1994 proposal to this program in which Burkholder and coinvestigator Parke Rublee proposed a plan for developing gene probes to help detect the dinoflagellate in estuarine waters. The final entry in this chapter is an article from *Discover* magazine, in which journalist Patrick Huyghe (1993) tells the story of the "killer algae" discoveries. You'll notice that the dinoflagellate was referred to as *Pfiesteria piscimorte* in early reports but was later formally named *Pfiesteria piscicida*.

New 'phantom' dinoflagellate is the causative agent of major estuarine fish kills

JoAnn M. Burkholder, Edward J. Noga*, Cecil H. Hobbs & Howard B. Glasgow Jr

Department of Botany, Box 7612, North Carolina State University, Raleigh, North Carolina 27695–7612, USA
* Department of Companion Animal and Special Species Medicine, North Carolina State University, 4700 Hillsborough Street, Raleigh, North Carolina 27606, USA

A WORLDWIDE increase in toxic phytoplankton blooms over the past 20 years[1,2] has coincided with increasing reports of fish diseases and deaths of unknown cause[3]. Among estuaries that have been repeatedly associated with unexplained fish kills on the western Atlantic Coast are the Pamlico and Neuse Estuaries of the southeastern United States[4]. Here we describe a new toxic dinoflagellate with 'phantom-like' behaviour that has been identified as the causative agent of a significant portion of the fish kills in these estuaries, and which may also be active in other geographic regions. The alga requires live finfish or their fresh excreta for excystment and release of a potent toxin. Low cell densities cause neurotoxic signs and fish death, followed by rapid algal encystment and dormancy unless live fish are added. This dinoflagellate was abundant in the water during major fish kills in local estuaries, but only while fish were dying; within several hours of death where carcasses were still present, the flagellated vegetative algal population had encysted and settled back to the sediments. Isolates from each event were highly lethal to finfish and shellfish in laboratory bioassays. Given its broad temperature and salinity tolerance, and its stimulation by phosphate enrichment, this toxic phytoplankter may be a widespread but undetected source of fish mortality in nutrient-enriched estuaries.

The alga, which represents a new family, genus and species within the order Dinamoebales (K. Steidinger, personal communication), was inadvertently discovered by fish pathologists[5] who observed sudden death of cultured tilapia (*Oreochromis aureus* and *O. mossambica*) several days after their exposure to freshly collected water from the Pamlico River. A small dinoflagellate increased in abundance before fish death, followed by a sharp decline unless additional live fish were introduced. Within 1–2 h of fish death, the sudden algal decline occurred because the cells produced resting cysts or non-toxic amoeboid stages and settled to the bottom of culture vessels (J.M.B., unpublished results).

The behaviour of the culture contaminant provided clues about its potential activity in estuarine habitat. We suspected that if low cell densities were lethal to fish in the wild, as in culture, then the alga would probably not be detected by routine monitoring efforts but might be found while a kill was in progress. We observed the dinoflagellate swarming in Pamlico water collected during a kill of nearly one million Atlantic menhaden (*Brevoortia tyrannus*; Table 1). Less than one day after the kill, few toxic vegetative cells remained suspended in the water although many fish carcasses still were present. In standard bioassays[6,7] maximum algal growth coincided with fish death, followed by rapid encystment (Fig. 1). This toxic dinoflagellate was found in abundance during several other major fish kills in the Pamlico and Neuse estuaries and in local aquaculture facilities (Table 1). Isolates from these events have been confirmed in laboratory bioassays as lethal to 11 species of finfish including commercially valuable striped bass (*Morone saxatilis*), southern flounder (*Paralichthys lenthostigma*), menhaden and eel (*Anguilla rostrata*).

The algal isolates all exhibit similar behaviour. Flagellated, photosynthetic toxic vegetative cells excyst after live fish or their fresh excreta are added to aquaria cultures under low light (Fig. 2*a*, *b*). The lag period for excystment ranges from minutes to days and increases with dormancy period or cyst age. The cells appear athecate but have thin cellulosic deposits (thecal plates, Fig. 2*c*)[8] beneath outer membranes that obscure the plate arrangement required for formal speciation[9]. This dinoflagellate completes its sexual cycle while killing fish. Vegetative cells produce poorly pigmented, anisogamous gametes; the male is smaller, and its extended longitudinal flagellum is five- to sixfold longer than the cell (Fig. 2*d*). The vegetative cells can also form large amoeboid stages (pigmented or colourless, 80–250 μm long with extended pseudopodia) with unknown role in the life cycle.

Gamete fusion results in planozygote formation followed by production of additional toxic vegetative cells if live fish are present. Without live fish, the planozygotes lose their flagella and form thick-walled cysts but remaining gametes continue to multiply especially under phosphate enrichment (batch cultures enriched with 50–400 μg PO_4^{3-} l^{-1} yielded significantly more gametes after 6 days than cultures with $PO_4^{3-} \leqslant 10$ μg l^{-1}; Student's *t* test, $P < 0.01$)[10]. Such stimulatory effects have not been observed with either nitrate or ammonium. A ciliated protozoan, *Stylonichia* sp.[11], is common in local estuarine waters and consumes both toxic vegetative cells and cysts. But a pigmented, toxic amoeboid stage of the alga, in turn, attacks the protozoan predator.

Each flagellated vegetative algal cell forms a peduncle or pseudopodium shortly after excystment[12,13]. The peduncle becomes fully extended during toxic activity (Fig. 3*a*), particularly at the optimal salinity of 15‰ (among tested salinities 5, 10, 15, 25 and 35‰; significantly higher vegetative cell production at 15‰, Students *t* test, $P < 0.01$)[10]. The lethal agent is an excreted neurotoxin (under analysis by D. Baden, personal communication). Water from which cells had been removed by gentle drop-filtration (0.22 μm-pore filters) induces neurotoxic signs in fish including sudden sporadic movement, disorientation, lethargy and apparent suffocation followed by death. The alga has not been observed to attack fish directly. It rapidly increases its swimming velocity to reach flecks of sloughed tissue from dying fish, however, using its peduncle to attach to and digest the tissue debris. Within several hours of fish death, toxic vegetative cells encyst in the presence or absence of fish carcasses (Fig. 3*b*–*d*). The cysts can be destroyed by dilute bleach. But after treatment with concentrated sulphuric acid or ammonium hydroxide, 35 days of desiccation, or nearly 2 years of dormancy, small percentages of the cysts have yielded viable toxic cells when placed in saline water (10–15‰) with live fish (J.M.B., unpublished results).

TABLE 1   Fish kills associated with the new toxic dinoflagellate, documented in North Carolina during 1991–92

Date	Location	Temperature (°C)	Salinity (‰)	Fish killed	Alga (cells ml^{-1})
1991					
May	Pamlico River	24	3	Menhaden	1,300
June	Pamlico River	29	10	Menhaden, others	1,100
Aug.	Pamlico River	31	12	Menhaden	600
Aug.	Pamlico River	30	8	Flounder walk	800
Sept.–Oct.	Neuse River	26	11	Menhaden, Blue crabs	1,200
Dec.	Taylor's Creek	15	30	Fish walk (flounder, eel, mullet, others)	~35,000
1992					
Jan.	Aquaculture	9	0	Striped bass	
Feb.	NC maritime museum (Newport River)	21	25	Fish spp.	
Feb.	National Marine Fisheries Service (Newport River)	15–21	25	Menhaden, others	

Locations on the Pamlico River followed a salinity gradient from 3 to 12‰ over an 8-km distance. Low dissolved oxygen (DO, 2.8 mg l^{-1}) was measured from the bottom water in the vicinity of the kill during June 1991. On all other dates, DO was ≥4.5 mg l^{-1} throughout the water column (R. Carpenter, K. Lynch and K. Miller, NC Dept. of Environmental Health and Natural Resources, personal communication). In each case from 1991, we observed abundant pigmented, flagellated vegetative cells in water samples taken during the kill. (Note, 'Walk' refers to cases in which stressed fish attempted to leave the water and beach before they died.) Water from fish kills in 1992 was collected within 1–4 days after fish death; hence, it was not possible to obtain data for cell abundances while the kills were in progress. Samples from a kill at a freshwater aquaculture facility in Aurora, NC contained ~40 vegetative cells ml^{-1} as well as cysts. Water from both kills in February 1992 (original source, the Newport River, 11 °C, 25‰) yielded toxic vegetative cells as well as gametes, planozygotes, amoeboid stages and cysts. We confirmed virtually identical morphology of these with previous isolates using scanning electron microscopy, and verified toxicity of each population using aquarium bioassays with tilapia and striped bass as test species.

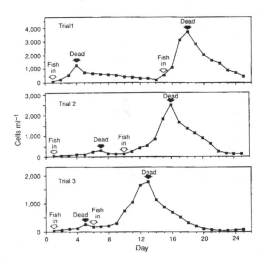

FIG. 1 Response of the dinoflagellate (toxic flagellated, vegetative stage) to *Oreachromis aureus* (length 3 cm, approximate age 40 days) in repeat-trial experiments. The trials were conducted as batch-culture aquarium bioassays (18° C, 10‰, 30 μEinst m^{-2} s^{-1}) which varied in the time interval between death of the first fish and addition of a second live fish (fish length 5 cm). The toxic vegetative stage attained greatest abundance just before fish death, followed by rapid decrease in abundance as the cells encysted or formed non-toxic, colourless amoebae and settled out. As we reduced the time interval without live fish, the second fish died more rapidly.

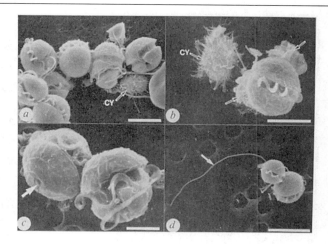

FIG. 2 Scanning electron micrographs of the toxic dino-flagellate. *a, b, c,* Several hours after a live tilapia was added to aquarium cultures (with environmental characteristics described in Fig. 1) after a 2-day period without fish. Most cysts (cy) had already produced flagellated vegetative cells (*a,* scale bar 10 μm; *b,* arrows indicate remains of cyst wall; scale bar, 8 μm); *c,* Cell showing the outline of plates or 'armour' (for example, arrow indicating an apical plate) beneath surficial membranes (scale bar, 5 μm); *d,* Male and female gametes (left and right, respectively). The presumed gametes were abundant ~1 day after addition of a live fish, when the fish began to exhibit neurotoxic signs of lethargy, inability to maintain balance, and apparent respiratory distress. In viewing live samples under light microscopy, we have observed such cells in fusion; note the extended length of the longitudinal flagellum on the male (arrow)(scale bar, 15 μm). Photos by C. Hobbs.

Laboratory bioassays and field collections have confirmed that this dinoflagellate is lethal to finfish across a temperature gradient from 4 to 28 °C[14–17], and across a salinity range of 2–35‰. It has also killed hybrid striped bass *Morone saxatilis × M. chrysops* in fresh water (0‰ salinity, in aquaculture facilities and in bioassays) with moderate concentrations of divalent cations (Table 1). Both native and exotic fishes are affected, suggesting stimulation by amino acids or other common, labile substance in fish excreta[18,19] (Table 1). Shellfish such as blue crabs (*Callinectes sapidus*, carapace width 8–10 cm) and bay scallops (*Aequipecten irradians*, shell width 4–5 cm) do not stimulate excystment or toxic activity. These shellfish remained viable for a 9-day test period while filtering low concentrations of the dinoflagellate (~50 cells ml⁻¹), although the scallop closing reflex slowed perceptibly. When placed in aquaria with dying finfish, however, blue crabs were killed within hours to several days, and scallops died within minutes.

Long-term data sets from Europe, Asia, and North America

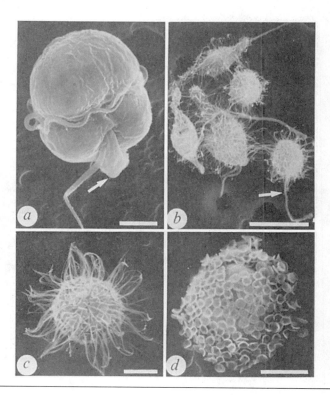

FIG. 3 Scanning electron micrographs of the dinoflagellate from aquarium cultures. *a,* While tilapia were dying. At its optimal salinity of 15‰, the flagellated vegetative cells assume a swollen appearance and the peduncle (arrow) becomes fully extended for use in saprotrophic feeding on bits of fish tissue (scale bar, 3 μm); *b,* Two hours after the tilapia died and were not replaced with live fish. Most vegetative cells had begun to form cysts; note that the longitudinal flagellum was still attached to one of the cells (arrow)(scale bar, 10 μm); *c,* One day after fish death, many completed cysts presumed to contain dinoflagellate cells were present. Note that the outer covering consists of scales with long bristles (scale bar, 3 μm); *d,* Nearly 1 month after fish death, the outer scales on most cysts had lost the bristle extensions (scale bar, 5 μm).

strongly correlate toxic phytoplankton blooms with increasing nutrient enrichment[20]. The Pamlico and Neuse Estuaries receive high anthropogenic loading of phosphorus and nitrogen[21,22]. Over the past century, cultural eutrophication has probably shifted the habitat to more favourable conditions for algal growth and toxic activity. Unlike other toxic phytoplankton, however, this dinoflagellate is ephemeral in the water column. The lethal flagellated vegetative cells move from the sediment surface to the water in response to lingering finfish which become lethargic from the toxin. The alga kills the organisms which stimulate it and then rapidly descends. Increasing reports worldwide describe unexplained kills involving fish that die quickly after exhibiting neurotoxic signs[23,24]. It is unlikely that this eurythermal, euryhaline dinoflagellate occurs only in these estuaries. We predict that with well timed sampling, this alga will be discovered at the scene of many fish kills in shallow, turbid, eutrophic coastal waters extending to geographic regions well beyond the Pamlico and the Neuse. □

Received 6 April; accepted 5 June 1992.

1. Hallegraeff, G. M., Steffensen, D. A. & Wetherbee, R. *J. Plankt. Res.* **10**, 533–541 (1988).
2. Robineau, B., Gagne, J. A., Fortier, L. & Cembella, A. D. *Mar. Biol.* **108**, 293–301 (1991).
3. Shumway, S. E. *J. World Aquaculture Soc.* **21**, 65–104 (1990).
4. Miller, K. H. *et al. Pamlico Environmental Response Team Report* (North Carolina Department of Environment, Health and Natural Resources, Wilmington, 1990).
5. Smith, S., Noga, E. J. & Bullis, R. A. *Proc. 3rd Int. Colloquium Path. Mar. Aquaculture* 167–168 (1988).
6. Ryther, J. H. *Ecology* **35**, 522–533 (1954).
7. Roberts, R. J., Bullock, A. M., Turner, M. K., Jones, K. & Tett, P. *J. mar. Biol. Ass. UK* **63**, 741–743 (1983).
8. Taylor, F. J. R. in *The Biology of Dinoflagellates* (ed. Taylor, F. J. R.) 24–92 (Blackwell, Boston, 1987).
9. Taylor, F. J. R. in *The Biology of Dinoflagellates* (ed. Taylor, F. J. R.) 723–731 (Blackwell, Boston, 1987).
10. Sokal, R. R. & Rohlf, F. J. *Biometry* 2nd edn (Freeman, San Francisco, 1981).
11. Thorp, J. H. & Covich, A. P. *Ecology and Classification of North American Freshwater Invertebrates*. 77 (Academic, New York, 1991).
12. Spero, H. J. *J. Phycol.* **18**, 357–360 (1982).
13. Gaines, G. & Elbrachter, M. in *The Biology of Dinoflagellates* (ed. Taylor, F. J. R.) 224–268 (Blackwell, Boston, 1987).
14. North Carolina Department of Environment, Health and Natural Resources. *1990 Algal Bloom Reports* (North Carolina Department of Environmental Management, Raleigh, 1990).
15. North Carolina Department of Natural Resources and Community Development. *1986 Algal Bloom Reports* (North Carolina Division of Environmental Management, Water Quality Section, Raleigh, 1987).
16. North Carolina Department of Natural Resources and Community Development. *1987 Algal Bloom Reports* (North Carolina Division of Environmental Management, Water Quality Section, Raleigh, 1988).
17. North Carolina Department of Natural Resources and Community Development. *1988 Algal Bloom Reports* (North Carolina Division of Environmental management, Water Quality Section, Raleigh, 1989).
18. Palenik, B., Kieber, D. J. & Morel, F. M. M. *Biol. Oceanogr.* **6**, 347–354 (1988).
19. Lagler, K. F., Bardach, J. E. & Miller, R. R. *Ichthyology* 268–272 (New York, 1962).
20. Smayda, T. J. in *Novel Phytoplankton Blooms* (ed. Cosper, E. M. *et al.*) 449–484 (Springer, New York, 1989).
21. Stanley, D. H. in *Proc. Symp. Coastal Water Resources* 155–164 (American Water Works Association, Wilmington, 1988).
22. Rudek, J., Paerl, H. W., Mallin, M. A. & Bates, P. W. *Mar. Ecol. Prog. Ser.* **75**, 133–142 (1991).
23. White, A. W. *Mar. Biol.* **65**, 255–260 (1981).
24. White, A. W. in *Proc. Int. Conf. Impact of Toxic Algae on Mariculture* 9–14 (Aqua-Nor 1987 Int. Fish Farming Exhib., Trondheim 1988).

ACKNOWLEDGEMENTS. We thank K. Lynch, K. Miller and R. Carpenter (NC Dept. of Environment, Health and Natural Resources), and N. McNeill and W. Hettler (National Marine Fisheries Service) for their help in obtaining water samples and V. Coleman, J. Compton, C. Harrington, and C. Johnson for laboratory assistance. This work was supported by the Albemarle-Pamlico Estuarine Study through the UNC Water Resources Research Institute, and the Department of Botany and the College of Veterinary Medicine at North Carolina State University.

CORRECTIONS

New 'phantom' dinoflagellate is the causative agent of major estuarine fish kills

JoAnn M. Burkholder, Edward J. Noga, Cecil H. Hobbs & Howard B. Glasgow Jr

Nature 358, 407–410 (1992)

IN this letter in the 30 July issue, Stephen A. Smith (Virginia-Maryland Regional College of Veterinary Medicine, Virginia Polytechnic Institute, Blacksburg, Virginia 24061-0442, USA), a former collaborator of the second author, was inadvertently omitted and should be added as a fifth author. The Office of Sea Grant (NOAA, US Department of Commerce, and the UNC Sea Grant College), and the North Carolina Agricultural Research Foundation should be acknowledged for their contributions to funding support.

Response of two zooplankton grazers to an ichthyotoxic estuarine dinoflagellate

Michael A.Mallin, JoAnn M.Burkholder[1], L.Michael Larsen[1] and Howard B.Glasgow,Jr[1]

Center for Marine Science Research, University of North Carolina at Wilmington, 7205 Wrightsville Avenue, Wilmington, NC 28403 and [1]Department of Botany, Box 7612, North Carolina State University, Raleigh, NC 27695-7612, USA

Abstract. The dinoflagellate *Pfiesteria piscicida* (gen. et sp. nov.), a toxic 'ambush predator', has been implicated as a causative agent of major fish kills in estuarine ecosystems of the southeastern USA. Here we report the first experimental tests of interactions between *P.piscicida* and estuarine zooplankton predators, specifically the rotifer *Brachionus plicatilis* and the calanoid copepod *Acartia tonsa*. Short-term (10 day) exposure of adult *B.plicatilis* to *P.piscicida* as a food resource, alone or in combination with the non-toxic green algae *Nannochloris* and *Tetraselmis*, did not increase rotifer mortality relative to animals that were given only non-toxic greens. Similarly, short-term (3 day) feeding trials using adult *A.tonsa* indicated that the copepods survived equally well on either *P.piscicida* or the non-toxic diatom *Thalassiosira pseudonana*. Copepods given toxic dinoflagellates exhibited erratic behavior, however, relative to animals given diatom prey. The fecundity of *B.plicatilis* when fed the toxic dinoflagellate was comparable to or higher than that of rotifers fed only non-toxic greens. We conclude that, on a short-term basis, toxic stages of *P.piscicida* can be readily utilized as a nutritional resource by these common estuarine zooplankters. More long-term effects of *P.piscicida* on zooplankton, the potential for toxin bioaccumulation across trophic levels, and the utility of zooplankton as biological control agents for this toxic dinoflagellate, remain important unanswered questions.

Introduction

The toxic estuarine dinoflagellate *Pfiesteria piscicida* (Dinophyceae; gen. et sp. nov.; Steidinger *et al.*, in preparation) was first observed in 1988 as a culture contaminant of unknown origin (Smith *et al.*, 1988; Noga *et al.*, 1993). Its lethal effects on a wide array of finfish and shellfish have since been documented at major fish kills and in laboratory trials (Burkholder *et al.*, 1992, 1995a). *Pfiesteria*-like species act as 'ambush predators' with chemosensory stimulation by live fish or their fresh secretions and tissues (Burkholder, 1993; Glasgow and Burkholder, 1993). Since its discovery in 1991 during a fish kill in the Pamlico Estuary, North Carolina, USA (Burkholder *et al.*, 1992), this dinoflagellate has been implicated as a causative agent of major fish kills in estuarine and coastal waters (Burkholder *et al.*, 1995a). *Pfiesteria*-like species apparently are widespread; recent investigations have demonstrated their presence at 'sudden-death' estuarine fish kill sites from the mid-Atlantic (Delaware and Chesapeake Bays) to both the Atlantic and Gulf Coasts of Florida, extending to the Alabama coast (Burkholder *et al.*, 1995a; Lewitus *et al.*, 1995).

Pfiesteria piscicida is the first known estuarine or marine dinoflagellate to have a complex life cycle with at least 19 flagellated, amoeboid, and encysted stages (Burkholder and Glasgow, 1995). 'Phantom-like' toxic flagellated vegetative cells (TFVCs, diameter ~9–20 μm), the most lethal known stage, develop from zoospores that emerge from resting cysts or are produced by other non-toxic

stages when live fish are detected, e.g. when a school of fish enters an estuarine
tributary to feed. The TFVCs swim up from the sediment in chemosensory
response to fish secreta; they excrete a potent neurotoxin which creates open
bleeding sores and causes sloughing of tissue in affected fish. The active TFVCs
consume small flagellated algae, protozoans and the sloughed bits of fish tissue
(Burkholder *et al.*, 1992, 1993). The fish are also narcotized by the toxin so that
they linger in the area. Consumption of small flagellated algae by TFVCs
(Steidinger *et al.*, 1995) while attacking fish results in a temporary, specialized
form of mixotrophy in which the chloroplasts are sequestered and retained for
short-term functional capacity (days). Hence, the TFVCs can utilize the small
algal flagellates' cleptochloroplasts as a 'stolen' means of photosynthesis (e.g.
Schnepf *et al.*, 1989; Fields and Rhodes, 1991; Schnepf and Elbrächter, 1992).

TFVCs produce gametes (diameter 5–8 μm) that are stimulated to complete
sexual reproduction (formation of planozygotes, diameter 10–60 μm) when the
prey begin to die. Upon fish death, the population may follow several possible
courses: (i) the cells transform into multiple, active, heterotrophic amoeboid
stages (length 5–250 μm; from TFVCs, gametes and planozygotes) which
consume the fish remains and prey upon other microorganisms; (ii) they revert
from sexual to non-toxic asexual forms (zoospores, diameter 5–9 μm),
especially in nutrient-enriched waters with abundant flagellated algal prey
(TFVCs and gametes); or (iii) they encyst under stressed conditions (TFVCs,
planozygotes, amoebae, and possibly gametes and zoospores; Burkholder *et al.*,
1992, 1995b).

Even extremely lethal cultures, fed live fish repeatedly, retain their toxicity
for <24 h when denied further access to fish prey (Burkholder *et al.*, 1992,
1993). Although *P.piscicida* is often ephemeral in the water column (hours)—
killing its prey and then either encysting or transforming to amoeboid stages and
following the carcasses down to the sediments or into shore—its toxic outbreaks
have been known to extend for days to weeks when schools of fish are available,
such as during autumn migrations of Atlantic menhaden (*Brevoortia tyrannus*
Latrobe) from shallow estuaries out to sea (Burkholder *et al.*, 1992).

The ciliated protozoan *Stylonichia* cf. *putrina* has been observed to consume
the TFVCs of *P.piscicida* without apparent adverse toxic effects (Burkholder *et
al.*, 1992). During prolonged feeding events, however, remaining planozygote
stages of the dinoflagellate can transform into large amoebae which in turn,
engulf the ciliate as predator becomes prey (Burkholder *et al.*, 1992, 1993).
Interactions between toxic stages of *P.piscicida* and other potential predators
are currently unknown, such as whether mesozooplankton or microzooplankton
aside from *Stylonichia* can consume or control it, or whether they are adversely
affected by its toxin. Other toxic dinoflagellates are used as food resources by
some zooplankters (Watras *et al.*, 1985; Turner and Tester, 1989), but they
reduce fecundity and survival in other zooplankton species or adversely affect
higher trophic levels through toxin bioaccumulation (White, 1980, 1981;
Huntley *et al.*, 1986).

The objective of this research was to determine the effects of *P.piscicida* on
the survival and fecundity of the common estuarine parthenogenic rotifer

Brachionus plicatilis Mueller, and the survival of an obligate sexually reproducing calanoid copepod, *Acartia tonsa* Dana. We expected that if the zooplankters were adversely affected by the toxic dinoflagelate, survival and/or fecundity would be depressed when TFVCs were presented as the available algal food supply, relative to the response of animals given non-toxic algae alone or in combination with *P. piscicida*.

Method

Culture isolates of *P. piscicida* were collected on 23 May 1991 from the Pamlico River Estuary near Channel Marker no. 9 at the mouth of Blount Bay in Beaufort County, North Carolina, during an active bloom of TFVCs while ~1 million Atlantic menhaden, southern flounder (*Paralichthys lethostigma* Jordan & Gilbert), hogchokers (*Trinectes maculatus* Block & Schneider) and spot (*Leiostomus xanthuris* Lacepede) were dying (Burkholder *et al.*, 1993). The cultures were maintained in a walk-in culture facility under 50 μE m^{-2} s^{-1} illumination (cool white fluorescent lamps) at 20°C with a 12:12 h light:dark (L:D) cylce in 40 l aerated aquaria filled with artificial seawater at 15‰ salinity (seawater derived by adding Instant Ocean salts to water from a well-water source on the grounds of North Carolina State University).

The dinoflagellate's TFVCs require an unidentified substance in fresh fish excreta (Burkholder *et al.*, 1992); hence, it was necessary to maintain cultures using live fish. We routinely fed the dinoflagellate tilapia (*Oreochromica mossambica* Peters, each 5–7 cm in length and washed thoroughly with deionized water) at a density of 15–20 fish day^{-1}, and removed all dead fish as live replacements were added. The cultures also contained the small blue–green algae, *Lyngbya* sp. (Cyanophyceae; mean cell diameter 2 μm, mean filament length 9 μm) and *Gloeothece* spp. (gelatinous colonial forms; mean cell biovolume 3 μm^3). Contact with culture water and aerosols has been associated with serious human health effects (Glasgow and Burkholder, 1994); therefore, all work was completed using full-face respirators with organic acid filters; disposable gloves, boots and hair covers; and protective clothing that was bleached after use to kill all dinoflagellate stages ($\geq$30% bleach).

Since some TFVCs in a given population are induced to transform to amoeba shortly after disturbance from transport or gentle pouring (more so when the TFVCs are pipetted), we minimized amoeboid transformations by completing all zooplankton feeding trials in the dinoflagellate culture room within close proximity of the stock culture tanks. In all treatments containing *P. piscicida*, the dinoflagellates were added by slow pouring to containers marked for the appropriate volume. Treatments were randomly distributed daily within four covered containers to reduce potential contamination from aerosol components of toxin from the stock dinoflagellate culture tanks, which were also covered but could not be tightly sealed.

The common estuarine rotifer, *B. plicatilis* (Lee *et al.*, 1985), was obtained from W.F. Hettler (NOAA Southeast Fisheries Center, Beaufort, NC). *Acartia tonsa*, an abundant estuarine zooplankter in the Albemarle–Pamlico estuarine system (Peters, 1968; Mallin, 1991), was cultured from samples that were taken from the Newport River Estuary near Beaufort, NC, mixed with cultures supplied by H.Millsaps (Chesapeake Biological Laboratory, Solomon, MD) where *P.piscicida* is also known to occur (Burkholder *et al.*, 1995a). Stock cultures of these organisms were maintained in a S/P Cryo-Fridge environmental chamber at 20°C, 15–20‰ salinity and a 12:12 L:D cycle. The calanoid copepod was grown on the centric diatom *Thalassiosira pseudonana* Hasle (Bacillario-phyceae; mean diameter and mean biovolume 10 μm and 490 μm^3, respectively; $n = 25$ cells), and the rotifers were maintained on a green algal culture containing *Nannochloris* sp. and *Tetraselmis* sp. [Chlorophyceae (Bold and Wynne, 1985); mean diameter and biovolume of *Nannochloris* = 3 μm and 9 μm^3, respectively; mean length and biovolume of *Tetraselmis* = 9 μm and 450 μm^3, respectively; $n = 25$ cells each; supplied by W.F.Hettler, NOAA]. Size ranges of all algal prey were well within that which is known to be grazed by each predator (for the copepod, adult and copepodite stages; Berggreen *et al.*, 1988). Algal cell densities and biovolumes were determined from samples preserved in acidic Lugol's solution (Vollenweider, 1974) with the Utermöhl method (Lund *et al.*, 1958), following the procedure of Burkholder and Wetzel (1989). Cells were quantified under phase contrast at 600× using an Olympus IMT2 inverted microscope.

Brachionus plicatilis was tested for survival and fecundity over a 9 day period as follows. Single rotifers (non-egg bearing) were placed in 15 ml of filtered test water (salinity 15‰, 21–22°C, taken from a fish culture without dinoflagellate contamination and filtered through Whatman 934AH filters with pore size ~1.5 μm) within 50 ml polyethylene cups. Water that had previously contained fish was used to facilitate comparison with dinoflagellate treatments, which required addition of *P.piscicida* in fish culture water to enhance toxic activity as long as possible, and to reduce the transformation of TFVCs to less toxic, inedible amoebae or inactive cysts. Each treatment included 10 replicate animals maintained separately. A designated 'unfed' treatment consisted of filtered fish water that contained only small blue–green algae [*Lyngbya* sp. and *Gloeothece* spp., ~2.40 × 10^5 μm^3 biovolume ml^{-1}; both non-toxic (Gorham and Carmichael, 1988)] and bacteria in the filtrate as potential prey items. *Brachionus plicatilis* is known to consume bacteria (Turner and Tester, 1992) and, hence, would not have been completely starved. In all treatments, both animals would also have had access to the blue–greens (~20% of the total available algal resources by biovolume), which are considered a poor food resource, especially for copepods (Fulton and Paerl, 1988; Gorham and Carmichael, 1988).

The first of three 'fed' treatments for the rotifer trials contained the dinoflagellate as a food resource, added at the fish culture density of ~2500 cells ml^{-1}, which is typical of phytoplankton densities in local estuaries during the summer growing season (Carpenter, 1971; Thayer, 1971; Mallin *et al.*, 1991). The toxic dinoflagellate culture was capable of killing three test tilapia at 3 h intervals; it was comprised of 88% TFVCs [mean diameter 10 μm, biovolume 510 ± 18 μm^3 (mean ±1 SE)], 10% zoospores and gametes (mean diameter

7 ± 2 µm, biovolume 195 ± 26 µm^3), and 2% amoebae (biovolume 410 ± 58 µm^3; $n = 25$ cells for each stage), for a total initial dinoflagellate biovolume of 9.33×10^5 µm^3 ml^{-1}. The *Nannochloris* component of the non-toxic green algal prey consisted of much smaller cells than the dinoflagellate; hence, we provided approximately equal biovolumes of toxic and non-toxic prey types in the designated treatments. The second 'fed' treatment consisted of the non-toxic green algal culture added to filtered fish water (3.38×10^5 µm^3 *Nannochloris* sp. $+ 4.92 \times 10^5$ µm^3 *Tetraselmis* sp. ml^{-1}, for a total of 8.30×10^5 µm^3 non-toxic greens ml^{-1}). The third 'fed' treatment was a 1:1 mixture by biovolume of *P.piscicida* (1250 cells ml^{-1}) and non-toxic greens. Treatments with the dinoflagellate alone or mixed with green algal food thus included full volume or half volume, respectively, as 'contaminated' filtered fish water from actively killing cultures—an important point since the lethal agents of *P.piscicida* is an exotoxin, with little residual toxic materials retained within the cells (H.B. Glasgow and J.M.Burkholder, unpublished data; D.Baden, U.Miami, unpublished data). Toxic filtrate from dinoflagellate cultures with fish (0.22 µm porosity filters) is known to be lethal to tilapia for 12–24 h, depending on the TFVC concentration (Burkholder *et al.*, 1992, 1993). On a daily basis, we used Palmer cells (Wetzel and Likens, 1991) to quantify algal cells from stock cultures 2 h before transferring animals to fresh food supply, and adjusted dilutions to ensure that the targeted algal densities were added throughout the experiment.

Rotifer tests were run for 9 days, with daily removal of each animal into a fresh test water/phytoplankton preparation. Mortality and the presence of eggs and live young were recorded daily at 17:00–20:00 h. After scanning a given replicate and removing the animal, the water and remaining phytoplankton were preserved with acidic Lugol's solution for quantification. From each treatment we randomly selected 4–6 replicate samples (excluding those that had contained dead animals) for analysis by the Utermöhl method, to quantify changes in the phytoplankton food supply during 24 h grazing periods.

Acartia tonsa was tested for survival in a similar manner over three repeat 3-day trials ($n = 10$ replicate animals maintained separately), with the following alterations. Test organisms were late-stage copepodites (CIV–CVI) of either sex, placed randomly into polyethylene cups as they were encountered. Non-toxic algal food was supplied as the centric diatom, *T.pseudonana*, of similar dimensions and biovolume as TFVCs of *P.piscicida*. The toxic dinoflagellate culture used in the copepod trials was capable of killing three small tilapia at 30 to 60 min intervals; the culture consisted of 93% TFVCs, 6% zoospores and gametes, and ~1% amoebae. Treatments for trials 1 and 2 included (i) 'unfed', without algae aside from small blue–greens; (ii) *T.pseudonana* at 2500 cells ml^{-1}); (iii) *P.piscicida* at 2500 cells ml^{-1}; and (iv) a mix of 1250 cells each of the diatom and the dinoflagellate. Trial 3 was conducted similarly, except that total cell densities were ~2000 ml^{-1}. Media volumes were maintained at 30 ml rather than the 15 ml volume used for the rotifer tests. Standard cladoceran toxicity bioassays use one animal per 15 ml water (Mount and Norberg, 1984; Horning and Weber, 1985). We increased the volume to 30 ml for the copepod to minimize containment effects. Volumes >15 ml for *B.plicatilis*, and >30 ml for

A.tonsa, would have imposed visual and manipulation problems when attempting to capture/transfer individual animals to fresh algal supply.

As for the rotifer feeding trials, the copepods were transferred daily into freshly prepared algal treatments. We also quantified changes in the phytoplankton food supply per 24 h grazing period similarly as for the rotifer bioassays. The selected diatom cell densities and biovolumes exceeded quantities necessary to achieve maximal feeding rates for adult *A.tonsa* (Houde and Roman, 1987; Paffenhöfer, 1988). After terminating bioassays, the guts of both *B.plicatilis* and *A.tonsa* were densely packed with both non-toxic algae and/or flagellated stages of *P.piscicida* when viewed in Lugol's-preserved material under light microscopy. Although we did not observe dinoflagellates consuming non-toxic algae in the mixed treatment for *B.plicatilis* or *A.tonsa*, it is likely that TFVCs also accounted for a portion of the non-toxic algal losses (Burkholder *et al.*, 1993; Burkholder and Glasgow, 1995).

Fisher's exact test (Horning and Weber, 1985) was used to test for significant differences in animal mortality among treatments within each trial. The mean number of young produced among all *N* animals exposed (regardless of time of mortality) is considered an approximate measure of an overall ecological effect of a toxicant upon a test organism (Hamilton, 1986). Thus, to analyze rotifer fecundity, we tested for differences in live young produced among treatments by ANOVA followed by the LSD. One-way ANOVA was also used to compare ingestion rates among treatments for both test animals (Statistical Analysis Systems Inc., 1985; significance level for all tests set at $\alpha = 0.05$).

Results

Bioassay trials with the rotifer, *B.plicatilis*, showed no mortality among any of the three algal food treatments (Table I). In the unfed treatment, 7 of 10 replicate animals died during the 9 day experimental period, and survival of unfed rotifers was significantly lower than in the other treatments (Fisher's exact test, $P = 0.0015$). Moreover, significantly fewer young were produced in the unfed treatment as compared with all algal treatments (Table I, Figure 1). There was no significant difference in rotifer fecundity when given *P.piscicida* or the mixed dinoflagellate/green algal food treatment, nor a significant difference in fecundity with non-toxic greens alone versus *P.piscicida* alone. The fecundity of *B.plicatilis* when given non-toxic greens + *P.piscicida* was marginally significantly greater than with non-toxic greens alone as a food source ($P = 0.056$). These data were supported by significantly greater ingestion rates after 1 day for animals given mixed food as opposed to either non-toxic or toxic algae, alone ($P = 0.05$; Table II). After day 6, however, rotifer fecundity in the non-toxic green algal treatment showed an increasing trend, whereas after day 7 fecundity in the dinoflagellate treatment decreased (Figure 1).

Brachionus plicatilis consumed $61 \pm 10\%$ (mean $\pm$ 1 SE) of the *Nannochloris* biovolume and $70 \pm 17\%$ of the *Tetraselmis* daily when fed non-toxic green algae, with consumption increasing to $98 \pm 2\%$ and $85 \pm 15\%$ of each green, respectively, in the mixed green/dinoflagellate treatment (Table III).

Table I. Mean survival and fecundity of *B.plicatilis* when exposed for 9 days to non-toxic green algae (greens) and/or *P.piscicida* (dino), including the SD and the LSD ranking of mean live young produced per rotifer ($n = 10$; $\alpha = 0.05$ for all treatments)

Treatment	Surviving adults	Total young	Young/adult	SD	LSD[a]
Unfed	3	4	0.4	1.0	A
Greens	10	88	8.8	5.1	B
Dino	10	104	10.4	2.8	BC
Greens + dino	10	131	13.0	3.2	C

[a]Means with the same letter were not significantly different.

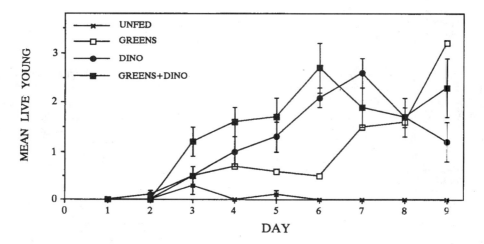

Fig. 1. Live young produced daily by *B.plicatilis* in each feeding treatment (means ± 1 SE).

Rotifer consumption of edible stages (TFVCs, gametes and zoospores) of *P.piscicida* was also higher in the mixed treatment, increasing from 53 ± 15% of the cells when given dinoflagellates alone to 84 ± 4% with mixed non-toxic/toxic algal food. After 24 h the percentage of inedible amoebae, transformed from TFVCs and gametes, was 20 ± 6% in the dinoflagellate-alone treatment and 47 ± 19% in the mixed algal treatment. Approximately one-third of the dinoflagellate cells available at the end of a daily feeding period were TFVCs, however, and more than half consisted of edible TFVCs, zoospores or gametes.

Trial 1 for *A.tonsa* showed no significant differences in survival among copepods fed any of the three algal treatments (Figure 2). Survival in the unfed treatment was significantly less than in all algal treatments ($P < 0.05$). In trial 2, there was no significant difference in survival between diatom-fed copepods and those fed *P.piscicida*, alone or in mixed toxic/non-toxic food supply (Figure 2). Copepods in all three 'fed' treatments survived significantly better than those in the unfed treatment ($P = 0.01$). In trial 3, the unfed copepods had significantly poorer survival than those given either dinoflagellate treatment (Figure 2).

Table II. Ingestion rates after the first 24 h (algal cells animal^{-1} h^{-1}, considering all trials) by the rotifer. *B.plicatilis*, and the copepod. *A.tonsa*, in bioassays with non-toxic algae, *P.piscicida* (flagellated stages) and mixed food supply ($n = 4$–6 randomly selected animals; data given as means $\pm$ 1 SE). Asterisks indicate *B.plicatilis* ingestion rates that were significantly higher in the mixed treatment than with one algal food source ($\alpha = 0.05$)

Treatment	Non-toxic algae	Dinoflagellate
Rotifer + dino	–	750 $\pm$ 210
Rotifer + greens		
Nannochloris	19 880 $\pm$ 3260	–
Tetraselmis	600 $\pm$ 150	–
Rotifer + mixed		
Nannochloris	31 900 $\pm$ 600	1180 $\pm$ 60*
Tetraselmis	730 $\pm$ 130	–
Copepod + dino	–	1240 $\pm$ 240
Copepod + diatom	1350 $\pm$ 320	–
Copepod + mixed	1240 $\pm$ 390	1070 $\pm$ 380

Table III. Consumption of algal prey after the first 24 h (as a percentage of the total biovolume initially available, considering all trials) by the rotifer. *B.plicatilis*, and the copepod, *A.tonsa* ($n = 4$–6 randomly selected animals; data given as means $\pm$ 1 SD)

Treatment	Non-toxic algae (%)	Dinoflagellate (flagellated; %)
Rotifer + dino	–	53 $\pm$ 15
Rotifer + greens	61 $\pm$ 10 *Nannochlorois*; 70 $\pm$ 17 *Tetraselmis*	–
Rotifer + mixed	98 $\pm$ 2 *Nannochloris*; 85 $\pm$ 15 *Tetraselmis*	84 $\pm$ 4
Copepod + dino	–	79 $\pm$ 16
Copepod + diatom	86 $\pm$ 20	–
Copepod + mixed	79 $\pm$ 25	68 $\pm$ 24

Copepods fed *P.piscicida* alone maintained significantly higher survival than those fed only diatoms ($P < 0.01$), whereas ingestion rates and survival were not significantly different among the non-toxic, toxic and mixed algal treatments ($P > 0.1$; Table II).

Acartia tonsa consumed ~80% of non-toxic diatom cells or toxic dinoflagellate cells when given either food source separately (Table III). Further, consumption of both non-toxic and toxic algae was comparable in the 'alone' or mixed algal treatment. After 24 h the inedible amoebae remained at <20% of the *P.piscicida* population in both dinoflagellate-alone and mixed algal treatments, with the remainder comprised of edible flagellated stages. Only 18% of the dinoflagellate cells were TFVCs, however, whereas >60% of the dinoflagellates were smaller and were probably either zoospores or gametes.

Although copepod survival was not impaired by the dinoflagellate during our short-term feeding trials, the presence of *P.piscicida* appeared to adversely affect the animals' behavior. In both dinoflagellate treatments, we observed that movement of *A.tonsa* was often extremely rapid and erratic, to the extent that some animals repeatedly collided head-first with the culture vessel walls. During trial 3, we attempted to quantify movement by counting the number of moves

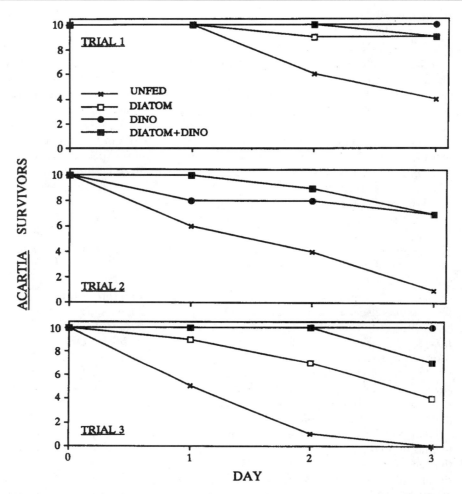

Fig. 2. Survival of *A.tonsa* for each feeding treatment over each of three experimental feeding trials. In trial 2, the copepod response in the diatom-only treatment was identical to that in the diatom + dinoflagellate treatment.

generated by use of swimming legs or antennae within a 30 s interval. For active animals, we estimated 29 ± 14 movements in treatments with the dinoflagellate. By comparison, copepods within the diatom-only treatment were 5- to 6-fold less active, and none were observed to collide with the vessel walls. Analysis of samples throughout the duration of the feeding trials did not indicate feeding impairment for *A.tonsa*, however, with similar ingestion rates and dinoflagellate losses daily within each trial. In trial 3, we extended the observations for an additional 4 days and found similar ingestion rates for *A.tonsa* among algal food treatments until day 7, when most animals probably had become stressed from handling and senescence.

Discussion

The lack of mortality among *B.plicatilis* in the *P.piscicida* treatments indicated that the toxin did not affect the rotifer. Most dinoflagellates had not encysted, as visual observations supported by cell counts confirmed that they were actively swimming in the test media after 24 h prior to water exchange. The rotifers actively consumed the dinoflagellate and benefited from this food resource, based on cell counts, observations of *P.piscicida* cells in the guts, and the high fecundity in the dinoflagellate-only treatment as compared to the low fecundity of unfed animals. In analogous behavior, the congeneric species *Brachionus calyciflorus* Pallas is capable of resisting the toxins produced by the blue–green alga *Microcystis aeruginosa* Kütz., and can utilize the blue–green as a supplementary nutritional source (Fulton and Paerl, 1987).

Since the fecundity of *B.plicatilis* in the two *P.piscicida* treatments was slightly elevated overall, relative to fecundity when given green algae alone, this dinoflagellate is probably a nutritious food source for the rotifer. The abundance of dinoflagellates in general has been positively correlated with zooplankton grazing rates in the Neuse River Estuary (Mallin and Paerl, 1994). Other researchers have noted the food value of, or grazing preference toward, various dinoflagellate species (Burkhill *et al.*, 1987; Uye and Takamatsu, 1990; Sellner *et al.*, 1991). For example, the copepods *A.tonsa*, *Oncaea venusta* Philippi and *Labidocera aestiva* Wheeler ingested *Gymnodinium breve* Davis cells in proportion to their availability over a broad range of concentrations without mortality or physiological incapacitation (Turner and Tester, 1989). *Centropages typicus* did not ingest *G.breve*. When offered a choice, however, *A.tonsa*, *L.aestiva* and *C.typicus* avoided *G.breve*, and instead consumed the non-toxic diatom, *Skeletonema costatum* (Greve.) Cleve.

Over the long term (weeks), the positive influence of *P.piscicida* on the fecundity of *B.plicatilis* may change, considering the decreasing trend shown in the fecundity data from days 7–9. This trend may be irrelevant in many field situations, however, since most *P.piscicida*-induced fish kills are <3 days in duration, and since the dinoflagellate often represents only 10–20% of the total available phytoplankton during toxic outbreaks (Burkholder *et al.*, 1995a).

The calanoid copepod *A.tonsa* also apparently found toxic *P.piscicida* cells a nutritious food source. Their survival was comparable to or higher than that of copepods fed a diatom diet, and their ingestion rates of diatoms and dinoflagellates were comparable when given either food resource alone or in combination. However, the animals' ability to detect and avoid the walls of the culture vessels was impaired in both the dinoflagellate-only and the dino-flagellate/diatom treatments, as evidenced by their extremely rapid and erratic movements and frequent, repeated collisions against the wall surfaces. In a natural habitat, such erratic behavior would be expected to increase the animals' susceptibility to visual predators, and might also impair their ability to detect and successfully mate with other individuals.

The utility of zooplankton as biocontrol agents for the suppression of toxic *P.piscicida* blooms remains an open question. The common estuarine zooplankters tested were able to consume the dinoflagellate's most lethal stage, and the rotifer was capable of reproducing successfully when TFVCs were its only food resource. Most zooplankton species, however, cannot reproduce quickly enough to control toxic or non-toxic short-term blooms. An additional question concerns whether the toxin, as yet uncharacterized, remains viable in zooplankton guts after consumption of the cells. If this is the case, then feeding by higher organisms on zooplankton in bloom areas may lead to bioaccumulation, with possible lethal or more insidious sublethal effects on their finfish and shellfish predators.

Acknowledgements

Funding support for this research was provided by the Office of Sea Grant, NOAA, US Department of Commerce (grant NA90AA-D-SG062) and the UNC Sea Grant College (project R/MER-17), the National Science Foundation (grant OCE-9403920), the UNC Water Resources Research Institute (grant 70136), the US Marine Corps–Cherry Point Air Station, the NCSU College of Agriculture & Life Sciences, the NC Agricultural Research Foundation (project 006034), the NC Agricultural Research Service (project 06188) and the Department of Botany at NCSU. Research assistance was contributed by J.Compton and F.Johnson. *Thalassiosira* cultures were provided by D. Kamykowski and R.Reed; *Acartia* cultures were provided, in part, by H.Millsaps, recommended by K.Sellner; and *Brachionus* and *Nannochloris* cultures were provided by W.Hettler. D.Rainer and B.MacDonald contributed sound counsel on laboratory safety precautions. We thank B.J.Copeland, J.Wynne and E.D.Seneca for their counsel and administrative support.

References

Berggreen,U., Hansen,B. and Kiørboe,T. (1988) Food size spectra, ingestion and growth of the copepod *Acartia tonsa* during development: implications for determination of copepod production. *Mar. Biol.*, **99**, 341–352.

Bold,H.C. and Wynne,M.J. (1985) *Introduction to Phycology*. Prentice-Hall, New York, pp. 476–512.

Burkhill,P.H., Mantoura,R.F., Llewellyn,C.A. and Owens,N.J.P. (1987) Microzooplankton grazing and selectivity of phytoplankton in coastal waters. *Mar. Biol.*, **93**, 581–590.

Burkholder,J.M. and Glasgow,H.B.,Jr (1995) Interactions of a toxic estuarine dinoflagellate with microbial predators and prey. *Archiv. für Protistenka*, in press.

Burkholder,J.M. and Wetzel,R.G. (1989) Epiphytic microalgae on natural substrata in a hardwater lake: seasonal dynamics of community structure, biomass and ATP content. *Arch. Hydrobiol.*, **83** (Suppl.), 1–56.

Burkholder,J.M., Noga,E.J., Hobbs,C.W., Glasgow,H.B.,Jr and Smith,S.A. (1992) New 'phantom' dinoflagellate is the causative agent of major estuarine fish kills. *Nature*, **358**, 407–410; *Nature*, **360**, 768.

Burkholder,J.M., Glasgow,H.B.,Jr, Noga,E.J. and Hobbs,C.W. (1993) *The Role of a New Toxic Dinoflagellate in Finfish and Shellfish Kills in the Neuse and Pamlico Estuaries.* Research Institute of the University of North Carolina, Raleigh, pp. 1–58.

Burkholder,J.M., Glasgow,H.B.Jr. and Hobbs,C.W. (1995a) Distribution and environmental conditions for fish kills linked to a toxic ambush-predator dinoflagellate. *Mar. Ecol. Proc. Ser.*, in press.

Burkholder,J.M., Glasgow,H.B.,Jr and Steidinger,K.A. (1995b) Stage transformations in the complex life cycle of an ichthyotoxic 'ambush predator' dinoflagellate. In Lassus,P., Arzul,G.,

Erard,E., Gentien,P. and Marcaillou,C. (eds), *Proceedings of the Sixth International Conference on Toxic Marine Phytoplankton*. Elsevier, Amsterdam, in press.

Carpenter,E.J. (1971) Annual phytoplankton cycle of the Cape Fear River estuary, North Carolina. *Chesapeake Sci.*, **12**, 95–104.

Fields,S.D. and Rhodes,R.G. (1991) Ingestion and retention of *Chroomonas* spp. (Cryptophyceae) by *Gymnodinium acidotum* (Dinophyceae). *J. Phycol.*, **27**, 525–529.

Fulton,R.S.,III and Paerl,H.W. (1987) Toxic and inhibitory effects of the blue-green alga *Microcystis aeruginosa* on herbivorous zooplankton. *J. Plankton Res.*, **9**, 837–855.

Fulton,R.S.,III and Paerl,H.W. (1988) Effects of the blue-green alga *Microcystis aeruginosa* on zooplankton competitive relations. *Oecologia (Berlin)*, **76**, 383–389.

Glasgow,H.B.,Jr and Burkholder,J.M. (1993) Comparative saprotrophy by flagellated and amoeboid stages of an ichthyotoxic estuarine dinoflagellate. In Lassus,P. (ed.), *Proceedings, Sixth International Conference on Toxic Marine Phytoplankton*. Institut Francais de Recherche Pour L'Exploitation de la Mer, Nantes, p. 78 (Abstract).

Glasgow,H.B.,Jr and Burkholder,J.M. (1994) Insidious effects on human health and fish survival by the toxic dinoflagellate *Pfiesteria piscimorte* from eutrophic estuaries. In *Proceedings of the 1st International Symposium on Ecosystem Health & Medicine*, Ottawa (Abstract).

Gorham,P.R. and Carmichael,W.W. (1988) Hazards of freshwater blue-green algae (cyanobacteria). In Lembi,C.A. and Waaland,J.R. (eds), *Algae and Human Affairs*. Cambridge University Press, New York, pp. 403–431.

Hamilton,M.A. (1986) Statistical analysis of the cladoceran reproductivity test. *Environ. Tox. Chem.*, **5**, 205–212.

Horning,W.B.,II and Weber,C.I. (eds) (1985) *Short-term Methods for Estimating the Chronic Toxicity of Effluents and Receiving Waters to Freshwater Organisms*. Report EPA/600/4-85/014. US Environmental Protection Agency, Cincinnati.

Houde,S.E.L. and Roman,M.R. (1987) Effects of food quality on the functional ingestion response of the copepod *Acartia tonsa*. *Mar. Ecol. Prog. Ser.*, **40**, 69–77.

Huntley,M., Sykes,P., Rohan,S. and Marin,V. (1986) Chemically mediated rejection of dinoflagellate prey by the copepods *Calanus pacificus* and *Paracalanus parvus*: Mechanism, occurrence and significance. *Mar. Ecol. Prog. Ser.*, **28**, 105–220.

Lee,J.J., Hutner,S.H. and Bovee,E.C. (1985) *An Illustrated Guide to the Protozoa*. Society of Protozoologists. Allen Press, Lawrence, KS.

Lewitus,A.J., Jesien,R.V., Kann,T.M., Burkholder,J.M., Glasgow,H.B. and May,E. (1995) Discovery of the "phantom" dinoflagellate in Chesapeake Bay. *Estuaries*, in press.

Lund,J.W.G., Kipling,C. and LeCren,E.D. (1958) The inverted microscope method of estimating algal numbers and the statistical basis of estimates by counting. *Hydrobiologia*, **11**, 143–170.

Mallin,M.A. (1991) Zooplankton abundance and community structure in a mesohaline North Carolina estuary. *Estuaries*, **14**, 481–488.

Mallin,M.A. and Paerl,H.W. (1994) Planktonic trophic transfer in an estuary: Seasonal, diel, and community structure effects. *Ecology*, **75**, 2168–2185.

Mallin,M.A., Paerl,H.W. and Rudek,J. (1991) Seasonal phytoplankton composition, productivity and biomass in the Neuse River Estuary, North Carolina. *Estuarine Coastal Shelf Sci.*, **32**, 609–623.

Mount,D.I. and Norberg,T.J. (1984) A seven-day life-cycle cladoceran toxicity test. *Environ. Tox. Chem.*, **3**, 425–434.

Noga,E.J., Smith,J.A., Burkholder,J.M., Hobbs,C.W. and Bullis,R.A. (1993) A new icthyotoxic dinoflagellate: cause of acute mortality in aquarium fishes. *Veterinary Record*, **1339**, 96–97.

Paffenhöfer,G.-A. (1988) Feeding rates and behavior of zooplankton. *Bull. Mar. Sci.*, **43**, 430–445.

Peters,D.S. (1968) A study of the relationships between zooplankton abundance and selected environmental variables in the Pamlico River Estuary of Eastern North Carolina. MS Thesis, North Carolina State University, Raleigh, NC, pp. 1–38.

Schnepf,E. and Elbrächter,M. (1992) Nutritional strategies in dinoflagellates. A review with emphasis on cell biological aspects. *Eur. J. Protistol.*, **28**, 3–24.

Schnepf,E., Winter,S. and Mollehauer,D. (1989) *Gymnodinium aeruginosum* (Dinophyta): A blue-green dinoflagellate with a vestigial, cryptophycean symbiont. *Plant Syst. Evol.*, **164**, 75–91.

Sellner,K.G., Lacoutre,R.V., Cibik,S.J., Brindley,A. and Brownlee,S.G. (1991) Importance of a winter dinoflagellate–microflagellate bloom in the Patuxent River Estuary. *Estuarine Coastal Shelf. Sci.*, **32**, 27–42.

Smith,S.A., Noga,E.J. and Bullis,R.A. (1988) Mortality in *Tilapia aurea* due to a toxic dinoflagellate bloom. In *Proceedings of the Third International Colloquium on the Pathology of Marine Aquaculture*, Glouster Point, VA, pp. 167–168 (Abstract).

Statistical Analysis Systems Inc. (1985) *SAS/STAT Guide for Personal Computers, Version 6.* SAS Institute, Inc., Cary, NC, pp. 1–378.

Steidinger,K.A., Truby,E.W., Garrett,J.K. and Burkholder,J.M. (1995) The morphology and cytology of a newly discovered toxic dinoflagellate. In Lassus,P., Arzul,G., Erard,E., Gentien,P. and Marcaillou,C. (ed.), *Proceedings of the Sixth International Conference on Toxic Marine Phytoplankton.* Elsevier, Amsterdam, in press.

Steidinger,K.A., Burkholder,J.M., Glasgow,H.B.,Jr, Truby,E.W., Garrett,J.K., Noga,E.J. and Smith,S.A. *Pfiesteria piscicida,* a new toxic dinoflagellate genus and species of the order Dinamoebales. *J. Phycol.,* in preparation.

Thayer,G.W. (1971) Phytoplankton production and the distribution of nutrients in a shallow unstratified estuarine system near Beaufort, N.C. *Chesapeake Sci.,* 12, 240–253.

Turner,J.T. and Tester,P.A. (1989) Zooplankton feeding ecology: Copepod grazing during an expatriate red tide. In Cosper,E.M., Bricelj,V.M. and Carpenter,E.J. (eds), *Novel Phytoplankton Blooms: Causes and Impacts of Recurrent Brown Tides and Other Unusual Blooms.* Springer-Verlag, Berlin, pp. 359–374.

Turner,J.T. and Tester,P.A. (1992) Zooplankton feeding ecology: bactivory by metazoan microzooplankton. *J. Exp. Mar. Biol. Ecol.,* 160, 149–167.

Uye,S. and Takamatsu,K. (1990) Feeding interactions between planktonic copepods and red-tide flagellates from Japanese coastal waters. *Mar. Ecol. Prog. Ser.,* 59, 97–107.

Vollenweider,R.A. (ed.) (1974) *A Manual on Methods for Measuring Primary Production in Aquatic Environments, 2nd edn International Biology Program Handbook 12.* Blackwell Scientific Publications, Oxford, pp. 1–213.

Watras,C.J., Garcon,V.C., Olson,R.J., Chisholm,S.W. and Anderson,D.M. (1985) The effect of zooplankton grazing on estuarine blooms of the toxic dinoflagellate *Gonyaulax tamarensis. J. Plankton Res.,* 7, 891–908.

Wetzel,R.G. and Likens,G.E. (1991) *Limnological Analyses,* 2nd edn. Springer-Verlag, New York.

White,A.W. (1980) Recurrence of kills of Atlantic herring (*Clupea harengus*) caused by dinoflagellate toxins transferred through herbivorous zooplankton. *Can. J. Fish. Aquat. Sci.,* 37, 2262–2265.

White,A.W. (1981) Marine zooplankton can accumulate and retain dinoflagellate toxins and cause fish kills. *Limnol. Oceanogr.,* 26, 103–109.

Received on March 14, 1994; accepted on September 22, 1994

NATIONAL SEA GRANT COLLEGE PROGRAM

STATEMENT OF OPPORTUNITY FOR FUNDING:
MARINE BIOTECHNOLOGY

DEADLINE: FEBRUARY 15, 1994

FIVE COPIES OF PROPOSAL AND SUMMARY REQUIRED

UNC SEA GRANT COLLEGE PROGRAM
BOX 8605, N.C. STATE UNIVERSITY
RALEIGH, NC 27696
(919) 515-2454

PLEASE CIRCULATE

NATIONAL SEA GRANT COLLEGE PROGRAM

STATEMENT OF OPPORTUNITY FOR FUNDING: MARINE BIOTECHNOLOGY

[Excerpts]

INTRODUCTION

The National Sea Grant College Program is accepting proposals in marine biotechnology for projects of one, two, or three years duration (with annual funding) to begin on or about August 1, 1994.

The primary emphasis will be on research opportunities although advisory and educational proposals also are eligible to compete for funds from an appropriation of $3.2 million. The maximum annual project budget that will be considered for single and multiple investigators is $75,000 and $500,000, respectively, in federal funds. As required by law, at least one third of the total cost of all projects must come from nonfederal matching funds. Industrial matching funds and other close interactions with industry are highly desirable. The National Sea Grant College Program has identified the following preferred areas for proposals:

Aquaculture
Biomaterials/Biosensors
Bioprocessing/Environmental Remediation
Policy/Economics
Education/Technology Transfer

Because of limited funding and widespread interest in marine biotechnology, proposals will be evaluated rigorously. Clear and complete proposals that meet the objectives of the National Sea Grant College Program as outlined in the enclosed materials will have the best chance of success.

BACKGROUND

The United States is the leader in research expertise in marine biotechnology. However, it faces strong competition in other countries that are moving ahead with national investment and planning in this field. Focused research in marine biotechnology in concert with commercial development offers the opportunity to provide scientific, economic, and social advancements. It will lead to new industries and new jobs and will help higher education to meet U.S. needs for scientists and technicians in an increasingly technical and competitive world. It will assist in reversing our trade deficit, which is $2.7 billion a year in seafood alone. It will lead to economic development and increased exports.

A national commitment to research and development in marine biotechnology will help respond to societal needs by (1) increasing the food supply through aquaculture, (2) developing new types and sources of industrial materials and processes, (3) opening new avenues to monitor health and treat disease, (4) providing innovative techniques to restore and protect aquatic ecosystems, (5) enhancing seafood safety and quality, and (6) expanding knowledge of processes in the world ocean.

A Sea Grant report, "Marine Biotechnology: Competing in the 21st Century," outlines a framework for research in support of marine biotechnology. Its three broad themes—Molecular Frontiers in the Ocean Sciences, Applications of Marine Biotechnology, and Marine Biotechnology and Society—encompass a range of research, education, and outreach needed to develop this field. (The report is available from the National Sea Grant Office and the thirty state and regional Sea Grant programs.)

The U.S. House of Representatives recently passed the "Marine Biotechnology Investment Act" which is designed to enhance the National Sea Grant College Program efforts in biotechnology to (1) expand the range and increase the utility of products from the oceans, (2) improve condition of marine ecosystems by developing substitute products that decrease the harvest pressure on living resources, (3) improve production of aquaculture, (4) provide new tools for understanding ecological and evolutionary processes, and (5) improve techniques for remediation of environmental damage.

The U.S. Senate is considering a companion bill to (1) expand the range and increase the utility of products from the oceans, (2) understand and treat human illness, (3) enhance the quality and quantity of seafood, (4) improve the stewardship of marine resources by developing and applying methods to restore and protect marine ecosystems, manage fisheries, and monitor biological and geochemical processes, and (5) to contribute to business and manufacturing innovations, create new jobs, and stimulate private sector investment.

In increasing the National Sea Grant College Program's budget by $3.2 million for fiscal year 1994, the U.S. Congress specified that the increase was to be used to expand research, education, and outreach in marine biotechnology. Focused research, education, and outreach in concert with commercial development offers promise of economic and social benefits. They will lead to new industries and new jobs and advance higher education to meet U.S. needs in an increasingly technical and competitive world.

Biotechnology may be defined as the application of scientific and engineering principles to provide goods and services through mediation of biological agents. The Federal Coordinating Council for Science, Engineering, and Technology, which identified marine biotechnology as an important field, defined biotechnology as "any technique that uses living organisms (or parts of organisms) to make or modify products, to improve plants or animals, or to develop microorganisms for specific use. The development of materials that mimic molecular structures or functions of living organisms is also included . . . [and] research involving recombinant DNA, DNA transfer techniques, macromolecular structure, cell fusion, bioprocessing, etc., and the infrastructure that supports that research." These broad definitions encompass more than DNA technology. Exclusive of agriculture, application of biotechnology in sewage treatment and water purification now comprises the largest sector in volume. Production of beer and spirits, cheese and other dairy products, baker's yeast, organic acids, and antibiotics follow in order of decreasing value. These traditional applications of biotechnology, which are based primarily on use of terrestrial organisms, are enormously important to the economy as well as human health and nutrition. For example, worldwide production of antibiotics by fermentation generates $10s of billions in sales annually.

While biotechnology is not new, developments in modern molecular biology indicate that it is still in its emerging phase. Many authorities expect biotechnology to be a primary basis of America's economic development and strength in the twenty-first century. Oceanic organisms harbor a major portion of the Earth's genetic resources (biodiversity), yet the large majority of marine organisms are not known well enough for their gene pool and biological processes to be accessible to those who develop and practice biotechnology in industry and academe. However, exploratory research shows the rich potential for exploiting the biochemical capabilities of marine organisms to provide models for new classes of pharmaceuticals, polymers, other chemical products, and new industrial processes as well as vaccines, diagnostic and analytical reagents, and genetically altered organisms for commercial use. Sea Grant has interests in the development of modern tools and technologies for aquaculture, resource management, seafood processing, bioprocessing, bioremediation, production of biomaterials and analogues, and controlling biofouling and biocorrosion.

The opportunities for advancing science and providing results directly supporting the development of marine biotechnology are broad. Because many of these developments will be realized only in the long-term it is essential to recognize that results will be enhanced through dissemination of information, through personnel mobility, and through cooperation and collaboration between universities and the private sector.

IMPROVED DETECTION OF AN ICHTHYOTOXIC DINOFLAGELLATE IN ESTUARIES AND AQUACULTURE FACILITIES

**A Research Proposal Submitted to
The National Sea Grant College Program
Marine Biotechnology Initiative**

by
JoAnn M. Burkholder and Parke A. Rublee

———————————————————

JoAnn M. Burkholder
Department of Botany
North Carolina State University

———————————————————

Parke A. Rublee
Department of Biology
University of North Carolina–Greensboro

———————————————————

Ernest D. Seneca, Head
Department of Botany
North Carolina State University

———————————————————

North Carolina State University

MARINE BIOTECHNOLOGY PROPOSAL: PROJECT SUMMARY

Title: Improved Detection of an Ichthyotoxic Dinoflagellate in Estuaries and Aquaculture Facilities

Principal Investigators:

JoAnn M. Burkholder, Associate Professor   *and*   Parke A. Rublee, Assistant Professor
Department of Botany, Box 7612   Department of Biology
North Carolina State University   UNC–Greensboro
Raleigh, NC 27695   Greensboro, NC 27412
Telephone: Telephone:
Time Devoted to Project: 15%   Time Devoted to Project: 25%

Project Period: <u>080194–073196</u>   **Budget Period:** <u>080194–073195</u>

Amount: $96,243 federal / $48,135 matching (NC Agricultural Research Service)

Objectives:

1) To refine rDNA probes, using cultures of the toxic dinoflagellate *Pfiesteria piscimorte,* to ensure high specificity in detecting flagellated, amoeboid, and encysted forms of this fish pathogen.
2) To test the probes to detect *P. piscimorte* in estuarine waters / sediments and aquaculture facilities.
3) To develop a "family" of probes that enable rapid, routine detection / quantification of intraspecific genetic strains of this toxic dinoflagellate at fish kill / ulcerative disease sites.
4) To apply fluorescent gene probes in screening for targeted chemosensory areas (portals of entry, attachment sites) for the dinoflagellate pathogen on/in fish prey.

Methodology:

State-of-the-art PCR amplification techniques for targeted sequence(s) of the dinoflagellate's rRNA genes from algal cultures will enable refinement of gene probes for use in routine detection of *P. piscimorte* in water / sediment samples from estuaries and aquaculture facilities. A fluorescent anti-sense oligoprobe will be developed and tested for rapid visual identification and localization of the alga in water, sediments, and fish tissues. Culture isolates from North Carolina estuaries and other fish kill sites in the mid-Atlantic and Southeast will be tested for intraspecific sequence variation, needed to develop a "family" of probes that vary in specificity depending on the targeted strain of *P. piscimorte*. These probes, in combination with fish bioassay experimental data, will also enable us to detect the presence of other toxic species that may closely resemble this dinoflagellate.

Rationale:

Unexplained "sudden-death" fish kills and ulcerative diseases have increased in North Carolina's estuaries over the past decade. The dinoflagellate *P. piscimorte* recently has been implicated as the causative agent of at least 30% of the major fish kills in these estuaries, and also has been tracked to fish kill sites on the mid-Atlantic Coast and Southeast. Despite its importance, the ephemeral toxicity, multiple cryptic flagellated and amoeboid stages, substantial size range (5–250 μm), and close resemblance of this pathogen to other non-toxic algae make reliable detection difficult in light microscopy analysis. Development of sensitive, highly specific gene probes for *P. piscimorte* will answer a critical need for its routine, early detection by regulatory staff and finfish / shellfish aquaculturists.

Some Suggested Reviewers:

Dr. Edward DeLong, Biol. Dept., Univ. Calif., Santa Barbara, CA 93106
Dr. Lucie Maranda, Dept. of Pharmacognosy & Env., URI, Kingston, RI 02881
Dr. Norman Pace, Inst. Molec. & Cell. Biol., Indiana Univ., Bloomington, IN 47405
Dr. John Paul, Dept. Mar. Sci., Univ. S. Florida, St. Petersburg, FL 33701
Dr. Karen Steidinger, FL DEP, FL Mar. Res. Inst., St. Petersburg, FL 33701
Dr. Susan Weiler, Dept. of Biol., Whitman College, Walla Walla, WA 99362

TABLE OF CONTENTS*

*The complete table of contents has been reprinted here to convey the full structure of the proposal package, but only the main body of the proposal argument has been reproduced.

TITLE: Improved Detection of an Ichthyotoxic Dinoflagellate in Estuaries and Aquaculture Facilities

PRINCIPAL INVESTIGATORS: JoAnn M. Burkholder and Parke A. Rublee

INTRODUCTION AND BACKGROUND

A. The Established Linkage between Sudden-Death Fish Kills, Unexplained Fish Disease, and a Toxic Ambush-Predator Dinoflagellate

A recent cosmopolitan rise in the frequency and spatial extent of toxic phytoplankton blooms suggests that these noxious species can significantly reduce our estuarine and marine fishery resources (White 1988, Shumway 1990, Robineau *et al*. 1991). Among the most notorious of toxic phytoplankton are the dinoflagellates that cause "red tides," resulting in the death or "poisoning" of many finfish and shellfish as well as humans (Steidinger & Baden 1984). The United States and other industrialized nations spend billions of dollars annually in research aimed toward prediction, mitigation, and control of toxic outbreaks (Shumway 1990, Culotta 1992). Despite these efforts, accumulating evidence suggests that toxic dinoflagellates may be increasing their activity and geographic range (Steidinger & Baden 1984, Smayda 1989, Hallegraeff 1993). Moreover, the past decade has yielded discoveries of previously unknown toxic species that have suddenly appeared in bloom concentrations, causing millions of dollars of damage to coastal fisheries and aquaculture industries in many U.S. coastal waters (Steidinger & Baden 1984, Smayda 1989, Shumway 1990).

Among the many estuaries in the United States with increasing incidence of unexplained ulcerative fish disease and "sudden-death" fish kills are the Pamlico, Neuse and New Estuaries in North Carolina (Miller *et al*. 1990, North Carolina Division of Marine Fisheries 1992). In the past three years a phytoplankter contaminant from these estuaries was implicated as the causative agent of approximately 30% of North Carolina's major estuarine fish kills (involving 10^3–10^9 fish; Burkholder *et al*. 1992a, 1993; Fig. 1). The dinoflagellate, *Pfiesteria piscimorte* (gen. et sp. nov.) Steidinger & Burkholder (Steidinger *et al*. 1994), represents a new family, genus and species. Unlike red tide dinoflagellates, its lethal forms are ephemeral in the water column and require live fish for toxic activity. When a school of finfish comes within detectable range, the dinoflagellate exhibits remarkable "phantom-like" behavior. The population swims up into the water from dormant cysts or amoebae on bottom sediments; the cells excrete a lethal toxin, complete sexual reproduction while killing the fish, and then rapidly descend back to the sediments (Burkholder *et al*. 1993). The toxin strips skin tissue from the fish and creates open bleeding sores; it also attacks the renal system, induces hemmorrhaging, suppresses the immune system, throws the nervous system into dysfunction, and apparently causes death by suffocation from muscle paralysis (Burkholder *et al*. 1993, E. Noga unpubl. data).

Toxic outbreaks of this "ambush-predator" dinoflagellate have been documented at salinities ranging from freshwater ($0^0/oo$ salinity) to full-strength seawater ($35^0/oo$), and across temperatures of 4–33°C (Burkholder 1993; Burkholder *et al*. 1992a, 1993). Further, the alga has proven lethal to all 28 species of native and exotic finfish and shellfish tested thus far (Burkholder *et al*. 1993; Table 1).

B. The Need for Highly Specific Molecular Probes to Detect This Fish Pathogen

Detection of *Pfiesteria piscimorte* (gen. et sp. nov.) demands well-timed sampling because the algal population displays rapid encystment without live fish, typically "disappearing" or settling out of the water

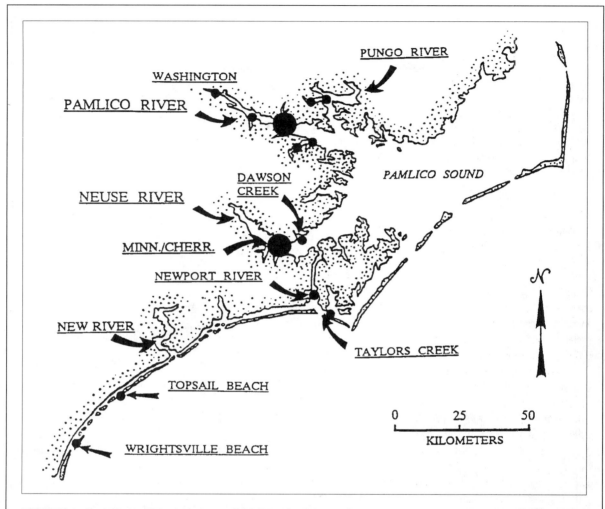

FIGURE 1. Locations where lethal stages of *Pfiesteria piscimorte* (gen. et sp. nov.) has been verified in North Carolina. Kill sites (documented in association with this toxic dinoflagellate in all labeled locations except the New River Estuary) are designated by blackened circles, with large circles representing sites where fish kills are most frequent with substantial area affected. The large circle on the Neuse River Estuary at Minnesott Beach / Cherry Point (MINN./CHERR.) designates the site with highest known fish loss; more than 1 billion Atlantic menhaden were killed, requiring bulldozers to clear the beaches over a 6-week period when the menhaden schools were moving out to sea.

column within only a few hours after fish death (Burkholder *et al.* 1992a). Efforts to establish the presence of *P. piscimorte* (gen. et sp. nov.) in fish kills also have been hampered because this small dinoflagellate resembles several other common nontoxic species so closely that it is difficult to reliably identify it in routine light-microscope analyses. The confusion is further compounded because the alga has a complex life cycle that includes rapid transformations (sometimes within minutes) among at least 15 different flagellated and amoeboid forms, many of which are amorphous and colorless and, hence, easily missed or mistaken for

TABLE 1. Species of finfish and shellfish that are known to be killed by *Pfiesteria piscimorte* (nov. gen et sp.)*

NATIVE ESTUARINE / MARINE SPECIES

American eel *(Anguilla rostrata)*
Atlantic croaker *(Micropogonias undulatus)*
Atlantic menhaden *(Brevoortia tyrannus)*
Bay scallop *(Aequipecten irradians)*
Black grouper *(Mycteroperca bonaci)*
Blue crab *(Callinectes sapidus)*
Channel catfish *(Icatalurus punctatus)*
Hogchoker *(Trinectes masculatus)*
Killifish (mummichog) *(Fundulus heteroclitus)*
Largemouth bass *(Micropterus salmoides)*
Littleneck clam *(Mercenaria mercenaria)*
Pinfish *(Lagodon rhomboides)*
Red drum *(Scianops ocellatus)*
Redear sunfish *(Lepomis microlophys)*
Southern flounder *(Paralichthys lethostigma)*
Spot *(Leiostomus xanthuris)*
Spotted sea trout *(Cynoscion nebulosus)*
Striped bass *(Morone saxatilis)*
Striped mullet *(Mugil cephalus)*
White perch *(Morone americana)*

EXOTIC (INTRODUCED) SPECIES

Clownfish *(Amphiprion percula)*
Goldfish *(Carrasius auratus)*
Guppie *(Poecilia reticulata)*
Hybrid Striped Bass *(Morone saxatilis × Morone chrysops)*
Mosquitofish *(Gambusia affinis)*
Tilapia *(Oreochromis aureus, Oreochromis mossambicus, Tilapia nilotica)*

* All 28 species of finfish and shellfish species tested thus far have proven susceptible to lethal effects of this ambush predator dinoflagellate (Burkholder *et al.* accepted <u>b</u>).

debris (Burkholder *et al.* 1993; research supported by a UNC Sea Grant mini-grant to PI JMB; Fig. 2, Plate 1). These considerations have led many U.S. coastal regulatory agencies to express the need for development of a species-specific "marker" that would enable rapid detection of this dinoflagellate, in its many stages, at sites of fish disease outbreaks or kills in their coastal waters and aquaculture facilities.

In the short time since its discovery in the Pamlico Estuary in 1992, *Pfiesteria piscimorte* (gen. et sp. nov.) has come to be regarded as a potentially widespread but typically undetected source of major fish mortality in nutrient-enriched estuaries (Burkholder *et al.* 1992a, 1993). Within the past two years, JMB and colleagues have tracked this organism to fish kill sites in eutrophic estuaries from the Delaware Bay south to Florida on the western Atlantic Coast, and as far west as Mobile Bay, Alabama on the Gulf Coast (Burkholder *et al.* 1993, Burkholder *et al.* submitted, Lewitus *et al.* submitted; Fig. 3). Scientists and regulatory staff on

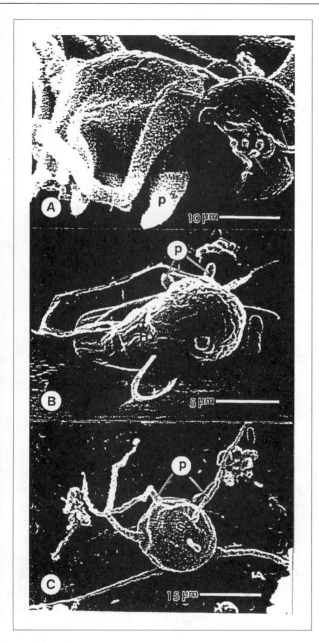

PLATE 1. Scanning electron micrographs of the ambush predator *Pfiesteria piscimorte* (gen. et sp. nov.), including (A) a large lobose amoeba [AM] with pseudopodia [p] and a TFVC [DI] beginning to form pseudopodia [p], but with the transverse flagellum [tf] still intact (× 3520); (B) More advanced state of TFVC conversion to an amoeba [ep = epitheca, hy = hypotheca] (× 2570); and (C) "Star" or filipodial amoeba stage [S] (formerly a TFVC) with long pseudopodia (× 1520).

the Atlantic, Gulf and Pacific Coasts have begun to hunt for the dinoflagellate at sites of ulcerative fish disease and sudden-death kill areas (Anonymous 1992; Kay 1992; Rensberger 1992a,b; Greer 1993; Huyghe 1993). **Methodology for rapid, reliable detection of this pathogen in advance of fish disease outbreaks**

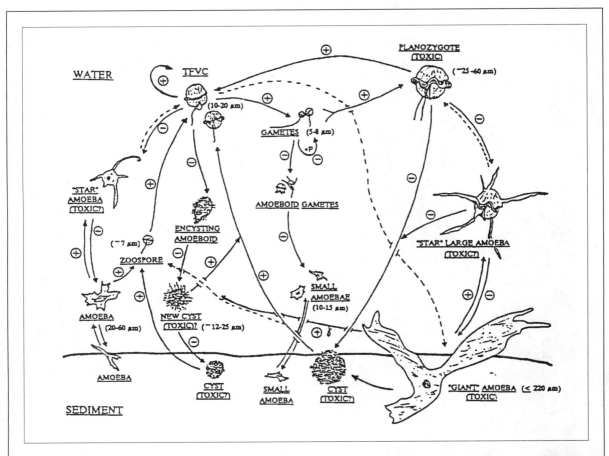

FIGURE 2. The life cycle of *Pfiesteria piscimorte* (gen. et sp. nov.), a representative toxic fish ambush-predator dinoflagellate (TFVC = toxic flagellated vegetative cell, the most toxic stage). Symbols indicate how the dinoflagellate appears in the presence (+) versus absence (–) of live finfish. Solid lines represent verified pathways in the complex life cycle; dashed lines indicate hypothesized additional pathways. "P" designates stimulation of amoeboid gamete production by phosphate enrichment. Stages known or suspected (?) to be toxic are also shown (from Burkholder *et al.* 1993).

and kills is critically needed in order to develop effective management strategies to predict its toxic activity and mitigate its acute and chronic effects on our estuarine fisheries.

C. Progress to Date on Gene Probe Development for *Pfiesteria piscimorte*

Within the past three years, both national and international workshops on toxic marine phytoplankton have identified an "imperative" for improved techniques to enable accurate, rapid detection and identification of harmful algae by regulatory staff on a routine basis (UNESCO 1991, Anderson *et al.* 1993). From a practical standpoint, the need for a sensitive highly species-specific technique to detect harmful species such as *Pfiesteria piscimorte* (gen. et sp. nov.) with multiple cryptic stages would be especially critical. Under a cur-

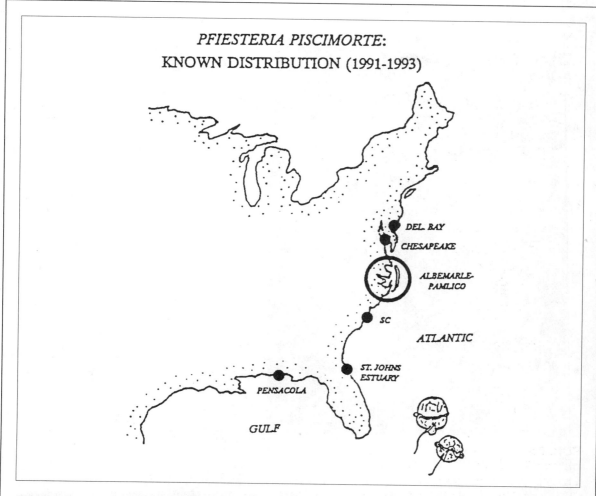

FIGURE 3. Locations where water samples have been confirmed to contain toxic stages of *Pfiesteria piscimorte* (gen. et sp. nov.) (data from PI JMB with H. Glasgow, B. Anderson, R. Lewis, K. Steidinger, and J. Landsberg); Mobile Bay, AL recently has been confirmed as an additional distribution site.

rent mini-grant from the UNC Sea Grant and the UNC Water Resources Research Institute (9/93–7/94), we are developing a gene probe to cultures of *P. piscimorte* (gen. et sp. nov.) that was isolated from the most concentrated known region of toxic activity by this organism—the Albemarle–Pamlico Estuarine System of North Carolina. At present we have successfully extracted DNA from *P. piscimorte* (gen. et sp. nov.), and have amplified a 1.0–1.1 kb fragment (Fig. 4) using "universal" primers to the small subunit ribosomal-DNA gene (16S-rDNA; Pace *et al.* 1986, Lynn & Sogin 1988, Britschgi & Giovannoni 1991, Manhart & McCourt 1992, Weisburg *et al.* 1991). The fragment has been ligated into the *BamHI-XhoI* restriction sites of the pBluescript II SK plasmid, which was then used to transform competent *Escherichia coli* DH5α cells with standard techniques (Sambrook *et al.* 1989, Ausubel *et al.* 1992). We now have available a total of 40 bacte-

FIGURE 4. Gell electrophoresis of PCR reaction of *Pfiesteria piscimorte* (gen. et sp. nov.) from the large lobose amoeba stage. Lanes 1 & 8, Lambda, *BstE*II DNA marker. Lane 2, positive control. Lanes 5 & 6, PCR products from the dinoflagellate (length ca 1.1 kb).

rial clones from 2 isolates of the dinoflagellate. We are set to begin sequencing these clones by the dideoxy chain termination method (Sanger 1977), while continuing to establish additional clones from other isolates. After the sequence has been determined, we will look for 17–34 base pair consensus regions within the 16S rDNA that are unique to this toxic alga, for use as species-specific gene probes.

The consensus sequence will be tested as a probe against samples of *P. piscimorte* (gen. et sp. nov.) and other algae. A radioactively end-labeled oligonucleotide probe corresponding to the sequence will be tested at high-stringency with Southern blots and later with dot/slot blots using preparations of the dinoflagellate's DNA in varying amounts. Salt and temperature concentrations will be adjusted so that the probe hybridizes specifically with samples containing the alga's DNA (high-stringency). Its specificity will be further tested by mixing samples of the dinoflagellate's DNA and nontarget DNA in known proportions, to determine whether the probe evokes a signal proportional to the amount of DNA from *P. piscimorte* (gen. et sp. nov.) in the mixed samples. We actually plan to test several probes under these conditions to identify optimal sequences for sensitivity and specificity among culture isolates representing different strains of this dinoflagellate. We are confident that we will have a radiolabeled probe that can be used on extracted DNA samples of cultured *P. piscimorte* (gen. et sp. nov.) by July 1994.

D. Goals and Hypotheses of the Proposed Research

notice bf [handwritten margin note]

The major goal of the proposed research is to refine state-of-the-art techniques for application of gene probes to enable routine detection of *Pfiesteria piscimorte* **(gen. et sp. nov.) in natural estuarine habitat and aquaculture facilities.** In this study we will extend application of gene probes to this toxic dinoflagellate for use on water and sediment field samples. We plan to determine the distribution of *P. piscimorte* (gen. et sp. nov.) in coastal systems and assess (a) genetic variability among populations; (b) whether this species is unique or representative of additional, as-yet-unknown toxic dinoflagellates, and (c) the distribution of "cyst banks" as it relates to nutrient conditions, especially to sites of anthropogenic nutrient input. We will also examine physical relationships (portals of entry / attachment) of this fish pathogen to its fish and invertebrate hosts, as indicative of sites of chemosensory attraction. Our research will focus on the following hypotheses:

1. Diverse races or strains within the species *Pfiesteria piscimorte* (gen. et sp. nov.) are abundant and widespread in eutrophic estuaries of North Carolina. Intraspecific genetic variability increases among populations in similar habitat of major estuaries from the mid-Atlantic to the Gulf Coast in the southeastern United States.
2. The unique primers that allow for PCR (polymerase chain reaction) amplification of targeted sequence(s) of rRNA genes in this toxic dinoflagellate can be used as a template from which other gene probes can be developed to enable detection of various strains within *P. piscimorte* (gen. et sp. nov.).
3. Sensitive, highly specific radiolabeled or fluorescent gene probes for *P. piscimorte* (gen. et sp. nov.), developed from cultured material, can be applied successfully for rapid detection of flagellated, amoeboid, and encysted stages of this fish pathogen—even in low abundance—in estuarine water column / sediment samples, and in aquaculture facilities.

OBJECTIVES

The overall objective of this research is to develop, test and refine application of gene probes for rapid, routine detection / quantification of the toxic dinoflagellate *Pfiesteria piscimorte* (gen. et sp. nov.), in all its stages, from natural estuarine waters and sediments and from aquaculture facilities. Specific objectives to accomplish this goal are as follows:

1. **Extend application of the gene probe(s) for this toxic dinoflagellate so that we can (a) detect it quickly during fish kills or disease outbreaks; (b) assay estuarine sediments for detection of ambient active (amoeboid) and dormant (encysted) populations; and (c) apply the probes as a basic tool** for further research on the ecology, toxicology, and developmental biology of this fish pathogen. [Years I–II]. Specifically, this work will require development in two directions:
    (i) **Gene probe** development for use as an extremely sensitive "field" detection tool to determine genetic variability in *P. piscimorte* (gen. et sp. nov.) populations from both the water column and the sediments, and for use as a basis for additional gene probes to related species. This gene probe will consist of the unique primers that allow for PCR amplification of the target sequence of rRNA genes, even if the rRNA genes are present in low copy number. The high quantity of DNA per dinoflagellate cell (ca 3–200 pg cell^{-1}; Triplett *et al.* 1993), relative to DNA content in other algae (average of 0.54 pg DNA cell^{-1}; Rizzo 1987) assures that substantial material can be obtained from relatively few cells of this organism.

(ii) **Fluorescent probe** development as an *"in situ"* probe for rapid visual identification and localization of the target species. This second probe will actually be an anti-sense **mRNA oligoprobe** with a fluorescent tag that binds to target mRNA, found in high copy number in the dinoflagellate's cells, to facilitate microscopic visualization (DeLong *et al.* 1989).

2. **Apply the probes for detection / quantification of this dinoflagellate from the water column and sediments of known sudden-death fish kill sites, and at major fish kill events** [Year I (pilot); Years II–III]. This will involve examining the intraspecific genetic variability in *P. piscimorte* (gen. et sp. nov.), a process that will also lead to detection of closely related species with similar behavior and appearance.

3. **Screen and identify targeted areas (portals of entry, sites of attachment) of the pathogen on/in fish prey,** of use in related research to identify the substance(s) from fish tissues that trigger toxin production, and to aid in understanding the pathology of this organism. The fluorescent probe will facilitate light-microscopy inspection of specific tissues. This application will also be of value in providing evidence for the presence of the organism in "individual" fish kill events.

4. **Determine utility of gene probes for this pathogen to the aquaculture industry** [Years II, III]. Our probes will enable identification of problematic periods for use of estuarine waters, and rapid determination of the degree of water source contamination by the toxic dinoflagellate. The probes will also be of use in discerning contaminated tanks or fish.

METHODOLOGY AND RATIONALE

A. Culturing This Unique Toxic Dinoflagellate (Burkholder; Objective 1; Years I–III)

Accomplishment of the central goal of this research—refinement of gene probes for *Pfiesteria piscimorte* (gen. et sp. nov.)—depends on successful maintenance of culture isolates for quality assurance in efficiency of extraction of the dinoflagellate's DNA from water and sediment samples. This organism is unique among toxic dinoflagellates in two major characteristics: (i) Unlike nearly all other known toxic dinoflagellates, it is animal-like rather than photosynthetic. As mentioned, it is capable of assuming a photosynthetic habit only by consuming other algae and digesting all but their chloroplasts, which it retains as cleptochloroplasts for extended periods (Steidinger *et al.* 1994); and (ii) *P. piscimorte* (gen. et sp. nov.) requires an as-yet-uncharacterized substance(s) from live fish at least daily to maintain toxic activity and complete its life cycle. These features mandate aeration of cultures, with best dinoflagellate growth when aeration is accomplished using box filters (Burkholder *et al.* 1993).

The challenge of safely culturing this alga demands considerable expense and effort. Because of the potential for impairment to human health from neurotoxic aerosols, all cultures are maintained in an isolated, quarantined modular facility with restricted access (Burkholder *et al.* 1993). Each aquarium must be separately housed within a completely contained, custom-designed isolation chamber that is vented (through HEPA filters) to the air outside the facility. As we collect isolates from sudden-death fish kill sites in North Carolina and other areas for testing with our gene probes, these isolation chambers will protect cultures from cross-contamination while affording safe working conditions. We have successfully maintained high concentrations ($\geq$ 1,000 to ca 90,000 cells mL^{-1}) of toxic flagellated stages and cysts using tilapia prey (*Oreochromis mossambica*; 3 fish, each 5–7 cm in length, added 1x or 2x daily per 9-L aquarium). The culture medium consists of 15⁰/₀₀-salinity water obtained by adding Instant Ocean salts to well water (NCSU School of Veterinary Medicine). To obtain unialgal cultures, fish are physically separated from dinoflagellate cells that have been isolated by hand (with micromanipulator at 400x) and inoculated into 0.45–0.80 μm-porosity dialysis tubing that allows passage of the toxin and finfish secreta / excreta.

Our modular culture facility includes a separate "cold" (uncontaminated) room for final cleaning of all aquaria and glassware with dilute bleach, a procedure which destroys even the encysted dinoflagellates without leaving toxic residues from the alga or the fixative. A larger adjacent room serves as the culture area, with facilities at one end for soaking contaminated labware and aquaria in strong bleach as initial steps in our cleaning procedures. We routinely use respirators (with HEPA + organic acid filters), disposable gloves, disposable boots, and clothing that is removed and treated with dilute bleach (0.05%) after use, as standard operating procedure. To de-contaminate aquaria, all reusable labware is soaked in dilute bleach for 12 hr to destroy the cells and cysts, and then is rinsed thoroughly with deionized water. All work in the culture facility further mandates a "buddy system" for additional safety assurance, with at least two people present at all times.

B. Rationale for Use of Genetic Probes

Recent advances in molecular biology have shown great promise in unraveling many basic and applied ecological questions. Brand (1988) reviewed the topic of genetic variation in phytoplankton and concluded that such variation, while not well understood, likely is important in determining the ecological "success" of algal species. Manhart & McCourt (1992) noted the potential value of molecular tools in clarifying phytoplankton speciation and species identifications. Within the past five years, an extensive literature on molecular biology of phytoplankton has become available, including studies of structural characteristics of algal nuclear, chloroplast, and mitochondrial genomes (e.g., Coleman & Goff 1991, Li & Cattolico 1992, Anderson *et al.* 1992, Reynolds *et al.* 1993), as well as information on ribosomal RNA genes (e.g., Rowan & Powers 1992, Scholen *et al.* 1993) and specific functional genes such as large subunit ribulose-1,5-bisphosphate carboxylase (*rbcL*; e.g., Pichard *et al.* 1993).

The potential value of such knowledge has been recognized by recent workshops to identify essential research needs for understanding and mitigating harmful algal blooms (UNESCO 1991, Anderson *et al.* 1993), where strong recommendations have called for development of immunologic and molecular probes to economically important toxic algal species. Such probes could provide rapid sample analysis while avoiding subjective visual identifications, made with light microscopy, that can be challenging even to scientists who are experienced in working with this dinoflagellate (Burkholder *et al.* 1993). Further, many of the difficulties associated with probe development (e.g., limited knowledge base, intensive laboratory effort, expense) have been resolved or reduced in recent years (e.g., significant advances in PCR technology and applications).

Our choice of gene probes over immunologic probes is based on several considerations. First, gene probes offer a high degree of specificity, and technology has improved to the point that development of such probes is realistic within the projected time frame. Second, gene probes have all the advantages of immunologic probes (e.g., specificity, sensitivity, capacity for linkage with radiochemical, fluorescent, or enzymatic markers), with the additional advantage of providing direct insights about genetic variability among isolates via the nucleotide sequence. Selection of rDNA probes over alternative gene probes to specific functional genes (e.g., enzymes) derives from their common use (e.g., DeRijk *et al.* 1992) as well as the state of our knowledge about the genetic structure of *Pfiesteria piscimorte* (gen. et sp. nov.). The rDNA probes have been developed and used extensively over the past decade as both intra- and interspecific markers, since ribosomal RNA is abundant in cells of all living organisms. Further, the DNA that codes for the RNA contains regions which are conserved across all kingdoms as well as variable regions that are unique at species / subspecies levels (Pace *et al.* 1986, Britschgi & Giovannoni 1991, Scholin *et al.* 1993). As a result, a large information base is available that includes both applications and rDNA sequences (e.g., GenBank, EMBL), and which spans a wide array of prokaryotic and eukaryotic organisms including dinoflagellates (e.g., Lenaers *et al.* 1989, Geraud *et al.* 1991, Fritz *et al.* 1991, Rowan & Powers 1992, DeRijk *et al.* 1992).

The use of probes to specific functional genes can offer insights about organism abundance and activity, as Pichard & Paul (1993) and Pichard *et al.* (1993) have shown with *rcbL* genes in photosynthetic algae. Our limited knowledge of *Pfiesteria piscimorte* (gen. et sp. nov.), however, makes it difficult to identify functional genes of optimal use as targets for such probes. For example, the organism "borrows" or retains chloroplasts from algal prey; aside from these "cleptochloroplasts," this dinoflagellate has none of its own (Steidinger et al. 1994). Hence, use of *rcbL* probes would be misleading. Further, the ability of this fish pathogen to transform rapidly among stages that are distinct in morphology and function suggests that targeting highly specific functional genes would be premature. Thus, at present we have focused our efforts toward developing "conservative" probes which we believe will have the most utility. In particular, with respect to aquaculture application of pretesting facilities for the presence of *P. piscimorte* (gen. et sp. nov.), this approach is less likely to lead to false negatives and subsequent economic loss. Ultimately, we hope to develop not only markers to specific strains of this dinoflagellate, but also to find markers that are unique to different stages in the life cycle and/or different functions of the organism—that is, unique signatures of mRNA expression to each stage (e.g., markers specific to flagellated, encysted, or amoeboid structures).

C. Research Plan for Extending Probe Applications

1. Fluorescent Probe Development (Burkholder/Rublee; required for Objectives 1-ii, 2–5; Years I–III; Table 2)—Immediately upon confirming suitability of the gene probe currently under development from cultured isolates, we will generate an anti-sense oligonucleotide probe tagged with a fluorescent marker (procedure of DeLong *et al.* 1989), using aminoethyl phosphate linkers and fluorescein dyes. This probe will first be tested for specificity and sensitivity against cultured dinoflagellate cells, followed by tests against water samples from fish kills or fish kill areas in which *Pfiesteria piscimorte* (gen. et sp. nov.) has been implicated as the causative agent.

2. Selection of Sediment Extraction Techniques (Rublee/Burkholder; required for Objective 3; Year I)—In sandy sediments, fluorescent probes should facilitate detection of the dinoflagellate's cysts as well as filipodial and lobopodial amoeboid stages, which tend to wrap their pseudopodia around particulate matter and tightly adhere when disturbed so that they are exceedingly difficult to detect by light microscopy. We anticipate that the amoebae will behave in this manner when they are sampled during or following distribution into aquarium cultures for bioassay of toxic activity in the presence of fish. The probes will be applied to water-column samples from the aquarium bioassays, of use as an aid in "ground truthing" our direct microscope counts.

We expect that detection of *Pfiesteria piscimorte* (gen. et. sp. nov.) in sediments with low cell abundance or high organic content will require use of more sensitive assays. We will probe these types of samples by extracting sediment DNA and using PCR amplification with our species-specific primers (e.g., Tsai & Olsen 1991, 1992; Moran *et al.* 1993). Since estuarine sediments vary in mineral and organic content (Rublee 1982; which may affect DNA extraction efficiency), we will first test defined sediments that have been seeded

TABLE 2. Timetable for completing the proposed research

PROJECT ACTIVITY	YEAR I	YEAR II	YEAR III
1. Culturing the dinoflagellate	————————————————————————————————		
2. Extending probe applications:			
A. Fluorescent probe development	——		
B. Development of sediment extraction methods	———		
C. Development of probes with varying sensitivity	—————————————————————————		
3. Field sampling effort:			
A. Analysis of natural "bloom" samples	– – –	——	——
B. Collection trips—North Carolina	– –—	——	——
C. Sampling from other locations (including network)	———	———	———
4. Probe application to detect the dinoflagellate on fish prey	– – —————— – – –		
5. Testing aquaculture sites		————————————————————	
6. Completion of reports and publications	———	———	———

with the dinoflagellate's cells, followed by tests of natural sediments using added cells as an "internal standard" for DNA extractions. The lysozyme-SDS-freeze-thaw extractions procedure of Tsai & Olsen (1992) will be followed by purification by electrophoresis (cf. Erb & Wagner-Döbler 1993, Herrick *et al.* 1993), to the lysozyme-hot phenol extraction method of Moran *et al.* (1993, omitting the DNAse digestion step). An alternative procedure, to be considered if the rDNA probes prove unsuitable, will be to assay directly for rRNA with anti-sense oligonucleotide primers using reverse transcriptase PCR (Moran *et al.* 1993).

3. Development of Probes that Differ in Sensitivity (Rublee/Burkholder; required for Objective 2; Year I [pilot]; Years II–III)—Within the past year, the discovery of a second Pfiesteria-like species (Landsberg *et al.* 1994) suggests that *Pfiesteria piscimorte* (gen. et sp. nov.) likely is one of a number of similar ambush-predator dinoflagellates (Burkholder 1993) that may play important roles in coastal ecosystems due to their varied life forms. An approach to screen for other such species which, like *P. piscimorte* (gen. et sp. nov.), may be cryptic with ephemeral stages, is to extend the utility of our probes by altering their specificity and developing a "family" of probes (*sensu* Anderson 1993). This will be accomplished either by (1) varying the conditions under which hybridization is allowed to occur (altering stringency), or (2) altering the nucleotide sequence in the probes themselves. We plan to test isolates as well as natural samples from kill sites under varying stringency conditions.

Improved selection of probe sequence will be achieved over time as we strengthen our understanding of the intraspecific genetic variability in *P. piscimorte* (gen. et sp. nov.) and related taxa. In particular, for all estuarine / aquaculture sampling efforts (see D, below), where we find dinoflagellates that appear visually similar to *P. piscimorte* (gen. et sp. nov.) but do not hybridize with probes under high-stringency conditions, we will reprobe using lower-stringency conditions as a test for other strains or closely related species (cf. Barkay *et al.* 1990). In either case, if we are able to isolate dinoflagellate cells from such samples, we will then use "universal" PCR primers to repeat the amplification, cloning, and sequencing process that was initially used to develop the *P. piscimorte* (gen. et sp. nov.) gene probes (see "Introduction and Background Information"). These steps will enable us to confirm new strains or species, and to develop additional probes of appropriate specificity.

D. Analysis to Verify Ability to Track the Pathogen in Natural Habitat
(Rublee/Burkholder; required for Objective 2; Year I [pilot], Years II–III)—

As a component of a related research effort (with separate funding support), over the next four years PI JMB will be intensively sampling the Minnesott Beach / Cherry Point region of the Neuse Estuary, which represents a major repeat-kill site for this dinoflagellate (Burkholder *et al.* 1993). In the proposed research, we plan to take advantage of that sampling effort in order to assess the utility of our probes in tracking toxic *Pfiesteria piscimorte* (gen. et sp. nov.) blooms. The water column will be sampled at biweekly to monthly intervals from April–October, which is the period associated with maximal growth and toxic outbreaks of this organism. We will check for increasing abundance of toxic stages using the Utermöhl technique (Lund *et al.* 1958) as in Burkholder & Wetzel (1989), and will verify our identifications using SEM. The data will be compared with / without application of our fluorescent probe to ground truth our direct counts by light microscopy. Fish kill events will be sampled more intensively (2-hr intervals on short-term kills; daily intervals during kills of longer duration [≥ 1 week]), extending for 1 week after the kill to track the dinoflagellate's decline. Analysis of the results will provide insights about the level of spatial / temporal resolution required to adequately sample such events. Live water samples from fish kills (parts D,E) will be screened for the presence of toxic stages using aquarium bioassays with our standard tilapia test species (Burkholder *et al.* 1992a). Tilapia is not endemic to North Carolina estuaries; it offers the advantages of constant availability, wide salinity tolerance, and certainty of no prior contamination by the alga.

E. Analysis of Natural Samples for Geographic Distribution
(Burkholder/Rublee; required for Objective 2; Year I [pilot-NC]; more intensive during Years II–III, including surveys of selected estuaries in DE, MD, GA, FL)—

An understanding of regional as well as local genetic variability among strains of *Pfiesteria piscimorte* (gen. et sp. nov.) is critical to development of highly specific gene probes that can be used by regulatory staff to detect this dinoflagellate in water samples on a routine basis. Our knowledge of intraspecific variability will be strengthened by the field component of the proposed research. Throughout this study we plan to collect natural water and sediment samples from sudden-death fish kills, known fish kill sites (from surveys of the Pamlico, Neuse and New River Estuaries in North Carolina; the Indian River in Delaware; the Choptank River in Maryland; the Savannah River in Georgia; and the St. Johns (Atlantic) and Pensacola (Gulf) areas in Florida that are suspected to harbor *Pfiesteria piscimorte* (gen. et sp. nov.). Samples will be obtained as follows:

1. **Bloom events during kills—We plan to continue to coordinate a network of state personnel and volunteer citizens** (established in 1991, with help from NC Division of Environmental Management and the NC Division of Marine Fisheries, and from concerned citizens groups such as the Neuse River Foundation and the Tar-Pamlico River Foundation) to help sample estuarine fish kills while in progress. We will also sample waters and sediments during fish kills ourselves whenever possible (i.e., whenever we are able to mobilize and complete the 3-hr drive to the North Carolina coast to sample the sudden-death fish kills while in progress). Water samples will be taken with an integrated water-column sampler (Neuse mean depth in the study area = 3.5 m); the upper 2 cm of sediment will be quantitatively sampled in two randomly positioned 0.1 m^2 quadrats per site with gentle siphoning guided by SCUBA divers (technique tested well at 26 sites in the Cherry Point area, October 1993).

   Our network has been highly successful in helping to establish the presence of the toxic dinoflagellate at kills during previous years. Based on data from May–Oct. 1991–1993, we expect that 8–12 major fish kills yr^{-1} (involving $\geq 10^3$ fish) will be linked to this organism (Burkholder *et al.* 1993). Following a kill in which *P. piscimorte* (gen. et sp. nov.) is found swarming in the water, with toxicity verified using aquarium bioassays, we will sample biweekly for 3 months to follow the dinoflagellate's population dynamics over time in both the water and sediments. Environmental variables to be sampled (as part of the separate funding support) will include temperature, salinity, pH, dissolved oxygen, total phosphorous, phosphate, total Kjeldahl nitrogen, nitrate, and ammonium, using techniques described in Burkholder *et al.* (1992b).

2. **Local surveys—We will collect replicate water and sediment samples from ca 50 representative estuarine sites in North Carolina** that previously have experienced fish kills (Neuse and Pamlico Rivers) or high incidence of unexplained ulcerative disease (Neuse, Pamlico, and New Rivers; Burkholder *et al.* 1993). Control sites will also be tested, using areas without incidence of fish kill / disease that are regarded as prime habitat for commercially important fisheries. These sites will span salinity gradients from rivers to sounds, and will include areas with/without wastewater discharge. Sampling of surveys sites in NC will begin during Year I, with major field effort planned for July–October (maximal kill / disease incidence) of Years II–III. These samples will be quickly assayed by our gene probe(s) for comparison with direct microscope counts, to confirm utility of the probe and strengthen our database about the distribution of *P. piscimorte* (gen. et sp. nov.) and abundance of flagellated / amoeboid stages in North Carolina estuaries.

   As part of this effort, we plan to examine correlations between the abundance of this pathogen and nutrient enrichment (P,N), to determine whether the dinoflagellate may be a useful biosensor of cultural eutrophication (as indicated by nutrient bioassays on some stages in culture [Burkholder *et al.* 1992], and by surveys of the New River Estuary that yielded significantly higher abundance of the toxic precursor stage in sewage outfall sites than in control areas without wastewater discharge [Burkholder & Glasgow unpubl. data; Fig. 5]). Accompanying environmental data will include the variables listed for analysis in the Minnesott Beach / Cherry Point area (1, above).

3. **Geographic distribution (Southeast region)—We will complete a sampling survey of sites in the selected estuaries of Delaware, Maryland, Georgia and Florida,** known or suspected to support populations of this dinoflagellate and close relatives. Culture isolates will be established from each location for tests against our gene probes. [Note that any research with the Pfiesteria-like species discovered by Landsberg *et al.* will be pursued in collaborative efforts led by our Florida Marine Research Institute colleagues, Drs. K. Steidinger and J. Landsberg.]

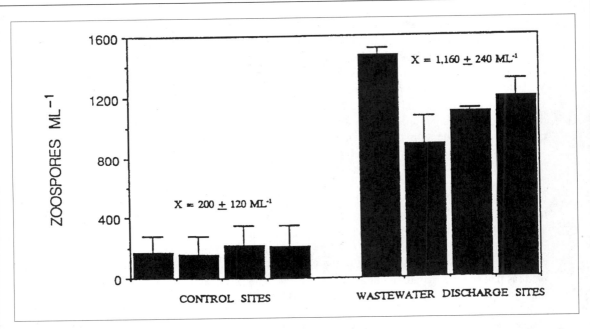

FIGURE 5. Abundance of the zoospore precursor stage to lethal TFVCs (see Fig. 2) from 4 control sites (upper Stones Bay, Swan Point, Ellis Cove, and Sneads Ferry Marina, in order of presentation) and 4 waste-water discharge sites (sampled within ca 100 m from the discharge point: Morgan Bay, mouth of French's Creek, lower Northeast Creek, and lower Stones Bay) at the end of the growing season in the New River Estu-ary, NC (duplicate samples per site; means ± 1 standard deviation). The data suggest that *Pfiesteria pisci-morte* (gen. et sp. nov.) may be of use as a biosensor of cultural eutrophication.

F. Localization of Sites with Maximal Chemosensory Attraction on Finfish
(Burkholder; required for Objective 3; Year II)—

In a related project, we have begun to screen fish tissues (e.g., gill, cloacal area, blood, mouth) for "swarming" activity or high incidence of attachment of dinoflagellate cells indicating obvious chemosensory attraction. This task has been approached using both light microscopy and scanning electron microscopy (SEM), but with only limited success. Light microscopy does not afford the high resolution needed to read-ily discern colorless amoeboid stages and small flagellates among fish debris. SEM greatly improves resolu-tion, but analysis is restricted to extremely small tissue areas. We anticipate that fluorescently labeled probe(s) will greatly facilitate screening of larger quantities of tissue by light microscopy.

Juveniles or young adults of three fish species (hybrid striped bass, tilapia, and mosquitofish *[Gambu-sia affinis]*), known to span a range of sensitivity to the toxins of this dinoflagellate (Burkholder & Glasgow unpubl. data), will be examined with the fluorescent probe(s) to discern sites / tissues that attract high concen-trations of toxic cells. In separate bioassays, each fish species will be exposed to an actively growing culture of *Pfiesteria piscimorte* (gen. et sp. nov.). At intervals between introduction and death of prey, individual ani-mals (each considered a "repeat trial" for the given species) will be sacrificed and gills, cloacal area, blood, mouth, and epidermis in localized regions of interest (e.g., for hybrid striped bass, the lymphatic canal region below the dorsal fin which typically fills with blood in response to the dinoflagellate's toxin) will be exposed to our fluorescent probe in the manner of DeLong *et al.* (1989). Time course analysis (1 hr, 2 hr, and 4 hr) of incubation with the fluorescent probe will be completed to determine the optimum exposure period for labeling dinoflagellate cells. Samples can then be screened for detection of the pathogen's flagellated / amoeboid stages among fish debris.

G. Aquaculture Site Testing
(Rublee and Burkholder; required for Objective 4; Year III)

This project component will be addressed using two approaches. First, we plan to obtain water samples and bottom scrapings in culture tanks used in finfish aquaculture at NMFS–Beaufort. For the past three years, PI JMB has counseled colleagues at this facility (Dr. A. Powell and Mr. W. Hettler, who are culturing southern flounder, Atlantic menhaden, spot *[Leiostomus xanthuris]*, bay scallops and other species) at this facility. The seawater supply is Taylors Creek, which is a known "hot spot" and kill site for *Pfiesteria piscimorte* (gen. et sp. nov.; Burkholder *et al.* 1993). In collaborative efforts we have learned that menhaden eggs do not hatch when lethal stages of the alga are abundant in the culture tank water (Burkholder *et al.* 1993). Moreover, southern flounder delay reproduction and develop large open, bleeding ventral sores when populations of toxic amoebae begin to grow on the tank floor. We will provide gene probe versus light microscope analyses of the samples as our probes come on line especially during years II–III of this effort. Secondly, through UNC–Sea Grant aquaculture extension specialists, we will obtain and similarly test samples from large-scale commercial operations for hybrid striped bass *(Morone saxatilis × Morone chyrsops),* a species which is known to be extremely susceptible to the dinoflagellate (Burkholder *et al.* 1993). Such sampling will enable additional ground truthing or quality assurance of our probes for an extended period of time, and also will provide a means to introduce this early detection technology into the aquaculture industry.

EXPECTED RESULTS

With bacterial clones containing rDNA fragments now established from two *Pfiesteria piscimorte* (gen. et sp. nov.) isolates, we have made significant progress toward developing a sensitive and reliable detection method for this toxic dinoflagellate. We expect to complete development of this probe for routine monitoring and detection by the fishing industry, as well as for use in basic ecological studies. Insights into the ecology, life history, and genetic character of this organism is a necessary prelude to the development of applications in medical research, and to developing management strategies to mitigate impacts of *P. piscimorte* (gen. et sp. nov.) on estuarine waters which serve as nursery areas and habitat for many commercially and recreationally important finfish and shellfish species.

Within the past year, one Ph.D. candidate has undertaken thesis research to examine nutritional controls on the population dynamics and transformations of *Pfiesteria piscimorte* (gen. et sp. nov.), and will be trained in development of gene probes. This graduate assistant will be continuing his dissertation experiments as a member of the project team for the proposed research. A Master of Science graduate research assistant will be conducting field population studies of this dinoflagellate as thesis research in this project, and she will also be trained in probe development. A second M.S. student and an undergraduate will be involved in gene probe development for the organism at UNC–Greensboro. At least four publications are planned per year from this research in internationally refereed journals. In other education-related activities, PI JMB has presented an average of more than 80 lectures and seminars per year in 1991–1993 about this toxic dinoflagellate and its effects on estuaries and aquaculture, including presentations to undergraduate students, graduate students, concerned citizens groups, regulatory staff, legislators, and scientists.

Research on this organism by PI JMB and her staff (1991–1993) has been translated into 37 languages and disseminated by the British Broadcasting Corporation, the Canadian Broadcasting Corporation, CNN, "Good Morning America," *Discover* Magazine, the *New York Times,* the *Washington Post,* and *Time* Magazine. There is strong public support in North Carolina, the United States, and other nations for rapid progress to be made in understanding and mitigating the effects of this organism on our fishery resources. To acquire this knowledge, techniques are critically needed to enable rapid, reliable, early detection of *P. piscimorte* (gen. et sp. nov.) prior to major ulcerative disease outbreaks or kill events. The central goal of the proposed research will be to meet this essential need by developing sensitive, highly specific gene probes for fish pathogen.

Education, graduate student training, and lecture / seminar presentations are important components of information transfer. To further aid in disseminating information about the availability of our gene probes for *Pfiesteria piscimorte* (gen. et sp. nov.), we plan to conduct demonstration / training workshops among regulatory agencies in the North Carolina Department of Environment, Health & Natural Resources (notably the Division of Environmental Management, Shellfish Sanitation, and Division of Marine Fisheries); the National Marine Fisheries Service–Beaufort, NC; and aquaculture specialists in UNC–Sea Grant extension program to inform them of the availability and power of this new technology as it is refined and tested. Staff members from these agencies have already expressed interest in participating in such workshops.

APPLICATION

This work, addressing several key targets identified by a national Sea Grant initiative for research in Marine Biotechnology, will yield applications of both immediate and long-term value. Development of a sensitive, highly specific diagnostic technique for this ichthyotoxic dinoflagellate pathogen will enable routine *in situ* monitoring of water quality in aquaculture. First, the probes can immediately be used to test for the presence of *Pfiesteria piscimorte* (gen. et sp. nov.) as a contaminant of aquaculture facilities, thereby preventing introduction of valuable fish stocks into ponds with high probability of later fish kills. The probes will enable early detection of the organism's presence so that steps can be taken (e.g., removal of fish so that ponds can be drained and treated with chlorox pellets, known to destroy all stages of the dinoflagellate including resistant resting cysts) so that toxic outbreaks in the ponds or tanks can be mitigated or avoided. In addition, the gene probes will facilitate detection of contaminated fish, leading to improved quarantine measures and improved quality assurance when fish and fish products are transported to/from other states—also an important consideration for human health, since this alga is known to be toxic to humans as well as fish (Burkholder *et al.* 1993; Burkholder & Glasgow unpubl. data and medical records; D. Baden, NIEHS Center Director & Rosenstiel Institute of U. Miami, unpubl. data on the neurotoxins of *P. piscimorte* [gen. et sp. nov.]).

In more long-term benefits, our research will enable more accurate assessment of environmental quality, as related to the presence of this dinoflagellate, in natural estuarine ecosystems. As we improve our knowledge of the distribution of the organism and its relation to nutrient enrichment, we will gain insights about controlling *Pfiesteria piscimorte* (gen. et sp. nov.) and mitigating its acute and chronic effects on commercially important fisheries. Additonal long-term value of this research relates to medical and biological research applications. Medical concerns include the potential for toxin accumulation in humans and, conversely, use of the toxin(s) in medical applications (cf. Carmichael 1994). In related efforts we are working with a colleague, Dr. Daniel Baden (also president of Chiral Corporation, Miami, FL, a firm that specializes in developing test kits for routine analysis of dinoflagellate toxins), to characterize the toxins produced by this alga. Once this is accomplished, we plan to join forces to develop rapid detection / quantification procedures for both the organism and its toxins in natural estuarine samples and aquaculture ponds. *P. piscimorte* (gen. et sp. nov.) also offers interesting features from a developmental biology standpoint; in response to environmental cues, its ability to transform rapidly among stages that differ dramatically in both form and function (Burkholder *et al.* 1993) suggest well-regulated developmental controls.

From an ecosystem standpoint, greater understanding of this abundant, widespread organism—with multiple amoeboid stages that previously were not even recognized as dinoflagellates—has already begun to change existing paradigms about the role of dinoflagellates in estuarine food webs (Mallin *et al.* submitted). Moreover, the data on dinoflagellate-related fish kills gained from our coordinated monitoring effort will contribute to a long-term data base that will be of value in correlating toxic outbreaks with environmental conditions, and evaluating the impacts of this toxic alga on our fisheries resources. Development of sensitive, highly specific gene probes for *P. piscimorte* (gen. et sp. nov.) will answer a critical need for routine, early detection of this toxic dinoflagellate by regulatory staff and finfish / shellfish aquacultrists.

REFERENCES

Anderson, D.A. & L. Hall. 1992. *State of Maryland Water Quality Monitoring Report for Fish Kills in 1990–1991.* Wye River Center, State of Maryland Department of Environmental Management.

Anderson, D.M. 1993. Identification of harmful algal species using molecular probes. In: *Proceedings, Sixth International Conference on Toxic Marine Phytoplankton.* Nantes, France, p. 22.

Anderson, D.M., S. Galloway & J.D. Joseph. 1993. *Marine Biotoxins and Harmful Algae: A National Plan.* Woods Hole Technical Report WHOI 93-02. NOAA, National Marine Fisheries Service.

Anderson, D.M., A.W. White and D.G. Baden. 1985. *Toxic Dinoflagellates.* New York, Elsevier Science Publishing Company.

Anderson, D.M., A. Grabher & M. Herzog. 1992. Separation of coding sequences from structural DNA in the dinoflagellate *Crypthecodinium cohnii. Molec. Mar. Biol.* 1:89–96.

Anonymous. 1992. "Phantom" algae are linked to mass fish deaths. *The New York Times,* 30 July, p.1.

Ausubel, F.M., R. Brent, R.E. Kingston, D.D. Moore, J.G. Siedman, J.A. Smith & K.H. Struhl. 1992. *Short Protocols in Molecular Biology.* 2nd ed. New York, John Wiley & Sons.

Barkay, T., M. Gillman & C. Liebert. 1990. Genes encoding mercuric reductases from selected gram-negative aquatic bacteria have a low degree of homology with *merA* of transposon Tn*501. Appl. Environ. Microbiol.* 56:1695–1701.

Brand, L.E. 1988. Review of genetic variation in marine phytoplankton species and the ecological implications. *Biol. Oceanogr.* 6:397–409.

Britschgi, T.B. & S.J. Giovannoni. 1991. Phylogenetic analysis of a natural marine bacterio-plankton population by rRNA gene cloning and sequencing. *Appl. Environ. Microbiol.* 57:1707–1713.

Burkholder, J.M. 1992. Phytoplankton and episodic suspended sediment loading: Phosphate partitioning and mechanisms for survival. *Limnol. Oceanogr.* 37:974–988.

Burkholder, J.M. 1993. Tracking a "phantom:" The many disguises of the new ichthyotoxic dinoflagellate. In: *Proceedings, Fifth International Conference on Modern and Fossil Dinoflagellates.* Zeist (The Netherlands), p. 22.

Burkholder, J.M. & R.G. Wetzel. 1989. Epiphytic microalgae on natural substrata in a hardwater lake: seasonal dynamics of community structure, biomass and ATP content. *Arch. Hydrobiol./Suppl.* 83:1–56.

Burkholder, J.M., R.G. Wetzel & K.L. Klomparens. 1994. Track SEM-autoradiography of adnate microalgae. In: *Periphyton Methods Manual,* by R.G. Wetzel (Ed.). Boston, Dr. W. Junk Publishers (in press).

Burkholder, J.M., H.B. Glasgow Jr. & C.W. Hobbs. Response of a ichtyotoxic estuarine dinoflagellate to gradients of salinity, light and nutrients. *Mar. Ecol. Prog. Ser.* (submitted).

Burkholder, J.M., H.B. Glasgow, E.J. Noga & C.W. Hobbs. 1993. *The Role of a New Toxic Dinoflagellate in Finfish and Shellfish Kills in the Neuse and Pamlico Estuaries.* Report No. 93-08, Albemarle–Pamlico Estuarine Study. Raleigh, North Carolina Department of Environment, Health & Natural Resources and U.S. Environmental Protection Agency–National Estuary Program, 58 pp. (3rd printing).

Burkholder, J.M., E.J. Noga, C.W. Hobbs, H.B. Glasgow, Jr. & S.A. Smith. 1992a. New "phantom" dinoflagellate is the causative agent of major estuarine fish kills. *Nature* 358:407–410; *Nature* 360:768.

Burkholder, J.M., K.M. Mason & H.B. Glasgow Jr. 1992b. Water-column nitrate enrichment promotes decline of eelgrass *Zostera marina* L.: Evidence from seasonal mesocosm experiments. *Mar. Ecol. Prog. Ser.* 81:163–178.

Carmichael, W.W. 1994. The toxins of cyanobacteria. *Sci. Amer.* 270:78–86.

Coleman, A.W. & J.L. Goff. 1991. DNA analysis of eukaryotic algal species. *J. Phycol.* 27:463–473.

Culotta, E. 1992. Red menace in the world's oceans. *Science* 257:1476–1477.

DeLong, E.F., G.S. Wickham & N.R. Pace. 1989. Phylogenetic strains: Ribosomal RNA-based probes for the identification of single cells. *Science* 243:1360–1363.

DeRijk, P., J-M. Neefs, Y. Vn de Peer & R. De Wachter. 1992. Compilation of small ribosomal subunit RNA sequences. *Nucleic Acids REs.* 20(Supp.):2075–2089.

Droop, M.R. 1974. Heterotrophy of carbon, pp. 530–559. In: *Algal Physiology and Biochemistry,* by W.D.P. Stewart (Ed.). Botanical Monograph Vol. 10. Berkeley, University of California Press, 989 pp.

Erb, R.W. & I. Wagner-Döbler. 1993. Detection of polychlorinated biphenyl degradation genes in polluted sediments by direct DNA extraction and polymerase chain reaction. *Appl. Environ. Microbiol.* 59:4065–4073.

Fritz, L., P. Milos, D. Morse & J.W. Hastings. 1991. *In situ* hybridization of luciferin-binding protein anti-sense RNA to thin sections of the bioluminescent dinoflagellate *Gonyaulax polyedra. J. Phycol.* 27:436–441.

Gaines, G. & M. Elbrachter. 1987. Heterotrophic nutrition, pp. 224–268. In: *The Biology of Dinoflagellates,* by F.J.R. Taylor (Ed.). Boston, Blackwell Scientific Publications, 785 pp.

Geraud, M.-L., M. Sala-Roviera, M. Herzog & M.-L. Soyer-Gobillard. 1991. Immunochemical localization of the DNA-binding protein HCs during cell cycle of the histone-less dinoflagellate protocista *Crypthecodinium cohnii* B. *Biol. Cell* 71:123–134.

Greer, J. 1993. Alien in our midst? Phantom algae suspected in Bay. *Marine Notes.* University of Maryland Sea Grant, March.

Hallegraeff, G.M. 1992. A review of harmful algal blooms and their apparent global increase. *Phycologia* 32:79–99.

Hallegraeff, G.M., D.A. Steffenson and R. Wetherbee. 1988. Three estuarine Australian dinoflagellates that can produce paralytic shellfish toxins. *J. Plankt. Res.* 10:533–541.

Hauser, D.C.R., M. Levandowsky, S.H. Hutner, L. Chunosoff & J.S. Hollwitz. 1975. Chemosensory responses by the heterotrophic marine dinoflagellate *Crypthecodinium cohnii. Microb. Ecol.* 1:246–254.

Herrick, J.B., E.L. Madsen, C.A. Batt & W.C. Ghiorse. 1993. Polymerase chain reaction amplification of naphthalene-catabolic and 16S rRNA gene sequences from indigenous sediment bacteria. *Appl. Environ. Microbiol.* 59:687–694.

Huyghe, P. 1993. A horrific predatory little plant is beating up on fish around the world. *Discover* Magazine, April.

Kay, J. 1992. Eerie killer algae may be stalking Bay fish. *San Francisco Examiner,* 17 Aug., p.1.

Landsberg, J.H., K.A. Steidinger & B. Blakesley. 1994. Fish-Killing dinoflagellates in a tropical aquarium. In: *Proceedings, Sixth International Conference on Toxic Marine Phytoplankton.* Amsterdam, Elsevier (in press).

Lenaers, G., C. Scholin, Y. Bhaud, D. Saint-Hilaire & M. Herzog. 1991. A molecular phylogeny of dinoflagellate protists (Pyrrhophyta) inferred from the sequence of 24S rRNA divergent domains D1 and D8. *J. Mol. Evol.* 32:53–63.

Levine, J.F., J.H. Hawkins, M.J. Dykstra, E.J. Noga, D.W. Moye & R.S. Cone. 1990. Species distribution of ulcerative lesions on finfish in the Tar-Pamlico Estuary, North Carolina. *Dis. Aquat. Org.* 8:1–5.

Lewitus, A.J., R.V. Jesien, T.M. Kana, J.M. Burkholder & E. May. Discovery of the "phantom" dinoflagellate in Chesapeake Bay. *Estuaries* (submitted).

Li, N. & R.A. Cattolico. 1992. *Ochromonas danica* (Chrysophyceae) chloroplast genome organization. *Mol. Mar. Biol. Biotech.* 1:165–174.

Lund, J.W.G., C. Kipling and E.D. LeCren. 1958. The inverted microscope method of estimating algal numbers and the statistical basis of estimates by counting. *Hydrobiologia* 11:143–170.

Lynn, D.H. & M.L. Sogin. 1988. Assessment of phylogenetic relationships among ciliated protists using ribosomal RNA sequences derived from reverse transcripts. *BioSystems* 21:249–254.

Mallin, M.A., J.M. Burkholder, L.M. Larsen & H.B. Glasgow Jr. Response of two zooplankton grazers to an ichthyotoxic estuarine dinoflagellate. *Mar. Ecol. Prog. Ser.* (submitted). [published 1995 in *J. Plankton Research.* 17:351–363]

Manhart, J.R. & R.M. McCourt. 1992. Molecular data and species concepts in the algae. *J. Phycol.* 28:730–737.

Miller, K.H., J. Camp, R.W. Bland, J.H. Hawkins, III, C.R. Tyndall and B.L. Adams. 1990. *Pamlico Environmental Response Team Report* (June–December 1988). North Carolina Dept. of Environment, Health, and Natural Resources. Wilmington, North Carolina, 50 pp.

Moran, M.A., V.L. Torsvik, T. Torsvik & R.E. Hodson. 1993. Direct extraction and purification of rRNA for ecological studies. *Appl. Environ. Microbiol.* 59:915–918.

North Carolina Division of Marine Fisheries. 1992. *Description of North Carolina's Coastal Fishery Resources.* Draft Report. Morehead City, North Carolina Department of Environment, Health and Natural Resources–Division of Marine Fisheries, 215 pp.

Pace, N.R., D.A. Stahl, D.J. Lane & G.J. Olsen. 1986. The analysis of natural microbial populations by ribosomal RNA sequences. *Adv. Microb. Ecol.* 9:1–55.

Pederson, B.H. & K.P. Anderson. 1992. Induction of trypsinogen secretion in herring larvae *(Clupea harengus)*. *Mar. Biol.* 112:559–565.

Pichard, S.L. & J.H. Paul. 1993. Gene expresion per gene dose: a specific measure of gene expression in aquatic microorganisms. *Appl. Environ. Microbiol.* 59:451–457.

Pichard, S.L., M.E. Fisher & J.H. Paul. 1993. Ribulose bisphosphate carboxylase gene expression in subtropical marine phytoplankton populations. *Mar. Ecol. Prog. Ser.* 101:55–65.

Rensberger, B. 1992a. Huge fish kills linked to slumbering algae. *The Washington Post,* 30 July, p.1.

Rensberger, B. 1992b. Look! In the water! It's a plant, it's an animal, it's a toxic creature. *The Washington Post,* 17 August, Science Feature.

Reynolds, A.E., B.L. McConaughy & R.A. Cattolico. 1993. Chloroplast genes of the marine alga *Heterosigma carterae* are transcriptionally regulated during a light/dark cycle. *Molec. Mar. Biol. Biotech.* 2:121–128.

Rizzo, P.J. 1987. Biochemistry of the dinoflagellate nucleus, pp. 143–171. In: *The Biochemistry of Dinoflagellates,* by F.J.R. Taylor (Ed.). Oxford, Blackwell Scientific.

Robineau, B., J.A. Gagne, L. Fortier and A.D. Cembella. 1991. Potential impact of a toxic dinoflagellate *(Alexandrium excavatum)* bloom on survival of fish and crustacean larvae. *Mar. Biol.* 108:293–301.

Rowan, R. & D.A. Powers. 1992. Ribosomal RNA sequences and the diversity of symbiotic dinoflagellates (zooxanthellae). *Proc. Natl. Acad. Sci.* 89:3639–3643.

Rublee, P.A. 1982. Seasonal distribution of bacteria in salt marsh sediments of North Carolina. *Est. coastal Shelf Sci.* 15:67–74.

Rublee, P.A. 1992. Community structure and bottom-up regulation of heterotrophic microplankton in arctic LTER lakes. *Hydrobiologia* 240:133–141.

Sambrook, J., E.F. Fritsch & T. Maniatis. 1989. *Molecular Cloning: A Laboratory Manual.* 2nd ed. New York, Cold Spring Harbor.

Sanger, F., S. Nicklen & A.R. Coulson. 1977. DNA sequencing with chain-terminating inhibitors. *Proc. Nat. Acad. Sci.* 74:5463–5467.

Scholen, C.A., D.M. Anderson & M.L. Sogin. 1993. Two distinct small-subunit ribosomal RNA genes in the North American toxic dinoflagellate *Alexandrium fundyense* (Dinophyceae). *J. Phycol.* 29:209–216.

Shimizu, Y. 1991. Dinoflagellates as sources of bioactive molecules, p.71. In: *Program and Abstracts of the Second International Marine Biotechnology Conference.* Baltimore, Society for Industrial Microbiology, October (Abstract).

Shumway, S.E. 1990. A review of the effects of algal blooms on shellfish and aquaculture. *Journal of the World Aquaculture Society* 21:65–104.

Smayda, T.J. 1989. Primary production and the global epidemic of phytoplankton blooms in the sea: A linkage?, pp. 449–484. In: *Novel Phytoplankton Blooms,* by E.M. Cosper, V.M. Bricelj and E.J. Carpenter (Eds.). Coastal and Estuarine Studies No. 35. New York, Springer-Verlag, 799 pp.

Steidinger, K.A. & D.G. Baden. 1984. Toxic marine dinoflagellates, pp. 201–262. In: *Dinoflagellates,* by D.L. Spector (Ed.). New York, Academic Press, 545 pp.

Steidinger, K.A., E.W. Truby, J.K. Garrett & J.M. Burkholder. 1994. The morphology and cytology of a newly discovered toxic dinoflagellate, 6 pp. In: *Proceedings, 6th International Conference on Toxic Marine Phytoplankton.* Amsterdam, Elsevier (in press).

Tsai, Y.-L. & B.H. Olsen. 1991. Rapid method for direct extraction of DNA from soil and sediments. *Appl. Environ. Microbiol.* 57:1070–1074.

Tsai, Y.-L. & B.H. Olsen. 1992. Rapid method for separation of bacterial DNA from humic substances in sediments for polymerase chain reaction. *Appl. Environ. Microbiol.* 58:2292–2295.

Triplett, E.L., N.S. Govind, S.J. Roman, R.V.M. Jovine & B.B. Prezelin. 1993. Characterization of the sequence organization of DNA from the dinoflagellate *Heterocapsa pygmaea* (*Glenodinium* sp.). *Molec. Mar. Biol. Biotech.* 2:239–245.

UNESCO. 1991. *Programme on Harmful Algal Blooms.* Workshop Report No. 80. Newport (RI), Intergovernmental Oceanographic Commission.

Weisberg, W.G., S.M. Barns, D.A. Pelletier & D.J. Lane. 1991. 16S ribosomal DNA amplification for phylogenetic study. *J. Bact.* 173:697–703.

White, A.W. 1988. Blooms of toxic algae worldwide: Their effects on fish farming and shellfish resources. In: *Proceedings of the International Conference on Impact of Toxic Algae on Mariculture.* Aqua-Nor '87 International Fish Farming Exhibition, August 1987, Trondeim, Norway, pp. 9–14.

Killer

TINY PREDATOR, HUGE PREY: MICROSCOPIC DINOFLAGELLATES ATTACK AND KILL FISH MORE THAN A MILLION TIMES THEIR SIZE. ONCE SATED, THEY SINK BACK TO THE RIVER'S BOTTOM TO AWAIT THEIR NEXT VICTIM.

BY PATRICK HUYGHE
Photographs by Andrew Moore

A horrific,
predatory little plant
is beating up
on fish around the
world.

Algae

ot long ago, one of the most bizarre creatures this world has ever seen showed up in North Carolina. Efforts to describe it have evoked the strangest comparisons: like grass feeding on sheep, said one scientist. And that's not stretching things much. The creature is in fact a tiny plant with a Jekyll and Hyde personality, one that preys on animal life a million times its size. It has caused massive kills of fish and other marine animals around the world. Though hordes of the little monsters could easily fit on the head of a pin, keeping a culture of them active, say its discoverers, requires about 15 fish a day. "It's really not like a normal culture," they say. No joke.

The discovery of this deadly little squirt has caused quite a commotion among marine ecologists; though they have long known the monster's relatives, no other members of the family have ever exhibited this upstart's horrific predatory behavior. The fish killer extraordinaire is a measly dinoflagellate, a kind of alga, or single-celled aquatic plant. Dinoflagellates are twilight-zone creatures: half-plant, half-animal. They produce chlorophyll, but they also move about, using their two flagella, or whiplike tails, to swim rapidly through the water. Dinoflagellates, along with diatoms and a dozen other forms of microalgae, serve as the very bottom of the food chain and, at 450 million to a billion years old, are among the most primitive forms of life on this planet.

Dinoflagellates have long been regarded as the bad boys of the phytoplankton community, though perhaps a little unjustly—so far, only 42 of some 2,000 marine species are known to be toxic. The oldest mention of their nasty behavior occurs in the Old Testament: the first plague visited upon Egypt is a blood red tide that kills the fish and fouls the water. Red tides—which can actually also be green, yellow, or brown— are caused by algae "blooms," excessive concentrations of dinoflagellates and other microalgae. The Red Sea itself may have been named for such a bloom.

More recently an increase of toxic blooms worldwide has alarmed marine ecologists. Not only do blooms appear to have proliferated during the last two decades, but species formerly thought harmless are now proving toxic. Their deadly appearances have led to immense revenue losses for the fishing industry and posed a threat to human health. A huge bloom in the Adriatic in 1989, for example, devastated the local tourist industry and the fishing community and caused $800 million in losses. An earlier toxic bloom, off the coast of Guatemala in 1987, caused one of the largest outbreaks of shellfish poisoning among humans in recent times. Although algal toxins can accumulate in shellfish without causing harm, when those shellfish are eaten by humans, serious illness and even death may occur. Almost 200 people fell ill during the Guatemalan outbreak; 26 died.

Researchers don't know for sure what is causing the increase in blooms. Some believe they may be a manifestation of a long-term cyclical trend. Others think that what we're seeing is the blooming of algal species that were always there in deep ocean waters but had never bloomed before. The new growth spurts, these researchers say, may be caused by natural factors, such as changes in currents or climate. Still others think the boom in blooms, especially in shallow waters and estuaries, is being triggered by an increase in nutrients via human sewage and agricultural runoff and that it signals the global deterioration of the marine environment. Whatever the cause, the discovery of the newest toxic dinoflagellate has done nothing but add a ghastly new twist to an old and increasingly perplexing problem.

The story properly begins one night early in 1988, with a massacre in Edward Noga's laboratory at North Carolina State University in Raleigh. Noga is a fish pathologist at the university's College of Veterinary Medicine. The scene was an aquarium filled with brackish water, and the victims were about 300 tilapias, African fish about two to three inches long; Noga was planning to use them for an immunology project. "All the fish were okay on Friday," recalls Noga, "and on Saturday they were all either dying or dead."

Noga and graduate student Stephen Smith immediately tried to trace the cause of the massive mortality. But they could find no pathogens on the skin or gills of the fish, and the cultures Smith took tested negative for bacteria. "The only thing that seemed unusual to me," recalls Noga, "was the presence in the water of an awful lot of dinoflagellates that seemed to have the same appearance. This suggested some kind of a bloom."

To test whether the dinoflagellates were responsible for the fish deaths, Noga and Smith set up five aquariums and stocked them with six fish apiece. Three of the tanks received a dose of a thousand dinoflagellates each; the two controls received none. Within 15 days blooms occurred in the three test tanks without discoloring the water, and 48 hours later all the fish in these tanks had died. It was a first: no dinoflagellate had ever been known to kill fish in an aquarium system before. The researchers then took water from a tank where fish were dying and filtered it, removing all bacteria, viruses, and dinoflagellates. When the water still proved lethal to 60 percent of the fish, they knew, says Smith, that "the organism was producing something very toxic to fish."

Noga, wanting the killer identified, began sending out cultures. But the dinoflagellates were so small (just 10 to 20 microns, or less than 4 to 8 ten-thousandths of an inch, in diameter), so nondescript, and so difficult to culture by ordinary

Quickly the creature filled the water with a lethal neurotoxin that paralyzed the fish, causing slow suffocation.

techniques that few scientists were interested in the little critters, and those who were couldn't come up with an answer. JoAnn Burkholder, an aquatic botanist then new to the North Carolina State campus, was no exception.

When Noga first contacted Burkholder, she expressed little interest in the creature, citing no special knowledge of dinoflagellates. But he insisted, and she finally agreed to take a look. The critters, she told him, "were just like three or four other nondescript little dinoflagellates common to North Carolina estuaries." She suggested that one of her part-time graduate students, Cecil Hobbs, might be interested in working on the ecology and life cycle of the mysterious "dino." Hobbs, a former high school biology teacher, had a personal interest in—and perhaps something of a vendetta against—dinoflagellates. His family's oyster beds in Sneads Ferry, North Carolina, had been wiped out by a red tide that hit the coast in 1987. The Hobbses' oysters weren't the only ones hit, of course. That bloom caused $25 million in losses to the local shellfish industry.

Making skillful use of a scanning electron microscope, Hobbs and Burkholder managed to get the first decent shots of the killer. Then they began to characterize the changes taking place in the dinoflagellate before, during, and after its kills. What they found had never been seen before. With no fish present, the creature simply sat in the sediment, encrusted in a hard, scaly, eggshell-like cyst. But when one or more fish began lingering overhead to feed, the creature shed its cyst, often within minutes. What emerged was a flagellated cell—the dinoflagellate stage by which this creature has become known. Quickly it filled the water with a lethal neurotoxin that paralyzed the fish, causing slow suffocation. In the face of impending death, the stunned fish leaned against the side of the aquarium, thrashing about as they struggled to get to the top. Then they dropped to the bottom, bumping their heads or falling on their tails.

Unlike most dinoflagellates, which move in a leisurely, winding sort of way, these made a beeline for their target—flecks of fish tissue stripped off by the toxin. They used a tonguelike absorption tube called a peduncle to attach themselves. As the frenzy increased, the dinos nearly doubled in size, the pedun-

cle itself becoming swollen and assuming different shapes. "In one picture," says Burkholder, "it looks like a huge hand with tentacles and a clawlike thing at the end. We call it our Darth Vader shot." In the presence of this tiny terror, fish are not long for this world. Although the time frame varies, death can occur in as little as 20 minutes.

Burkholder and Hobbs think the attack is actually signaled by something excreted by the fish, though precisely what, they don't yet know. The massacre ends, they think, as the amount of excreta in the water diminishes. The dinos then do one of two things. In less than a minute they can create and encase themselves in a new cyst and drop to the bottom

JoAnn Burkholder found that massive fish kills in North Carolina's Pamlico River were being caused by a new genus of dinoflagellate

to await more prey. This is apparently a protective measure; the cyst marks a dormant stage spurred by a lack of food or some other modification in their environment—a sudden change in water temperature, or turbulence from a storm. But alternatively, if there is no environmental stress and food (fish tissue) still remains, the dinos bizarrely enter instead a stage in which they transform themselves into amoebas. They shed their flagella, lose some of their toxicity, stop photosynthesizing, and become more animallike; once their rapid transformation is complete, they continue to feed on the fish bits at a more leisurely pace.

Burkholder has discovered that even after lying dormant for two years the dinoflagellate can still kill its prey, although it may then take about six weeks for the toxic cells to emerge from their cysts (whereupon, if no fish are present, they simply return to the cyst stage to await another passage of fish). While in the cyst stage, though, the dinos look absolutely harmless. Indeed, they fooled everyone—the fish, obviously, but also the scientists who had seen these cysts before but had never seen in them the face of a killer. Because these cysts so resembled those of a completely different kind of alga, a harmless chrysophyte, to anyone examining the water after a fish kill it looked as if no trace of a villain remained.

Hobbs and Burkholder had been working with the dinoflagellates for about a year when another wave of die-offs struck

Noga's aquariums, killing a thousand fish. "I'm not sure how his tanks got contaminated," says Hobbs, "but it was probably because we were all using the same refractometer" (an instrument used to measure the water's salinity). "We thought we could contain this thing in one room," adds Noga, "but we were very much mistaken. This is worse than any other infectious agent I've ever dealt with."

Hobbs and Burkholder went on to discover just how deadly their creature was. They found that the dinoflagellates could kill fish in anything from fresh water to full-strength seawater, though optimal growth occurred in water of midrange salinity. The researchers also learned that no fish was immune to the dinos' toxin, although some, like striped bass, were more susceptible than others.

Burkholder and Noga now knew enough about the dino to search for it in the wild. But since the killer was too elusive to be caught by haphazard sampling, they enlisted help from North Carolina state biologists in hopes of nailing the phantom while a kill was in progress. "They'd had a lot of sudden-death kills in the past decade or so with no explanations," recalls Burkholder. "During 1988, the best-monitored year, they found 88 fish kills in the local Pamlico River between May and October. Their reports noted the fish sometimes exhibited neurotoxic symptoms and acted panicky. They told me that in some of the kills the fish were actually trying to get out of the water and onto the beach before they died; they called it a flounder, or crab, walk."

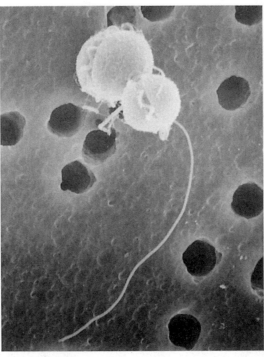

During its feeding frenzy, a dino can nearly double in size; it uses a long, tonguelike tube to attach itself to a flake of fish.

nos left in the water. Their retreat had been as rapid as their blitzkrieg attack.

Burkholder and Hobbs now definitely had their hands on the killer. But they had yet to tag it with an identity. The following October, however, they managed to take a photograph of it that revealed the faint outline of two cellulose structures sitting beneath surface membranes on the dinoflagellate cells. They suspected that these could be "armored plates" that serve as a protective covering for the cell when it is active. The presence or absence of such plates is one of the ways botanists classify a dinoflagellate species. Until this photo, the researchers had assumed the killer belonged among the "naked" species, which lack such armor.

At the end of the month, at an international toxic-phytoplankton conference in Rhode Island, Burkholder announced her findings: that this particular dino had plates, that the species went into a cyst stage, and that it hunted fish as prey. The audience, Burkholder recalls, was stunned. Most of the researchers there had assumed that phytoplankton toxins played a more passive role, as chemical deterrents to predators. But if Burkholder was right, this "plant" was a hunter armed with the equivalent of poison-tipped arrows. Burkholder's findings started a scramble; though somewhat doubtful, researchers from the United States and 14 other countries with similar sudden-death fish kills returned home to hunt for the phantom alga in their local waters.

A few months later, in January 1992, the super-villain struck again—but this time its target was human. Howard Glasgow Jr., a research associate of Burkholder's, was kneeling next to a drain on the floor of Noga's lab, where the dinoflagellate experiments were being conducted. He was carefully rinsing out a ten-gallon aquarium that he planned to use for his upcoming project on the phantom dinos when he began to realize that he was not thinking properly.

Thoughts began rolling through his mind at a terrific pace; but when he reached to pull himself up, his hand seemed to take forever to move. Sensing trouble, Glasgow decided to leave the chamber, but his steps turned into something of a moon walk. "I don't know if it was really slow, or if I was thinking fast," he recalls, "but something was drastically wrong." Glasgow recov-

In May 1991 the phantom killer was finally caught with its peduncle on the goods. Kevin Miller, a state environmental technician, was sampling menhaden (a marine fish of the herring family) from an estuary off the Pamlico when a squall blew up. Forty minutes later, after seeking shelter in a protected bay, he returned to the sampling site to find the fish behaving abnormally. Then they started dying—by the millions. He took water samples and sent them off to Burkholder and Hobbs, who confirmed their toxicity and determined that the dinoflagellate responsible was the same one they'd had in culture for three years. And as expected, samples taken at the site the next day showed almost no diered 15 minutes later and now remembers the experience as more euphoric than frightening. But the aerosol effects of toxin produced by dinoflagellates can be serious. For example, whenever the red tide produced by another dino, *Gymnodinium breve*, hits the coast of Florida, a dangerous aerosol is released for several days. During that period health warnings are posted telling people with heart and respiratory conditions to stay off the beach. "The alga will come in on wave action," explains Burkholder, "and when the cell fragments become airborne in sea spray, they can elicit sneezing, coughing, and severe asthmatic reactions."

Meanwhile the search for the killer's identity went on. Burkholder, unsuccessful in her attempts to strip the outer membranes from the cells, turned for help to a colleague, Karen Steidinger, chief of research at the Marine Research Institute of Florida's Department of Natural Resources and an expert in dinoflagellates. After several unsuccessful attempts on her own, in June 1992 Steidinger, using 100-proof alcohol, finally succeeded in stripping the stubborn membranes off the dino to reveal the armored plates beneath. With the plates' existence finally confirmed, Steidinger and Burkholder were able to verify that their killer represented a new genus and a new species. They have proposed calling it *Pfiesteria piscimorte* (the genus name was picked to honor the late dino specialist Lois Pfiester of the University of Oklahoma and also because Burkholder liked the name's echo of both *feast* and *cafeteria*; the species name means "fish killer").

Over the next few months Burkholder and Glasgow continued to discover just how bizarre the new dinoflagellate really is. Its life cycle consists of more than 15 stages—it's a veritable metamorphosing monster. The dino lives most of its life as an amoeba; in one toxic "giant" amoeboid stage it grows to be nearly 20 times the size of the flagellated toxic cells. The stage when the dino leaves its cyst to attack fish is actually quite ephemeral, appearing only when fish are present. The researchers also found that this stage is the only time the creature sexually reproduces; they again suspect that something in fish excreta triggers the reproductive urge.

Never before had marine dinoflagellates shown such stages. "I've been working with marine dinoflagellates for 29 years now," Steidinger says, "and I've never seen anything like it before. It has the most diverse life cycle I've ever seen. And this can't be the only species out there. There have to be others."

Since the 1991 kill of menhaden off the Pamlico River, Burkholder and her students have confirmed the presence of the dinoflagellate in 11 other fish kills in North Carolina. Outside the state, the phantom killer has been found in Delaware in the Indian River. Anecdotal evidence suggests it was also responsible for a kill in Maryland in the Wye River, a tributary of the Chesapeake. They expect that many other kills previously labeled "unknown" may now be explained.

While researchers the world over are looking for the elusive dino in their own waters, others are trying to identify

Howard Glasgow, seen through an aquarium used to hold the killer algae, was briefly affected by too close an encounter with their toxin.

the toxin. Glasgow, for his part, is searching for the algal aphrodisiac—what it is in fish excreta that stimulates dinoflagellates to burst out of their cysts, feed on fish, and reproduce sexually. Burkholder is working on discovering the kind and concentration of organic and inorganic nutrients that stimulate the phantom dino's growth. She's already found that waters with excessive amounts of phosphorus and nitrogen seem to favor the organism. Last but not least, she hopes to find natural predators as a way to eventually control the dinoflagellate. One candidate, a microscopic animal called a rotifer that is a natural predator in estuaries, looks promising. In dino-infested waters, Burkholder has seen the guts of rotifers packed with dinoflagellates.

But finding an able opponent for the dino may not be easy. Burkholder laughs nervously as she recalls the day an undergraduate came running into her office with the news that all the dinoflagellates in a fish tank were dying. When Burkholder looked under the microscope she was surprised to find a protozoan, which had somehow gotten into the tank, busily consuming both the dinos and their cysts as the dinos themselves were attacking and killing the fish. But once the fish had died, the remaining dinoflagellates started to circle the protozoan. Then, like something from a grade B science fiction movie, a few of those dinoflagellates formed into giant toxic amoebas, which—you guessed it—completely engulfed the protozoan. ⒟

11

Research on Supernova Remnants

REYNOLDS AND COLLEAGUES

A distinctive feature of research in astrophysics is the complementarity of two types of work: astronomical observations and theoretical modelling. In this chapter we include examples of both types of research conducted by Dr. Stephen P. Reynolds, an astrophysicist at North Carolina State University. The first document is a research letter published in 1994 in *Monthly Notices of the Royal Astronomical Society,* in which Reynolds and his coauthors propose a hypothesis about the origins of a particular supernova remnant (SNR), supported by reports of satellite observations of the SNR. The second is a full-length research report published in 1990 in the *Astrophysical Journal,* in which Reynolds and coauthor Michael Fulbright generate and compare model maps of SNR radiation, testing them against observational data. This report was later cited in the final document in this chapter, a 1994 grant proposal submitted to NASA in which Reynolds and co-investigators Kasimierz Borkowski and John Blondin proposed to develop a new theoretical model of supernova dynamics to be used in interpreting satellite data. The proposal is accompanied by an excerpt from NASA's program description.

X-ray evidence for the association of G11.2 − 0.3 with the supernova of 386 AD

Stephen P. Reynolds,[1]★ Maxim Lyutikov,[2]★ Roger D. Blandford[2]★ and Frederick D. Seward[3]★

[1] Physics Department, North Carolina State University, Raleigh, NC 27695-8202, USA
[2] California Institute of Technology, 130–33, Pasadena, CA 91125, USA
[3] Harvard–Smithsonian Center for Astrophysics, Cambridge, MA 02138, USA

Accepted 1994 August 17. Received 1994 August 16; in original form 1994 May 4

ABSTRACT

We present *ROSAT* observations of the radio-bright supernova remnant G11.2 − 0.3 which support the hypothesis that it is the remnant of the historical supernova SN 386 AD. The PSPC data imply a post-shock temperature of $kT = 2.4 \, (+2.3, -1.0) \, \mathrm{keV}$ (95 per cent confidence limits), equivalent to a shock speed of 1400 km s^{-1}, and an absorbing column density $N_H = 1.3 \, (+0.4, -0.2) \times 10^{22}$ cm^{-2}, for a homogeneous, equilibrium plasma model. For an assumed age of 1607 yr, these data imply a distance of 9 kpc and a peak apparent magnitude of about 2 mag, consistent with a Type Ia supernova.

Key words: supernovae: individual: SN 386 AD – supernova remnants – X-rays: ISM.

1 INTRODUCTION

Clark & Stephenson (1977) describe the historical records reporting the supernova of 386 AD, during the Chin dynasty. Little information is available, except the asterism (locating the object to within about 15°) and the fact that the supernova disappeared after about 3 months (not for seasonal reasons). They judge the uncertainty in the duration to be about 1 month; they also conclude, from lack of Chinese observations of Mira, that to be noticed an object would require an apparent magnitude ≲ 1.5 mag, consistent with a supernova at a distance $d \sim 5$ kpc. The positional and distance information led Clark & Stephenson to associate SN 386 AD with the Galactic plane supernova remnant G11.2 − 0.3, which has $d \gtrsim 5$ kpc from H I observations (Radhakrishnan et al. 1972), discounting weak negative-velocity absorption (Becker, Markert & Donahue 1985). Radhakrishnan et al. (1972) report no absorption at velocities corresponding to larger distances, but no appreciable emission occurs at such velocities either, so there is no upper bound on the distance to be obtained from H I data. Although recognizing the positional and distance agreement, Clark & Stephenson rate the identification as only possible because the remnant's inferred diameter was too large for the empirical diameter–time relation (Clark & Caswell 1976).

★ E-mail: Stephen_Reynolds@ncsu.edu; lyutikov@tapir.caltech.edu; rdb@tapir.caltech.edu; fds@asc.harvard.edu

Subsequent, high-resolution radio studies (Downes 1984; Becker et al. 1985; Green et al. 1988) of G11.2 − 0.3 show a quite regular, circular shell of mean diameter 4.4 arcmin, ranking seventh in surface brightness at 1 GHz among Galactic shell remnants. In fact, the surface brightness and structure have been used to argue for a *smaller* age than 1607 yr (Downes 1984). It seems unlikely that a second supernova, within a few hundred years of SN 386 AD, should occur in the same region of the sky and go unnoticed by the Chinese. We shall henceforth adopt the hypothesis that G11.2 − 0.3 is the remnant of SN 386 AD, and examine the consequences.

2 OBSERVATIONS AND DISCUSSION

As part of a programme studying the relation between radio and X-ray structures in young supernova remnants, we have observed G11.2 − 0.3 with the *ROSAT* PSPC on 1993 April 3, for about 5300 s, receiving about 3800 counts. The image is shown in Fig. 1, smoothed with a Gaussian of 20-arcsec FWHM. It closely corresponds with the radio image, down to individual features in the shell. Spectra from the remnant as a whole, from the central region, and from the brightest shell region to the south-east are indistinguishable. A fit of the entire spectrum with a single-temperature model (Raymond & Smith 1977) assuming cosmic abundances produces a reduced χ^2 value of 2.1. The plot of N_H–T parameter

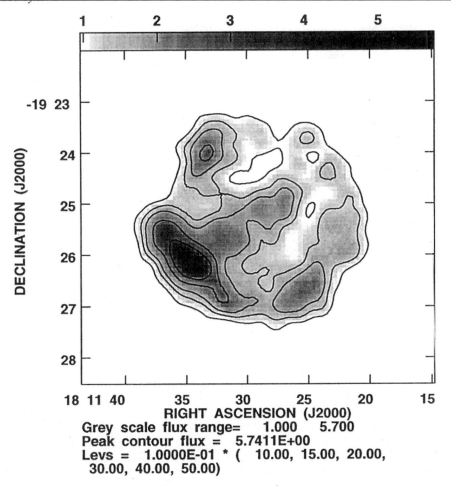

Grey scale flux range=   1.000   5.700
Peak contour flux = 5.7411E+00
Levs =   1.0000E-01 * (10.00, 15.00, 20.00,
30.00, 40.00, 50.00)

Figure 1. PSPC image of G11.2 − 0.3, smoothed with a Gaussian of 20-arcsec FWHM.

space is shown in Fig. 2; the best fit is at $kT = 2.4$ keV and $N_H = 1.3 \times 10^{22}$ cm^{-2}, based on 21 channels between 0.4 and 2.4 keV. This is not a bad fit for such an oversimplified model. An inhomogeneous model based on Sedov dynamics (see below) furnishes a similar temperature for the outer rim. The accuracy of these measurements is limited by the *ROSAT* spectral response which is of low energy resolution and only strong below ~ 2 keV, where the continuum is raised substantially by unresolved lines. The presence of significant signal above 2 keV cannot be produced by lower temperatures, however, so this temperature is strictly only a lower bound. It corresponds to a Rankine–Hugoniot shock speed $v_s = (16kT/3\mu m_p)^{1/2} \gtrsim 1400$ km s^{-1} assuming electron–ion equilibration ($\mu = 0.6$ for cosmic abundances). Here m_p is the proton mass and μm_p the mean mass per particle. Incomplete electron heating (Draine & McKee 1993) would also result in a larger shock velocity. Adopting the Sedov relation, for a blast wave of radius R_s and age t, of

$R_s = 2.5 v_s t$, we obtain lower bounds of $R_s \gtrsim 5.8$ pc and $d \gtrsim 9$ kpc. [Note that both values will be reduced if the shock is less decelerated, as will be the case before the Sedov phase is attained. We argue below, however, that estimates of the swept-up mass vindicate the Sedov assumption. Unless the swept-up mass is significantly less than that ejected, $R_s \propto t^{0.4}$ is a good approximation, as observed in Tycho's supernova remnant (SNR) (Tan & Gull 1985), although Tycho still shows evidence of a reverse shock.]

We may now use the observed column density to estimate the optical extinction to the supernova. Adopting the empirical relation $A_V = (4.8 \pm 0.4) \times 10^{-22} N_H$ (Gorenstein 1975; Bohlin, Savage & Drake 1978) and the average peak magnitudes for supernovae of Types Ia, Ib and II of $M_V \sim M_B = -19.1 \pm 0.7, -17.5 \pm 0.4$ and -16.5 ± 1.5 mag, respectively (van den Bergh & Tammann 1991; Cappellaro et al. 1993) (adopting $H_0 = 70 \pm 20$ km s^{-1} Mpc^{-1}), we predict peak apparent magnitudes $V \simeq (1.9 \pm 1.1) + 5 \log(d/9$ kpc) for

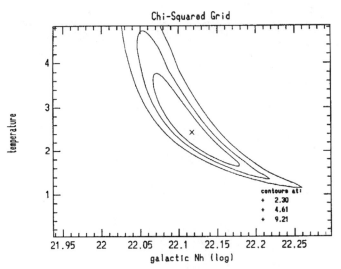

Figure 2. Plot of $N_H - T$ parameter space with contours of constant χ^2 for fits to the spectrum of G11.2 − 0.3. The three contours are the 68, 95 and 99 per cent confidence levels.

Type Ia, and $(3.5 \pm 1.0) + 5 \log(d/9 \text{ kpc})$ and $(4.5 \pm 1.8) + 5 \log(d/9 \text{ kpc})$ for Types Ib and II, where the stated errors include the uncertainty in A_V (including both the error in N_H and the error in the conversion), the uncertainty in the Hubble constant, and the intrinsic dispersions in the properties of observed supernovae (Cappellaro et al. 1993), combined in quadrature. Only a Type Ia supernova appears to be consistent with the historically inferred visual peak magnitude of ~ 1.5 mag.

This conclusion is predicated upon G11.2 − 0.3 being in the Sedov phase, and we can substantiate this by fitting the observed X-ray count rate. First, concentrating upon the 1.4–2.4 keV band, and allowing for absorption, we infer a total X-ray luminosity of $4 \times 10^{35}(d/9 \text{ kpc})^2 \text{ erg s}^{-1}$. Adopting the homogeneous model with an emissivity of $10^{-23}(n_e/1 \text{ cm}^{-3})^2 \text{ erg cm}^{-3} \text{ s}^{-1}$ (where n_e is the post-shock electron density) and a volume filling factor of $f = 0.25$ (appropriate for an adiabatic shock with a compression ratio of 4), we obtain a pre-shock hydrogen density $n_0 = 1.1 \text{ cm}^{-3}$, reasonable at a Galactocentric radius of ~ 2 kpc, and the explosion energy $E = 10^{51} \text{ erg}$. Alternatively, a spectral fit to the data based on a Sedov dynamical model with equilibrium ionization and cosmic abundances leads to a density $n_0 = 2.2 \text{ cm}^{-3}$, a post-shock temperature $kT = 1.9 \text{ keV}$, a remnant radius $R_s = 5 \text{ pc}$, a distance $d = 7.8 \text{ kpc}$, and an explosion energy $E = 10^{51} \text{ erg}$. The estimate for the swept-up mass, $M = 23 \text{ M}_\odot$, is large enough to vindicate our assumption of Sedov dynamics. The agreement between these two different and quite simple models seems to be reasonably good. The absence of a well-defined edge in the radio image (Green et al. 1988) might suggest an earlier evolutionary stage, by analogy with Cas A, but our understanding of electron acceleration at shock waves is sufficiently incomplete that we

are prepared to over-rule this argument in favour of the X-ray inferences.

In summary, the PSPC observations of G11.2 − 0.3 give us a lower limit to the shock temperature, and the absorbing column density. We can then derive a self-consistent model by adopting Sedov dynamics and assuming that G11.2 − 0.3 is the remnant of SN 386 AD with an age of 1607 yr. The derived peak visual magnitude and decay time are compatible with a Type Ia supernova. Two future observational studies may support or refute these arguments. First, the *ASCA* X-ray satellite with its superior energy resolution and high-energy sensitivity should provide a more accurate estimate of the post-shock temperature which should reduce the uncertainty in the distance. We note that a significantly larger temperature would lead to an unreasonably large explosion energy, and a significantly lower temperature would permit a Type Ib or II supernova. Secondly, the radio (Downes 1984; Becker et al. 1985; Green et al. 1988) and, especially, the *Einstein* HRI images (Downes 1984; Becker et al. 1985) exhibit a local emission peak near the centre of the remnant, suggestive of a 'composite' SNR which would require the presence of a central neutron star. If future radio or X-ray observations were to discover a central pulsar at this location, we would have to consider the possibility of a non-thermal component in the X-ray spectrum, although in our data this central emission only contains ~ 20 per cent of the counts. Alternatively, further corroboration of the historical association of G11.2 − 0.3 with SN 386 AD, as well as that of RCW 86 (Clark & Stephenson 1977) or MSH 15−52 (Thorsett 1992) with SN 185 AD, should aid in an understanding of the transition from young to middle-aged remnants, with implications for the statistics of SNRs and supernovae and for the physics of strong shock waves.

ACKNOWLEDGMENTS

We thank Rudolf Danner and Shri Kulkarni for advice. Support under NASA contracts NAG 5-2177 and NAG 5-2212 is gratefully acknowledged.

REFERENCES

Becker R. H., Markert T., Donahue M., 1985, ApJ, 296, 461
Bohlin R. C., Savage B. D., Drake J. F., 1978, ApJ, 224, 132
Cappellaro E. et al., 1993, A&A, 268, 472
Clark D. H., Caswell J. L., 1976, MNRAS, 174, 267
Clark D. H., Stephenson F. R., 1977, The Historical Supernovae. Pergamon, Oxford
Downes A., 1984, MNRAS, 210, 845
Draine B. T., McKee C. F., 1993, ARA&A, 31, 373
Gorenstein P., 1975, ApJ, 198, 95
Green D. A., Gull S. F., Tan S. M., Simon A. J. B., 1988, MNRAS, 231, 735
Radhakrishnan V., Goss W. M., Murray J. D., Brooks J. W., 1972, ApJS, 24, 49
Raymond J. C., Smith B. W., 1977, ApJS, 35, 419
Tan S. M., Gull S. F., 1985, MNRAS, 216, 949
Thorsett S. E., 1992, Nat, 356, 690
van den Bergh S., Tammann G., 1991, ARA&A, 29, 363

BIPOLAR SUPERNOVA REMNANTS AND THE OBLIQUITY DEPENDENCE OF SHOCK ACCELERATION

Michael S. Fulbright and Stephen P. Reynolds

North Carolina State University
Received 1989 June 19; accepted 1990 January 15

ABSTRACT

Young, adiabatic-phase shell supernova remnants (SNRs) generate a radio synchrotron flux which implies a greater energy density in relativistic particles or magnetic field than can be accounted for by shock compression of ambient cosmic rays and magnetic field. It is likely that diffusive shock acceleration produces the observed relativistic electrons in at least some young remnants. We have generated synthetic radio maps of model SNRs based on several classes of assumptions regarding both shock acceleration and magnetic field evolution. We have not considered turbulent amplification of magnetic field. For these models to replicate the pronounced bipolarity in observed remnants, we find that the efficiency of the acceleration process must depend upon the obliquity angle θ_{Bn} between the shock normal and the magnetic field (assumed uniform). We calculate azimuthal intensity ratios to enable a quantitative comparison with observations. Isotropic acceleration models cannot produce ratios as large as observed in five of a sample of nine young remnants. In our simple parameterization, "quasi-perpendicular" ($\theta_{Bn} \sim 90°$) models can produce observed ratios but predict a smaller median ratio than seen in the sample, and, for one remnant, they are only marginally consistent with observations. However, for small angles between the external magnetic field and the line of sight, "quasi-parallel" models ($\theta_{Bn} \sim 0°$) produce images unlike any observed remnants and predict a much larger median ratio than seen in the sample. Our version of "quasi-parallel" acceleration is distinctly less favored by observations.

Subject headings: hydromagnetics — nebulae: supernova remnants — particle acceleration — radiation mechanisms — shock waves

I. INTRODUCTION

Supernova remnants can serve as prime laboratories for the study of the acceleration of energetic electrons and the turbulent amplification of magnetic field. Adiabatic-phase remnants, in which the maximum shock compression is a factor of 4, must exhibit one or both of these processes, since such remnants are far too bright in synchrotron radio emission for the emissivity to result from simply compressing ambient magnetic field and cosmic-ray electrons. However, little agreement has been obtained on basic questions such as the relative importance of generation of electrons versus magnetic field, or the nature of electron acceleration, whether diffusive shock acceleration (first-order Fermi) or turbulent acceleration in the remnant interior (second-order Fermi). We shall address some of these questions by calculating synthetic images of synchrotron emission from young supernova remnants (SNRs).

In at least some young remnants, strong arguments can be made that the dominant location of electron acceleration is the shock wave. For the remnant of SN 1006, high-resolution radio observations (Reynolds and Gilmore 1986) show a very sharp outer edge over much of the remnant circumference, with a rapid rise in brightness strongly suggestive of acceleration at the shock wave. Furthermore, an optical filament exhibiting high-velocity Hα emission, probably from charge exchange on shocked partially ionized material (Chevalier, Kirshner, and Raymond 1980), lies at the edge of the radio emission rather than considerably beyond it as would be expected if the radio emissivity appeared in the remnant interior. In more general terms, Dickel *et al.* (1989) propose a model for radio emission from young SNRs in which interior turbulence resulting from

the shock overrunning small inhomogeneities generates magnetic field. However, they point out that electron acceleration by the same mechanism would both take too long and result in a curved spectrum contrary to observations. Thus there is good reason to believe that shock acceleration is the dominant producer of fast electrons in at least some adiabatic-phase SNRs.

Kesteven and Caswell (1987) and Roger *et al.* (1988) have recently called attention to a fairly common morphology among SNRs that they term "bilateral symmetry" or "barrel-shaped": shell remnants bright on two opposing limbs, with lower brightness regions between them around the shell. They discuss possible reasons for this systematic property, including asymmetries in the surrounding medium and in the ejection process. The sample of objects represented in their work, however, includes both historical remnants such as SN 1006 (G327.6 + 14.6), which certainly has an adiabatic shock wave, and much older, larger, fainter remnants such as G296.5 + 10.0, in which the likely mechanism for synchrotron emission is just compression of ambient material. It is likely that mechanisms for "barrel-shaped" morphology, or, as we shall describe it, "bipolarity," differ for remnants in these two very different stages of evolution, as discussed by Roger *et al.* (1988). We intend to investigate mechanisms for bipolarity in young remnants, under the assumption that shock acceleration physics may be responsible. We note here that some remnants show bipolar structure in X-rays as well; SN 1006 is a good example (Pye *et al.* 1981). Since the X-rays result from shock-heated thermal material, perhaps dominantly from ejecta in the youngest remnants, the explanation of X-ray bipolarity seems considerably more difficult than that of radio bipolarity. The possibility of a link between magnetic-field geometry and the efficiency of heating of electrons in collisionless shocks may be a fruitful line of investigation.

The most obvious mechanism to produce a bipolar structure is the compression by a factor of 4 of magnetic field where it is perpendicular to the shock normal, compared to no amplification where it is parallel (van der Laan 1962; Whiteoak and Gardner 1968). However, this mechanism can produce only a limited amount of azimuthal modulation of intensity. Roger *et al.* (1988) and Leckband, Spangler, and Cairns (1989) point out that another possible mechanism for producing bipolar structure in shell remnants is a systematic dependence of the efficiency of shock acceleration on the obliquity angle θ_{Bn} between the shock normal and the external magnetic field, if the field is assumed to be fairly well ordered on the scale of the remnant diameter. Shock acceleration theorists are divided on whether quasi-parallel ($\theta_{Bn} \sim 0°$) or quasi-perpendicular ($\theta_{Bn} \sim 90°$) shocks are more efficient in accelerating electrons. The quasi-parallel geometry seems more adapted to classical diffusive shock acceleration, (see reviews such as Drury 1983 or Blandford and Eichler 1987) while quasi-perpendicular geometry allows the so-called shock drift mechanism (Pesses, Decker, and Armstrong 1982; Decker and Vlahos 1985, among others) in which electric fields along the shock front accelerate particles, a process which can be considerably more rapid (Jokipii 1987). Leckband, Spangler, and Cairns (1989) attempted to study this issue by examining the limb-to-center ratios of SNRs, inferring the direction of the external magnetic field for each remnant using a model of the galactic magnetic field, and comparing with model calculations for profiles of SNRs. They could not come to a definite conclusion.

We shall address the same problem using a somewhat different technique. We propose to answer several well-defined questions about bipolar structure in young SNRs. What is the maximum bipolarity that can be produced if relativistic electrons are produced isotropically behind the shock? If this is inadequate to explain observed SNRs, what can be said about models exhibiting quasi-perpendicular or quasi-parallel favoritism? We shall generate synthetic images of SNRs for each set of physical assumptions ("model"), and for various values of the aspect angle ϕ between the magnetic field outside the remnant and our line of sight. In a quasi-parallel model, remnant emission occurs in two bright polar caps; in a quasi-perpendicular model, equatorial barrels result. To discriminate among these, an ensemble average over aspect angles must be considered. This frees us from the necessity of determining the aspect angle of a given remnant.

The approach of generating synthetic images to compare with observed ones is, we believe, an extremely powerful one. We intend to investigate many related problems with extensions of this work. For instance, while observed azimuthal brightness variations contain information about obliquity dependence of particle acceleration, radial variations depend on the evolutionary history of particle acceleration, including such crucial but undetermined facts as the dependence of electron acceleration efficiency on shock velocity for very strong shocks. Polarized-flux images, beyond the scope of the present paper, also contain a great deal of information which can be extracted in a similar way. These studies should allow us to make better use of the enormous amount of morphological information collected over the past two decades by hard-working SNR observers.

II. MODEL ASSUMPTIONS

The SNRs modeled are assumed to be young and expanding adiabatically, thus described by the Sedov (1959) analytic self-similar solution for the dynamics of a point explosion in a homogeneous medium. It is assumed that the interstellar medium surrounding the SNR has a constant density and is permeated by a constant magnetic field with a strength of 3 μg. A strong shock is assumed, so the compression factor will be 4. Many of the details of our modeling procedure are similar to those described in Reynolds and Chevalier (1981).

For remnants in which the reverse shock has not yet reached the center, the outer regions will be well described by Sedov dynamics, but the synchrotron emissivity in the ejected material will be negligible because the remnant has expanded sufficiently that the stellar magnetic field, which accounts for the magnetic field inside the contact discontinuity between shocked interstellar medium and shocked ejecta, has been greatly diluted. To take account of this, as well as to avoid the unphysical results the Sedov solution produces as one approaches the explosion center, we assume that the material inside a radius $r = R_{crit}$ has a synchrotron emissivity of zero. We have identified R_{crit} with the current radius of the ejecta (i.e., the contact discontinuity). The models we illustrate below have been calculated with $R_{crit} = 0.8$, or 80% of the remnant's total radius. For the material outside $r = R_{crit}$, we must calculate the synchrotron emissivity j_ν by the relation (Pacholczyk 1970)

$$j_\nu = c_j K B_\perp^{(s+1)/2} \nu^{(1-s)/2} \text{ ergs s}^{-1} \text{ cm}^{-3} \text{ Hz}^{-1} \text{ sr}^{-1} ,$$

where c_j is a constant depending upon the value of s, $B_\perp$ is the component of magnetic field perpendicular to the line of sight, and K and s are related to the relativistic electron energy distribution by

$$N(E) = K E^{-s} \text{ cm}^{-3} \text{ ergs}^{-1} .$$

We assume the remnant expands into a region of constant magnetic field. While observational data on length scales of 1–20 pc are scarce, the galactic field is not expected to show strong variations on these scales (Heiles 1976). In all cases, it is assumed that the shock compresses and possibly disorders the external field. The thermal matter is assumed to be highly ionized, and therefore the magnetic field is frozen in the material. When a fluid element's dimensions are modified, the magnetic field inside the element will change so that the magnetic flux through the element's surface remains constant. When a fluid element is radially compressed, the component of the element's magnetic field perpendicular to the shock direction is amplified by the compression factor, which in these calculations is 4. After the element is shocked it begins to expand, with different expansion factors for radial and tangential dimensions, thus causing different evolution of radial and tangential components of magnetic field (B_r and B_t, respectively). This was originally pointed out by Duin and Strom (1975) and elaborated in Reynolds and Chevalier (1981):

$$B_r \propto \frac{r_i^2}{r^2} ,$$

and

$$B_t \propto \frac{\rho(r)r}{r_i} ,$$

where r_i is the radius at which the fluid element under consideration was originally shocked.

Determining the current value of the magnetic field at a point behind the shock requires several steps. First one must find the initial shocked value of the magnetic field by taking the tangential component of the external magnetic field, multiplying it by 4, and adding it to the radial component, which is left unchanged. Using the Sedov description of the explosion, one can calculate how long ago the element we are currently interested in was shocked. This information allows us to determine the expansion of the element since shocking and the decay of the magnetic field. We describe this scenario for the magnetic field evolution as the ordered case.

Using the ordered model of the magnetic field, however, leads to model SNRs with fractional polarization approaching 70%, whereas the maximum observed fractional polarization is much lower (typically 10%–15%). This implies that the shock somehow disorders the magnetic field. To model the effect of disordering the magnetic field, the following approach was taken. The immediate postshock field strength was calculated as in the ordered field case, and the randomized field was determined by simply choosing a new, randomly oriented magnetic field vector with the same magnitude. The subsequent evolution of the magnetic-field components was as in the ordered case. This prescription amounts to assuming the magnetic field is disordered on scales as small as the separation of our integration points. The polarization resulting from this model is, of course, very small, but that result is closer to observed remnants than the high polarization of the totally ordered field case. In any case, the degree of order of the magnetic field makes only small differences to the total-intensity image. (The polarization images will be discussed in a later paper.)

Current shock acceleration theory cannot predict the dependence of the accelerated relativistic electron energy density u_r on either the speed of the shock or the angle between the shock normal and the external magnetic field θ_{Bn}. We have chosen to model u_{r2}, the relativistic electron energy density just behind the shock, as proportional to the postshock pressure. This situation is what one would intuitively mean by "constant efficiency." We assume that the accelerated particles do not diffuse from the fluid element in which they were accelerated, subsequently evolving adiabatically as the element expands. Using only these assumptions, we have the isotropic particle acceleration case. As mentioned above, however, it is possible that the acceleration efficiency depends in some way on the angle θ_{Bn}. To model quasi-parallel acceleration, we multiply the above u_{r2} by $\cos^2 \theta_{Bn2}$, where θ_{Bn2} is the angle between the shock normal and the *postshock* magnetic field. [The two angles are simply related: $\sin^2 \theta_{Bn2} = (\cot^2 \theta_{Bn}/16 + 1)^{-1}$.] One justification for this can be found in the description of particle injection by Eichler (1981) and Edmiston, Kennel, and Eichler (1982) in which the likelihood of a particle downstream recrossing the shock to be further accelerated is proportional to $\cos^2 \theta_{Bn2}$. Similarly, to model quasi-perpendicular acceleration we multiply the above u_{r2} by $\sin^2 \theta_{Bn2}$, though there is no equivalent justification for this expression. As mentioned above, in the shock drift mechanism, only the acceleration rate is expected to differ from that in the quasi-parallel case, and the rate is completely unimportant for the relatively low-energy electrons we observe in SNRs. So the physical justification for a quasi-perpendicular predominance would probably occur in the physics of the injection process. In models which have a disor-

dered magnetic field, we use the postshock magnetic field before disordering to calculate θ_{Bn2}, because the disordering process is assumed to take place over a longer time scale than the acceleration, which occurs in the close proximity of the shock.

III. METHOD OF CALCULATION

Calculation of the synthetic radio maps of the model supernova remnants was straightforward. The observer measures the polarization components of the incoming radiation in terms of a Cartesian two-dimensional system (a, b) in the plane of the sky. The radio map is simply the projection of the surface of the supernova remnant onto the plane of the observer's two-dimensional frame of reference. The basic equations governing the production and transfer of synchrotron radiation, including Faraday rotation, are (in terms of Stokes parameters I, Q, U, and V) (Pacholczyk 1970)

$$\frac{dI^{(a)}}{ds} = \beta U^{(ab)} + j_v^{(1)} \sin^2 \chi + j_v^{(2)} \cos^2 \chi ,$$

$$\frac{dI^{(b)}}{ds} = -\beta U^{(ab)} + j_v^{(1)} \cos^2 \chi + j_v^{(2)} \sin^2 \chi ,$$

$$\frac{dU^{(ab)}}{ds} = -2\beta I^{(a)} + 2\beta I^{(b)} - (j_v^{(1)} - j_v^{(2)}) \sin 2\chi ,$$

where $I^{(a)}$ and $I^{(b)}$ are the intensities of the radiation polarized along the respective axis, and $U^{(ab)}$ is a measure of the intensity of the circularly polarized radiation. The other variables are χ, which is the angle between the component of the magnetic field perpendicular to the observer's line of sight and the a axis, and β, which is the Faraday rotation rate (rad pc^{-1}). To generate a synthetic radio map of the remnant, one integrates this system of equations along lines of sight spaced evenly over a grid centered on the origin of the ab system. A map of arbitrary resolution can be generated, since the resolution of the map is roughly the sampling frequency used for the grid. At each point along the line of sight one calculates $j_v^{(1)}$ and $j_v^{(2)}$ using the synchrotron emissivity j_v for the volume element:

$$j_v^{(1)} = \frac{1}{2} j_v \left(1 + \frac{s+1}{s+7/3} \right) ,$$

$$j_v^{(2)} = \frac{1}{2} j_v \left(1 - \frac{s+1}{s+7/3} \right) ,$$

where $s = 2|\alpha| + 1$, and α is the spectral index for the remnant (typically $\alpha \sim -0.5$), and $j_v^{(1)}$ and $j_v^{(2)}$ are, respectively, the synchrotron emissivities for radiation linearly polarized perpendicular and parallel to the perpendicular projection of the magnetic field.

Although the models include Faraday rotation, initially we have concentrated on studying the map of total intensity $I = [(I^{(a)})^2 + (I^{(b)})^2]^{1/2}$ of each remnant. We have therefore let $\beta = 0$, effectively disabling Faraday rotation. Faraday rotation can only alter the plane of polarization; it cannot change the intensity map of a supernova remnant. For this reason, the results of these initial models are independent of Faraday rotation. Again, polarization results will be discussed in a later paper.

We first calculated maps at a resolution of 128 by 128 pixels, then convolved these maps to lower resolutions. The original map was sampled densely enough so that after convolving it with a Gaussian of the same size as the desired resolution we had a map which more closely represented what a real instrument would have generated than the original map.

We attempted to quantify the degree of "bipolarity" by using a measure we call the azimuthal intensity ratio A, essentially the ratio of maximum to minimum intensity around the shell of emission. To calculate A from a model, we first find the point on the map where the intensity is greatest. Then along the contour of points at the same radius as the point of maximum intensity, we find the point with the lowest intensity. The ratio of the maximum to the minimum is the value of A for this contour. To get an idea of the range of possible values, we also calculated A on circular contours inside and outside of the original contour where the peak had dropped to 90% of its maximum value. We found, using this measure, that the various models studied produced significantly different results, possibly enabling one to use this method as a diagnostic in studying actual remnants to determine the nature of magnetic field amplification and relativistic particle acceleration. Of particular importance is the wide variation in A both between models and for different values of aspect angle ϕ for the same model. This is important because measuring the value of A for a real object is often somewhat uncertain, for instance because the shell cannot be followed unambiguously over a full 360°.

It is easier to work with azimuthal slices of the model images if one first converts the map from a Cartesian coordinate system to a polar coordinate system. Circles in the Cartesian plane become lines in the polar plane. Figure 1a shows a sample map in the Cartesian plane, and Figure 1b shows the same map in the polar plane. In the polar plane, consecutive vertical lines represent concentric circles in the Cartesian plane. To calculate A, one takes the ratio of the maximum to the minimum value on the vertical slice that also contains the maximum intensity pixel on the entire map. The upper and lower values of A we quote in Table 1 were determined along the slices on opposite sides of the peak slice, where the maxima fall to 90% of the map's peak intensity.

Because the synchrotron emissivity depends upon the component of the magnetic field perpendicular to the line of sight, as the inclination of the field external to the remnant is varied the radio map produced will also change. When the external magnetic field is parallel to the observer's line of sight ($\phi = 0°$), then one predicts that the emissivity will be lowest, and because the system is symmetric about the magnetic field the remnant will be axially symmetric. In this case $A = 1$. When the external magnetic field is in the plane of the sky, then the emissivity should attain its maximum value. In this case, A should be maximum. Thus it is necessary to calculate models at many aspect angles to obtain an idea of how remnants embodying the magnetic field and relativistic particle acceleration assumptions in question would appear. Then, using assumptions about the large-scale structure of the galactic magnetic field, one can weight each aspect angle by its likelihood and from this generate statistical predictions about the ensemble of observed remnants.

IV. RESULTS

Figures 2, 3, and 4 show images calculated from each of the three models, convolved to 32 beams per remnant diameter. (After model arrays were calculated, they were imported into AIPS [Astronomical Image Processing System], where all

TABLE 1

AZIMUTHAL INTENSITY RATIOS A FOR MODELS

Aspect angle ϕ	A_{90}[a]	A_{32}[b]	A_{10}[c]
Isotropic Model			
0°	1.1 (1.0, 1.0)	1.0 (1.0, 1.0)	1.0 (1.0, 1.0)
15°	1.1 (1.1, 1.1)	1.1 (1.1, 1.1)	1.1 (1.1, 1.1)
30°	1.3 (1.3, 1.3)	1.2 (1.2, 1.2)	1.2 (1.2, 1.2)
45°	1.7 (1.6, 1.7)	1.7 (1.6, 1.7)	1.6 (1.6, 1.6)
60°	2.6 (2.5, 2.6)	2.5 (2.3, 2.5)	2.3 (2.1, 2.3)
75°	4.6 (3.7, 4.7)	4.2 (3.4, 4.4)	3.2 (2.7, 3.5)
90°	6.3 (4.5, 6.8)	5.6 (4.0, 6.0)	3.7 (3.0, 4.3)
Quasi-Parallel Model			
0°	1.2 (1.1, 1.1)	1.0 (1.0, 1.0)	1.0 (1.0, 1.0)
15°	1.3 (1.1, 1.1)	1.2 (1.1, 1.2)	1.1 (1.1, 1.2)
30°	2.8 (2.5, 6.6)	3.5 (2.2, 6.8)	2.7 (1.8, 3.8)
45°	21 (10, 38)	16 (8.1, 25)	8.2 (5.5, 11)
60°	67 (36, 120)	51 (31, 79)	24 (18, 31)
75°	460 (260, 670)	280 (190, 350)	76 (60, 93)
90°	$>10^5$ ($>10^5$, $>10^5$)	2000 (1800, 2200)	160 (140, 170)
Quasi-Perpendicular Model			
0°	1.0 (1.1, 1.0)	1.0 (1.0, 1.0)	1.0 (1.0, 1.0)
15°	1.1 (1.1, 1.1)	1.1 (1.1, 1.1)	1.1 (1.1, 1.1)
30°	1.3 (1.3, 1.3)	1.3 (1.3, 1.3)	1.3 (1.3, 1.3)
45°	1.8 (1.8, 1.8)	1.8 (1.8, 1.8)	1.7 (1.7, 1.7)
60°	3.1 (2.9, 3.1)	3.0 (2.7, 3.0)	2.6 (2.5, 2.7)
75°	7.4 (5.1, 7.8)	6.4 (4.6, 6.9)	4.2 (3.3, 4.9)
90°	15 (6.9, 21)	11 (5.8, 14)	5.3 (3.8, 6.8)

NOTE.—The numbers in parentheses give the values of A measured along inner and outer 90% radii, respectively, as described in § III.

[a] For a resolution of 90 beams per diameter.
[b] For a resolution of 32 beams per diameter.
[c] For a resolution of 10 beams per diameter.

image manipulation was performed.) All images are normalized gray-scale maps of total intensity, with contours spaced at every 10% of the peak value. The different images for each model type are maps which result from using different values of ϕ. Models were calculated where ϕ varied from 0° to 90°, at intervals of 15°. As will be discussed, the resolution of the map affects the degree of bipolarity observed. Table 1 lists the values of A at 90, 32, and 10 beams per remnant diameter, and these values are plotted in Figures 5a–5c.

At the resolution of 90 beams per diameter, the three models have distinctly different A versus ϕ curves. In the isotropic particle case, A varies from 1 to 6.3. In the quasi-parallel case, A varies from 1 to basically infinity, while in the quasi-perpendicular case, A varies from 1 to 15. The value of A also depends on the resolution of the map. One expects as the remnant becomes totally unresolved that $A \rightarrow 1$. The dependency of A on resolution is shown in Figure 6 for the quasi-perpendicular model with $\phi = 90°$. The dotted line on this graph shows the values obtained for A when the alternative slices were used.

One question we asked was whether A depended upon the value of $R_{\rm crit}$ used. As a test, the totally random magnetic field, isotropic particle acceleration model with the external magnetic field at 90° was used. Maps with $R_{\rm crit} = 0.4, 0.6, 0.8,$ and 0.9 were generated. The value of A varied by less than 5% over this range of $R_{\rm crit}$. One reason for this is that the maximum intensity occurs near the remnant's periphery, so in all these cases the lines of sight which contain both the peak and minimum pixels do not intersect the sphere with radius $R_{\rm crit}$.

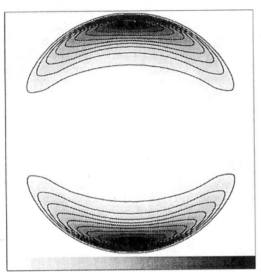

FIG. 1a

Only if R_{crit} were greater than about 95% of the remnant's radius would it start to influence the value of A. Such a large value would imply a remnant barely out of the free-expansion stage, that is, probably younger than any remnant in the galaxy (with the possible exception of Cas A). Another reason that the value of A has little dependence on R_{crit} is that the synchrotron emissivity j_v decreases as a high power of the thermal matter density ρ, and the value of ρ drops quickly with radius. By decreasing the value of R_{crit} and thereby increasing the depth at which the lines of sight penetrate the remnant, one does little in increasing the total intensity because the newly contributing material has a very low emissivity compared to the rest of the line of sight. The weak dependence on R_{crit} also means that the particular dynamics used are not critical to the results. A more careful calculation, for instance using Chevalier's (1982) two-shock similarity solution, should not differ dramatically from that shown here in the values of A produced. (Radial variations are another matter.)

V. DISCUSSION

The important qualitative conclusion to be drawn from Figures 2, 3, and 4 is that different simple parameterizations of the variation of electron acceleration efficiency with shock

FIG. 1b

FIG. 1.—(a) A typical model image, in this case a "quasi-parallel" model with aspect angle 90°, shown at a resolution of 32 beams per diameter. Contours on this and all subsequent contour plots are at intervals of 10% of the peak intensity. (b) The same image transformed with a Cartesian-to-polar coordinate transformation about the center. Abscissa is radius; ordinate is azimuthal angle. The range of A shown in Fig. 6 corresponds to values obtained taking vertical slices on either side of the slice through the peak, along the radii where the maximum is 90% of the map peak.

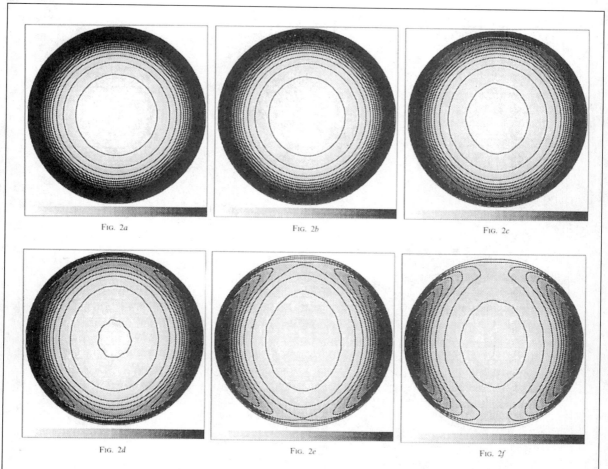

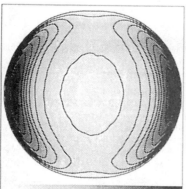

Fig. 2.—Sequence of images for the "isotropic" (acceleration efficiency independent of θ_{Bn}) models, at a resolution of 32 beams per diameter. (a) Aspect angle $\phi = 0°$; (b) $\phi = 15°$; (c) $\phi = 30°$; (d) $\phi = 45°$; (e) $\phi = 60°$; (f) $\phi = 75°$; (g) $\phi = 90°$.

obliquity produce images differing considerably in appearance. In particular, the simple quantitative observable A shows very different behavior with aspect among the various models, as shown by Table 1. This wide difference in predicted values of this quantity makes the discrimination among models plausible, even though observed remnants are typically considerably nonspherical, irregular, and patchy. However, if A can be determined for a set of remnants even to within a factor of 2, a model discrimination may be possible.

The results of the previous section can be compared with observed radio images of supernova remnants both individually and statistically. The null hypothesis that no obliquity dependence of electron acceleration is necessary can be firmly rejected either way. This hypothesis apparently was first stated by van der Laan (1962), elaborated by Whiteoak and Gardner (1968) among many others, and repeated by Leckband, Spangler, and Cairns (1989). In none of this work was the quantitative extent of bipolarity producible by this model examined. In fact, a trivial calculation shows that for $\phi = 90°$, a point on the remnant's "equator" will have an emissivity larger than at the pole by a factor $4^{1-\alpha}$, suggesting a maximum value of A of about 9 for $\alpha = -0.6$. But radio observations of the remnant of SN 1006 with a resolution of 90 beams per diameter (Reynolds

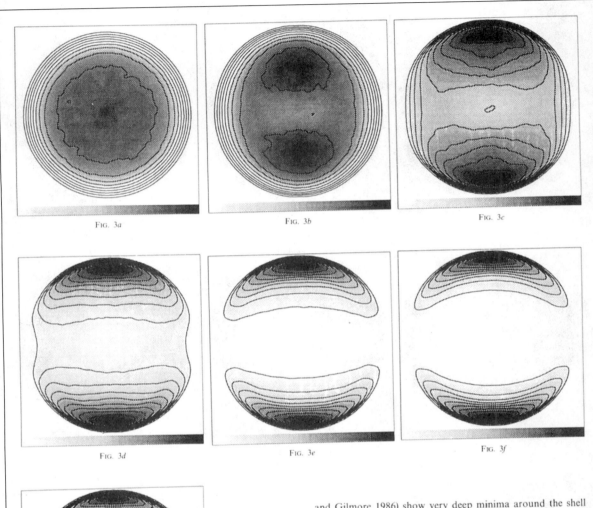

FIG. 3a   FIG. 3b   FIG. 3c

FIG. 3d   FIG. 3e   FIG. 3f

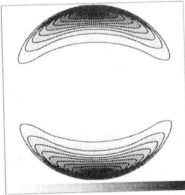

FIG. 3g

FIG. 3.—Sequence of images for the quasi-parallel models, at a resolution of 32 beams per diameter. (a) Aspect angle $\phi = 0°$; (b) $\phi = 15°$; (c) $\phi = 30°$; (d) $\phi = 45°$; (e) $\phi = 60°$; (f) $\phi = 75°$; (g) $\phi = 90°$.

and Gilmore 1986) show very deep minima around the shell between the bright eastern and western lobes; we estimate from that map a value of $A \sim 15$ at least. Since the maximum value of A producible by the isotropic model is less than 7 at this resolution, this model is completely inconsistent with SN 1006. We consulted other published work to compare it with a wider range of remnants.

We have selected a sample of bright shell remnants to which to apply these results, from the catalogue of Green (1988). In order to select objects most likely to be in the adiabatic phase, a surface-brightness minimum criterion was set: $\Sigma > 0.32$ Jy arcmin^{-2} at 1 GHz. This selects the brightest 20% of galactic remnants. A more physically significant cutoff results from demanding that the mean remnant emissivity, proportional to Σ/D where D is the remnant diameter, be at least 5000 times the mean galactic background emissivity, or $\Sigma/D > 0.01$ Jy arcmin^{-2} pc^{-1}. This value results from using the data of Beuermann, Kanback, and Berkhuijsen (1985) on the galactic synchrotron background: a value of $T_b \sim 300$ K toward the Galactic center at 408 MHz, with an assumed line-of-sight

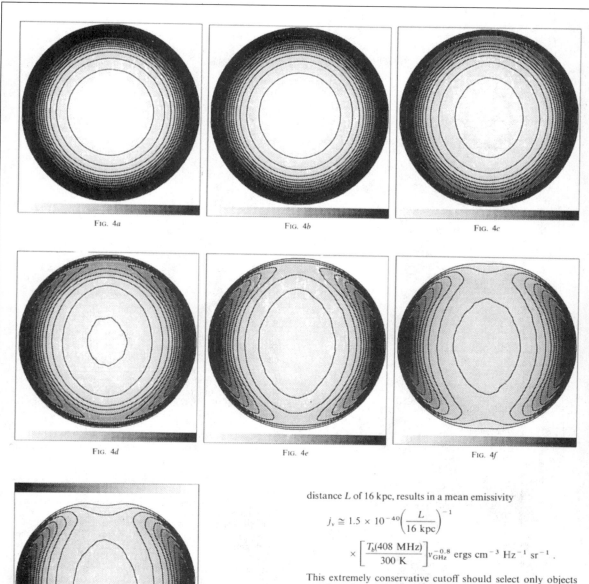

Fig. 4a   Fig. 4b   Fig. 4c

Fig. 4d   Fig. 4e   Fig. 4f

Fig. 4g

Fig. 4.—Sequence of images for the quasi-perpendicular models, at a resolution of 32 beams per diameter. (a) Aspect angle $\phi = 0°$; (b) $\phi = 15°$; (c) $\phi = 30°$; (d) $\phi = 45°$; (e) $\phi = 60°$; (f) $\phi = 75°$; (g) $\phi = 90°$.

distance L of 16 kpc, results in a mean emissivity

$$j_\nu \cong 1.5 \times 10^{-40}\left(\frac{L}{16 \text{ kpc}}\right)^{-1}$$

$$\times \left[\frac{T_b(408 \text{ MHz})}{300 \text{ K}}\right]\nu_{\text{GHz}}^{-0.8} \text{ ergs cm}^{-3} \text{ Hz}^{-1} \text{ sr}^{-1} .$$

This extremely conservative cutoff should select only objects certain to be producing some of their own synchrotron emissivity by means other than compression. For a spectral index of -0.5, a compression ratio of over 30 is required for this emissivity increase. For objects with unknown distances, we evaluated the criterion assuming a distance of 20 kpc. These two standards left a list of 25 objects, of which four are Crab-like, therefore presumably pulsar-driven and to be excluded. (We note that the criteria are severe enough that if SN 1006 were not known to be in the adiabatic phase from its small age, it would not qualify for the sample. It will in fact be left out of the statistical discussion below.) The best available published maps of the remaining 21 objects were scrutinized for bipolar structure.

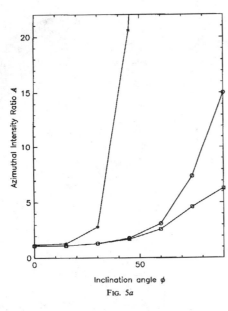

Fig. 5a

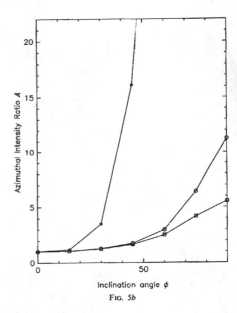

Fig. 5b

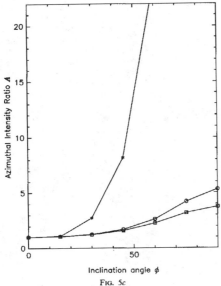

Fig. 5c

Fig. 5.—Azimuthal intensity ratio *A* for all three models. *Squares*: isotropic; *circles*: quasi-perpendicular; *stars*: quasi-parallel. (*a*) Resolution of 90 beams per diameter. (*b*) Resolution of 32 beams per diameter. (*c*) Resolution of 10 beams per diameter.

Of these 21, four are not well resolved and had to be excluded. Two others have very peculiar morphologies not consistent with a simple shell. The remaining 15 all have well-resolved shell structure. The criteria for bipolarity used were the appearance of two maxima around the shell, separated by roughly half the circumference (100°–260°), differing in intensity by no more than a factor of 2. By this standard, eight remnants were bipolar, and seven were roughly "monopolar," that is, with one major shell maximum and a minimum roughly across from it. (Monopolar remnants are discussed with reference to gradients in external magnetic field in a related work; Reynolds and Fulbright 1990.) The eight bipolar objects are listed in Table 2, along with SN 1006, with the values of *A* deduced from the maps in the quoted references. For G1.9 + 0.3, no contour plot was published; the limit for *A* quoted was deduced from a gray-scale image. We estimate that the values for *A* are accurate to about 20%—quite good enough for comparison with the models. The typical resolution of the maps was adequate to support the conclusions we draw below.

Table 2 shows that five of the nine bipolar remnants (including SN 1006) show values of *A* too large to be accounted for by the compression of magnetic field and isotropic particles model, for any resolution and orientation. Thus this picture can be emphatically ruled out. (Higher resolution should only increase observed values of *A*.) Young supernova remnants show azimuthal brightness variations that demand anisotropic processes producing relativistic particles (or magnetic field) in their interiors. We lean strongly to the point of view that it is the particle distribution that accounts for the bipolarity; if magnetic field is generated internally by turbulent amplification, it is difficult to imagine a process that could produce large-scale modulation of the strength of such a field in remnants that are roughly circularly symmetric. Small-scale irregularities in the surrounding medium can cause fluctuations in the synchrotron brightness (Dickel *et al.* 1989), but systematic trends such as bipolarity do not arise. A cylindrically symmetric external medium could cause bipolarity but ought to retard the position of the shock where the density is higher, causing bright regions to have smaller radii. This is not observed.

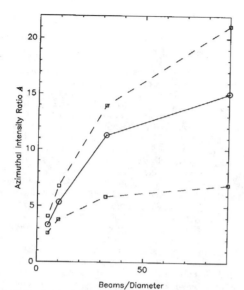

Fig. 6.—Resolution dependence of A for quasi-perpendicular models. The range of values shown at each resolution corresponds to measuring A around radii on either side of the radius of the map maximum, as described under Fig. 1b, and it gives an idea of the variation likely to occur in a real, not-quite-circular remnant.

take it, will produce images unlike any observed whenever $\phi \lesssim 15°$. If the galactic magnetic field is largely in the plane, roughly one-sixth of observed remnants should have $\phi < 15°$, so seeing no center-brightened pulsarless remnants (out of about 140) argues strongly against this picture. In addition, for aspect angles between 15° and 30°, remnants have two smallish maxima atop a uniform "halo" or plateau, showing no structure describable as a "shell." This morphology is not shown by any of the nine objects of Table 2, nor by any of the 15 other objects illustrated in Kesteven and Caswell (1987). Neither the center-brightened nor the double peak-halo structure seems to be observed in supernova remnants, while one or the other should occur one-third of the time among remnants in which shock acceleration is the dominant source of relativistic electrons. Again, if the galactic magnetic field is in the plane, the median value of ϕ should be about 45°, so the median value of A for a randomly oriented distribution should be 16 (for a resolution of 32 beams per diameter), while the figures of Table 2 indicate a median of about 8.

The simple quasi-perpendicular model is consistent with observations, with the possible exception of G332.4−0.4 whose value of A (deduced from a contour plot) is marginally too high for its resolution. A determination of A from the actual radio data could clarify this point. The quasi-perpendicular model can produce large values of A for high resolutions: for 90 beams per diameter, the resolution of the map of SN 1006 by Reynolds and Gilmore (1986), the model gives $A(\phi = 90°) = 15$. For 32 beams per diameter, the median value is only 1.7, compared with the sample median of 8; but the latter applies to remnants observed with varying resolutions. It is somewhat troubling that one would deduce most of the large-A objects in Table 2 to have $\phi \sim 90°$, but with a sample this small, a firm statement cannot be made. For no aspect angle is a model produced that is qualitatively unlike observed objects.

We regard the argument against θ_{Bn}-independent acceleration to be conclusive. The statistical arguments are, of course, weaker. One source of possible bias can be ruled out, however; the radio spectral luminosity of all models calculated varies only over a range of a factor of 2, so selection effects based on morphology/brightness correlations are not a problem. Both quasi-parallel and quasi-perpendicular models do a poor job reproducing the median value of A of the sample of eight remnants, but the statistical significance of this result is quite weak.

We now consider the two simple parameterizations of quasi-parallel and quasi-perpendicular acceleration processes. First, note from Figure 3 that for all aspect angles less than about 15°, quasi-parallel models are center-brightened. No center-brightened supernova remnant is observed with a spectral index steeper than $\alpha = -0.45(S_\nu \propto \nu^\alpha)$ (though several cases are ambiguous). The few center-brightened objects are assumed to owe this morphology as well as their flat spectra to energy input from a central pulsar, telling us nothing about shock acceleration. Thus our simple description of quasi-parallel acceleration, in which the efficiency of electron acceleration is proportional to the likelihood that a given thermal electron behind the shock can scatter back upstream to over-

TABLE 2

Observed Azimuthal Intensity Ratios

Remnant	Surface Brightness (1 GHz) (Jy arcmin^{-2})	A	Resolution (beams per diameter)	Reference
G0.0+0.0 (Sgr A E)	15	2.5	48	1
G43.3−0.2 (W49 B)	4.0	>10	18	2
G4.5+6.8 (Kepler)	2.7	~10	60	3
G41.1−0.3	2.5	4.3	26	4
G11.2−0.3	1.8	3	30	4
G1.9+0.3	0.53	>10	36	5
G39.2−0.3	0.48	1.3	8	6
G332.4−0.4	0.44	8	10	7
G327.6+14.6 (SN 1006)	0.027	15	90	8

Note.—SN 1006 does not qualify for the sample using the surface-brightness criterion described in the text, but its known age and expansion proper motion indicate that it is in the adiabatic phase.

References.—(1) Ekers et al. 1983; (2) Pye et al. 1984; (3) Matsui et al. 1984; (4) Becker, Markert, and Donahue 1985; (5) Green and Gull 1984; (6) Caswell et al. 1982; (7) Caswell et al. 1980; (8) Reynolds and Gilmore 1986.

The fortuitous congregation of seven of the eight objects in the first 45° of galactic longitude is unfortunate and may bias the sample toward particular values of ϕ. We hope to extend our sample to fainter objects in the future, and we also hope to encourage future observers to report A from their data, since its determination from contour plots is necessarily crude.

It must be noted that while the theoretical justification for our quasi-parallel parameterization is thin, that for our quasi-perpendicular model is nonexistent (though it is the same parameterization as used by Leckband, Spangler, and Cairns 1989). We simply chose a simple functional description by analogy with the quasi-parallel case. While Jokipii (1987) and others have argued that quasi-perpendicular "shock drift" acceleration takes place much more rapidly than quasi-parallel diffusive acceleration, the electrons we observe in radio supernova remnants have energies of typically 1–10 GeV, so low that they are essentially instantaneously accelerated, and difference in rates is unobservable. After the (unknown!) injection process has inserted electrons into the mechanism with gyroradii of several times those of thermal ions, there is no current justification for a dependence of acceleration efficiency on shock obliquity. We intend in future work to explore possibilities for any such dependence; any candidates will have to pass the test of producing model images consistent with observed remnants.

The reader will have noticed that we draw considerably firmer conclusions than do Leckband, Spangler, and Cairns (1989), despite using a rather similar approach. We have tried to do a careful job calculating the model intensities and have chosen for study a quantitative measure, the azimuthal intensity ratio A, that appears to be a better discriminant than the limb-to-center ratio. Since we find that some models are limited in the range of values of A they can produce, while others produce images unlike any observed remnants, we are freed from relying on either a complex statistical treatment or a model of the galactic magnetic field, which itself requires that distances to the remnants be known. In addition, our surface-brightness selection criteria may result in a more homogeneous sample of objects.

VI. CONCLUSIONS

Our initial attempt to use model images to investigate the physics of nonthermal radiation from supernova remnants has led to several useful conclusions:

1. The three model cases considered produce plots of azimuthal intensity ratio A versus external magnetic field inclination angle ϕ which are quite distinct.

2. The isotropic particle acceleration case systematically produces model images in which A is too low to explain observed objects. We therefore conclude that the shock acceleration efficiency of electrons must vary over the shock.

3. Using the quasi-parallel case, which has some theoretical justification, models with A varying from 1 to infinity are possible. But these models are unlikely to be correct because they produce images which are center-brightened or otherwise at odds with observations, for a significant range of aspect angles.

4. Finally, the quasi-perpendicular case, which lacks the physical foundation of the quasi-parallel case, produces objects in which A is greater than in the isotropic case, and the images it produces are not center-brightened. However, the predicted median value of A for a randomly oriented sample of remnants is considerably smaller than that observed for our sample of nine objects. One observed object (G332.4−0.4) may have a value of A too large (for its resolution) to be produced by the quasi-perpendicular models, but higher resolution studies will be necessary to confirm this.

Several qualifications of these results must be listed. First, the value of A for a model is highly resolution dependent; as the resolution decreases, the value of A will as well. Second, we have shown only that a particular simple prescription describing quasi-parallel favoritism is inconsistent with observations, not that no model favoring quasi-parallel geometry is possible. However, we anticipate that any such model will have to pass the tests described here, of being able to describe observed objects without predicting chimerical ones. We stress that none of the results obtained here depend on knowing distances to individual remnants. The only effect of distance information in the entire analysis is in the mean-emissivity lower cutoff for the sample of Table 2, as described in § V. We expect the synthetic-images approach to be a powerful way to continue to investigate the variation of the efficiency of shock acceleration with the angle between the shock normal and the external magnetic field.

We would like to thank D. Ellison and R. Jokipii for useful conversations, and K. Funk for assembling an invaluable collection of reference works on radio SNR images. S. P. R. acknowledges partial support from NASA grant NAG 8-731 and NSF grant AST-8817567; M. S. F. acknowledges summer support from NSF grant DMR 88-03370.

REFERENCES

Becker, R. H., Markert, T., and Donahue, M. 1985, *Ap. J.*, **296**, 461.
Beuermann, K., Kanback, G., and Berkhuijsen, E. M. 1985, *Astr. Ap.*, **153**, 17.
Blandford, R. D., and Eichler, D. 1987, *Phys. Rept.*, **154**, 1.
Caswell, J. L., Haynes, R. F., Milne, D. K., and Wellington, K. J. 1980, *M.N.R.A.S.*, **190**, 881.
———. 1982, *M.N.R.A.S.*, **200**, 1143.
Chevalier, R. A. 1982, *Ap. J.*, **258**, 790.
Chevalier, R. A., Kirshner, R. P., and Raymond, J. C. 1980, *Ap. J.*, **235**, 186.
Decker, R. B., and Vlahos, L. 1985, *J. Geophys. Res.*, **90**, 47.
Dickel, J. R., Eilek, J. A., Jones, E. M., and Reynolds, S. P. 1989, *Ap. J. Suppl.*, **70**, 497.
Drury, L. O'C. 1983, *Rept. Progr. Phys.*, **46**, 973.
Duin, R. M., and Strom, R. G. 1975, *Astr. Ap.*, **39**, 33.
Edmiston, J. P., Kennel, C. F., and Eichler, D. 1982, *Geophys. Res. Letters*, **9**, 531.
Eichler, D. 1981, *Ap. J.*, **244**, 711.
Ekers, R. D., van Gorkom, J. H., Schwarz, U. J., and Goss, W. M. 1983, *Astr. Ap.*, **122**, 143.
Green, D. A. 1988, *Ap. Space Sci.*, **148**, 3.
Green, D. A., and Gull, S. F. 1984, *Nature*, **312**, 527.
Heiles, C. 1976, *Ann. Rev. Astr. Ap.*, **14**, 1.

Jokipii, J. R. 1987, *Ap. J.*, **313**, 842.
Kesteven, M. J., and Caswell, J. L. 1987, *Astr. Ap.*, **183**, 118.
Leckband, J. A., Spangler, S. R., and Cairns, I. H. 1989, *Ap. J.*, **338**, 963.
Matsui, Y., Long, K. S., Dickel, J. R., and Greisen, E. W. 1984, *Ap. J.*, **287**, 295.
Pacholczyk, A. G. 1970, *Radio Astrophysics* (San Francisco: Freeman).
Pesses, M. E., Decker, R. B., and Armstrong, T. P. 1982, *Space Sci. Rev.*, **32**, 185.
Pye, J. P., Becker, R. H., Seward, F. D., and Thomas, N. 1984, *M.N.R.A.S.*, **207**, 649.
Pye, J. P., Pounds, K. A., Rolf, D. P., Seward, F. D., Smith, A., and Willingale, R. 1981, *M.N.R.A.S.*, **194**, 569.
Reynolds, S. P., and Chevalier, R. A. 1981, *Ap. J.*, **245**, 912.
Reynolds, S. P., and Fulbright, M. S. 1990, in *Proc. 21st Internat. Cosmic-Ray Conf. (Adelaide)*, **4**, 72.
Reynolds, S. P., and Gilmore, D. M. 1986, *A.J.*, **92**, 1138.
Roger, R. S., Milne, D. K., Kesteven, M. J., Wellington, K. J., and Haynes, R. F. 1988, *Ap. J.*, **332**, 940.
Sedov, L. I. 1959, *Similarity and Dimensional Methods in Mechanics* (New York: Academic Press).
van der Laan, H. 1962, *M.N.R.A.S.*, **124**, 179.
Whiteoak, J. B., and Gardner, F. F. 1968, *Ap. J.*, **154**, 807.

MICHAEL S. FULBRIGHT: Department of Astronomy and Steward Observatory, Tucson, AZ 85721

STEPHEN P. REYNOLDS: Department of Physics, North Carolina State University, Box 8202, Raleigh, NC 26795-8202

ASTROPHYSICS THEORY PROGRAM

**NASA Research Announcement
Soliciting Proposals
for Theory in Space Astrophysics
for period ending
November 29, 1993**

**NRA 93-OSSA-06
Issued: August 27, 1993**

**Office of Space Science
National Aeronautics and Space Administration
Washington, DC 20546**

[Introduction]

This NASA Research Announcement (NRA) solicits proposals for theoretical research with ultimate relevance to the interpretation of data from Space Astrophysics observations. The program, which was initiated 9 years ago, consists of research projects with a duration of 3 years.

The NASA Astrophysics Theory Program is intended to support efforts to develop the basic theory needed for NASA's Space Astrophysics program. This program seeks to address theoretical problems in Space Astrophysics of a wider scope than those typically studied within any particular subdiscipline of Space Astrophysics, or studied as a part of a data analysis program. Participation in this program is open to all categories of organizations including educational institutions, profit and nonprofit organizations, NASA Centers, and other Government agencies. Proposals may be submitted at any time during the period ending November 29, 1993. The proposal deadline will be adhered to strictly, and proposals received after November 29, 1993 will be held for the next review cycle which will commence in 1996. Proposals will be evaluated by scientific peer review panels with the goal of announcing selection by the middle of December 1993, and with availability of funds tentatively scheduled for January 1994 (subject to the NASA budget cycle).

X-RAY EMISSION AND DYNAMICS OF SUPERNOVA REMNANTS

A PROPOSAL TO THE
NATIONAL AERONAUTICS AND SPACE ADMINISTRATION'S
ASTROPHYSICS THEORY PROGRAM

In response to NRA 93-OSS-06

Proposing Organization: North Carolina State University

Principal Investigator: Stephen P. Reynolds, Department of Physics

December 7, 1994

Stephen P. Reynolds
Associate Professor of Physics, North Carolina State University
Principal Investigator

Kazimierz J.B. Borkowski
Research Associate, Department of Astronomy, University of Maryland
Co-Investigator

John M. Blondin
Assistant Professor of Physics, North Carolina State University
Co-Investigator

TABLE OF CONTENTS*

* The complete table of contents has been reprinted here to convey the full structure of the proposal package, but only the main body of the proposal argument has been reproduced.

I. INTRODUCTION

Supernovae and their remnants are called upon to perform a wide range of duties in the galactic ecosystem. They seed the interstellar medium (ISM) with heavy elements; dump kinetic energy into turbulence; accelerate cosmic rays, which affect ionization levels in molecular clouds with profound consequences for astrochemistry and star formation; heat the "hot phase" of the interstellar medium; and perhaps trigger star formation by compressing clouds on the point of instability. Understanding these diverse responsibilities demands a detailed knowledge of the physics of supernova remnants (SNRs), from the gross hydrodynamics of the blast wave (and reverse shock, at early times) to the nonthermal physics of the shock waves.

X-ray emission provides perhaps the most direct and fundamental knowledge about the physics of SNRs. Its diagnostic values are many: X-ray intensity depends quadratically on the gas density, X-ray spectra are sensitive to electron temperatures, gas abundances and SNR ages, X-ray morphology presents us with information on the SNR structure (e.g., see Canizares 1990 for a brief introduction to X-ray emission from SNRs). X-ray evidence has provided some of the most fundamental information we have on supernovae and remnants: total energy of SN events, amount and composition of ejected material, supernova subtype: information essential to many areas of astrophysics beyond the study of SNRs for their own sake. This information is extracted from observations, however, in a highly model-dependent fashion. The current analysis of X-ray spectra relies heavily on one-dimensional self-similar solutions such a point blast explosion (Sedov) model or (more infrequently) on one-dimensional hydrodynamical simulations, coupled with nonequilibrium ionization calculations. An account of early work on X-ray emission from young SNRs is given by Chevalier (1988), with references to papers which describe the basic strategy for modeling X-ray observations in use today.

The major problem with using X-ray observations as the diagnostic of hot shocked plasmas has been the modest spectral and spatial resolution of X-ray satellites. This problem is being addressed by new observatories such as ROSAT and ASCA. These new high-quality X-ray observations mark a beginning of a new era in SNR research, to be developed further with the anticipated launch of AXAF in 1999. ROSAT observations of SNRs are just beginning to be analyzed, and spectral observations are being gathered by the ASCA satellite. Initial results indicate that these new data are providing qualitatively new information which in many cases cannot be explained within the framework of existing theoretical models. Once these data are analyzed and interpreted, a new, more complete picture will emerge of SNRs, their progenitors, and their interactions with the circumstellar and interstellar medium.

To take advantage of this observational breakthrough, intensive theoretical efforts in X-ray emission modeling are necessary. The spectral imaging data provided by ASCA will require multi-dimensional hydrodynamic simulations coupled with non-equilibrium ionization calculations. The authors of this proposal have developed just such a tool, which allows the calculation of X-ray spectra from multidimensional hydrodynamic simulations. One of this proposal's main objectives is to use this unique tool to generate a new class of detailed theoretical predictions of the X-ray morphology and spectrum of SNRs at various evolutionary phases, and to interpret new observations of young, X-ray bright SNRs with the help of these predictions.

However, X-ray emission alone is not a perfect tracer of SNR dynamics. Absorption effects and uncertainties about electron temperatures, or even the possibility of strongly non-Maxwellian electron distributions, can impair the diagnostic power of X-rays. Many of these problems can be addressed by combining radio information with X-ray information. While many remnants show radio emission beyond the confines of X-rays, as in W49 B (Moffett & Reynolds 1994) and other middle-aged remnants, there are no examples of X-rays found in the absence of radio emission; radio emission thus more reliably indicates at least the outer shock wave. Turbulence in remnant interiors probably amplifies magnetic field, so that one might expect synchrotron enhancements at the location of such turbulence which would allow dynamically unstable regions to be identified in observed objects. In addition, nonthermal effects, for instance collisionless heating of electrons at the shock, may influence X-ray morphology and spectra. Such collisionless heating may depend on the obliquity angle Θ_{Bn} between the shock normal and upstream magnetic field. The remnant of SN1006, with apparently lineless power-law X-ray spectra from some areas, and a bipolar morphology resembling the radio, provides an excellent example. The efficacy of processes such as electron heat conduction may depend on the degree of small-scale order in the magnetic field, which can be investigated with radio polarization information. Full use of X-ray spectra and images of SNRs requires concomitant study of radio emission.

Much is unknown about the physics responsible for producing radio emission, for instance the relative importance of shock acceleration and second-order ("stochastic") acceleration in producing the relativistic electron populations we infer, the detailed mechanisms involved in injection of thermal electrons into the shock acceleration mechanism, and the possible importance of magnetic-field amplification in determining SNR radio morphology. To use the radio information as efficiently as possible requires expertise in both radio observations and in the modeling of strong shocks and of turbulence.

Current models for analyzing X-ray spectra of SNRs often rest on extremely simple dynamical substrates which cannot accurately describe all facets of real SNR evolution. For example, it is still common practice to use Sedov-Taylor models when interpreting the X-ray emission of SNRs, despite the fact that many, if not most, SNRs are not well described by such a model. Other similarity solutions exist to describe the global dynamics of SNRs in the early two-shock phase (the "self-similar driven wave", or SSDW; Chevalier 1982), and such solutions have been used in the interpretation of X-ray emission. Furthermore, many young SNRs may actually be in a transition phase between these self-similar solutions, for which time-dependent hydrodynamic simulations must be used (e.g., Band & Liang 1988). More recently, advances have been made using computational hydrodynamics to study the evolution of SNRs in multidimensions, addressing the role of dynamical instabilities (Chevalier, Blondin, & Emmering 1992), and the interaction of SNRs with a non-spherical circumstellar medium (Borkowski, Blondin, & Sarazin 1992). However, a similar problem exists in the study of SNR dynamics; to advance the understanding of SNR dynamics, one tends to make simplistic assumptions about the emissivity of the gas in order to use, say, SNR X-ray images for comparison with hydrodynamic models.

It is time to develop a combined approach, putting the most advanced techniques of each of these three areas—X-ray emission, synchrotron emission, and global dynamics—into one research effort. We shall propose below to combine our well-developed programs in each of these areas to make such progress. We shall bring together sophisticated, multidimensional hydrodynamic simulations and advanced modeling of non-equilibrium ionization and X-ray emissivity to produce model X-ray images, to confront published and new SNR observations at X-ray wavelengths. Our model calculations will allow the production of spectra integrated over all or part of an image, and images integrated over arbitrary bandpasses, to compare with any satellite imaging or spectral information. Where necessary, we shall also produce model radio images, which can further constrain the process of abstracting astrophysically interesting information from observations. Without such a combined approach of individually state-of-the-art techniques, the next generation of understanding of SNRs, and all the galactic processes that depend on them, cannot be achieved.

The proposed research is timely—it is new observations of SNRs with the ASCA satellite, and to a lesser degree with the ROSAT satellite, which have been the primary cause of the current renaissance of SNR studies. We bring to this field the only tool capable of addressing the spectral imaging data being collected by the ASCA satellite, and one well able to meet the theoretical challenge of the AXAF mission.

We summarize the thrust of our proposed work in the following questions:

1. What role do instabilities play in mixing ejecta and interstellar material, forming clumps in the ejecta, generating turbulence, amplifying magnetic fields, and in creating the observed X-ray and radio morphology of various SNRs?
2. What are the chemical abundances in the shocked ambient medium and in the shocked SN ejecta as a function of position within the remnant? The latter question is of paramount importance in view of the SN role in enriching the interstellar medium (ISM) in heavy elements, and must be understood for the correct interpretation of spatially resolved spectra.
3. What is the nature of supernova progenitors and of their surroundings, as inferred from the geometry and dynamics of individual SNRs?
4. What are the major outlines of the hydrodynamical evolution of a supernova remnant as it interacts first with circumstellar material, then with possibly inhomogeneous interstellar medium? How do multidimensional treatments differ from one-dimensional (1D) analytic results? Are there distinctive X-ray, optical, or radio signatures of these evolutionary stages which will allow us unambiguously to identify the stage of any observed SNR?
5. How is the heating of electrons (producing X-ray emission) related to the acceleration of electrons (producing radio emission)? What information can radio observations provide on collisionless heating of electrons, an important issue for X-ray modeling? Where are energetic particles accelerated, and with what efficiency?

These questions must be addressed to attack such fundamental issues as the total energy input from SNRs into the ISM, and the proportions introduced as turbulent motions, thermal energy, fast particles, and magnetic field; the origin of cosmic rays; the microphysics of collisionless shock waves; the statistics of SN and SNRs, and the use of remnants to diagnose the state of the circumstellar medium (CSM) and ISM. Thus they reach beyond the particulars of understanding SNRs for their own sake, and are important for a wide range of astronomical applications: studying the Galactic energy budget, the nature of the ISM, the properties of SN progenitors, and the properties of cosmic rays.

II. PROJECT DESCRIPTION

A. Background

A complete understanding of SNR dynamics must include a self-consistent account of both the thermal plasma and relevant aspects of the physics of strong shock waves, both because of the influence of those aspects on the dynamical evolution, and because proper interpretation of X-ray and radio observations of SNRs, and hence testing of theory against those observations, cannot be accomplished without such understanding. We describe below the current state of research in X-ray and radio emission from SNRs and in the global dynamics of SNRs.

1. X-Ray Emission of SNRs

X-ray observations of tenuous hot plasmas have been used extensively to study physical conditions and chemical abundances in a variety of astrophysical objects, including SNRs and clusters of galaxies. This is possible because X-ray spectra of such objects are dominated by numerous emission lines, whose analysis can yield a wealth of information about the X-ray emitting gas. When such systems are in an equilibrium state (kinetic equilibrium and collisional ionization equilibrium), their X-ray emission properties depend only on the current physical state of the system, which may be deduced from the observations. The ages of young SNRs are much too short to allow them to achieve full ionization or kinetic equilibrium by the present time, implying that the present-day X-ray emission from these systems depends on their history, and coupling their emission and dynamics.

The previous decade has seen the development of methods to account for departures from equilibrium ionization and equipartition in simple, one-dimensional hydrodynamical models. X-ray spectra of young SNRs are now routinely analyzed with the help of one-dimensional nonequilibrium ionization models (Itoh 1977; Gronenschild and Mewe 1982; Shull 1982; Hamilton, Sarazin, and Chevalier 1983; Nugent et al. 1984; Hughes and Helfand 1985; Hamilton, Sarazin, and Szymkowiak 1986a, b; Itoh, Masai, and Nomoto 1988; see Kaastra & Jansen 1993 for recent developments in numerical techniques). Morphological analysis of SNR images is at a much less developed stage, though some work has been done (e.g., Pye et al. 1981; Reid, Becker, and Long 1982; Seward, Gorenstein, and Tucker 1983; Petre et al. 1982; Matsui et al. 1984; Hughes 1987; Jansen et al. 1988; Matsui, Long, and Tuohy 1988). It is straightforward to compute the radial brightness distribution in one-dimensional models (Kaastra & Jansen 1993 illustrate this procedure for Cas A SNR). While one-dimensional nonequilibrium ionization models have been adequate to interpret older X-ray observations of young SNRs in most cases, this certainly does not hold true for new high-quality observations gathered with the ASCA and ROSAT satellites. A good example is provided by Cas A where X-ray emitting material expands asymmetrically (Mushotzky 1994), suggesting a ring-like distribution of the SN ejecta and/or of the ambient CSM. A preliminary analysis of these ASCA data on Cas A reinforces the conclusion about the inadequacy of the current modeling efforts—it is impossible to model these data in the framework of existing nonequilibrium ionization models (Mushotzky 1994). This failure is most likely related to the use of vastly oversimplified, one-dimensional models for Cas A SNR. It seems improbable that these models can properly describe the complex geometry of this remnant. One-dimensional models are hardly adequate for an initially spherically-symmetric distribution of the ambient medium and the SN ejecta because the gas in SNRs is often subject to various hydrodynamical instabilities. This is one reason why one-dimensional models fail completely to account for the morphology of young SNRs: they produce thin X-ray emitting shells which have never been seen in any SNR. In general, one must calculate the ionization and kinetic state of the plasma in two- or three-dimensional hydrodynamical simulations in order to determine the X-ray emission. X-ray properties depend sensitively on the details of the flow, because of the quadratic dependence of the X-ray emissivity on gas density.

While multidimensional ionization calculations have been computationally prohibitive in the past, the situation is changing rapidly due to the availability of fast computers, vast data storage, and efficient algorithms. It is now possible to match rapid advances in X-ray astronomy with multidimensional hydrodynamics and ionization calculations. Our new technique for accomplishing this task, described in § IIB, is fully developed, and has already been applied to X-ray data on Kepler's SNR (Borkowski, Sarazin, & Blondin 1994). This novel technique will allow us to investigate a number of fundamental problems which had been intractable until now because of their multidimensional nature. The solution of these problems is necessary in order to interpret new X-ray observations of SNRs.

X-ray emission is dependent on the extent and location of electron heating. Indeed, the very observation of X-rays from SNRs directly implies the thermalization of electrons in strong shock waves. However, this thermalization process is not well understood. (Electron heating in all X-ray emission models, including our numerical code, is usually parameterized in a simple fashion, and then deduced by comparison with observations.) First, the collisionless shocks in SNRs directly heat ions only, leaving electrons to gain energy only through relatively slow Coulomb collisions. For young SNRs the Coulomb collision time scale is long enough that electrons become heated relatively slowly, and may never reach the ion temperature (Itoh 1977; Cox and Anderson 1982; Masai et al. 1988). In fact, for shock velocities greater than 2000 km s^{-1} as inferred for the historical SNRs, the ion temperature should be $3m_p v_s^2 /16k \sim 10^8$ K, which is not unambiguously observed. Furthermore, the X-ray morphology of many SNRs is puzzling. While some work has been done on modeling the fairly common situation of a radio shell but filled-center X-ray morphology (White and Long 1991), invoking evaporation by saturated conduction of small dense clouds in the ISM, the odd bipolar morphology of SN 1006 in X-rays (Pye et al. 1981), mirroring the radio structure (Reynolds and Gilmore 1986), seems to demand an influence in electron heating also manifested in nonthermal electron acceleration. Such an influence may be the operation of plasma instabilities that depend on the shock obliquity (Cargill and Papadopoulos 1988). The understanding of the physics of electron heating is important for the interpretation of X-ray images of SNRs and their comparison with dynamical simulations, as well as an important problem in plasma astrophysics in its own right.

2. Radio Emission of SNRs

SNR shocks are commonly assumed to produce the Galactic cosmic rays (see, e.g., Blandford and Eichler 1987 for a review) by first-order Fermi acceleration, though many details of the process are not well understood (Jones and Ellison 1991). The process must be intrinsically nonlinear, as SNRs must put 5%–10% or more of their energy into cosmic rays to explain the observations, so the cosmic rays can influence the shock structure, through both broadening the velocity profile (pre-accelerating the upstream fluid, in an SNR frame of reference) and changing the bulk fluid's adiabatic index and hence compressibility. Furthermore, the particles probably generate the Alfvén waves from which they also scatter, and the waves can damp into the thermal fluid. Many authors have considered these issues, separately and in combination, in the context of SNRs.

Our chief concern will be with the use of radio synchrotron radiation from SNRs to trace locations of shocks and of interior turbulence, and to give information on shock properties which may be relevant for understanding electron heating. However, in spite of the central position held by electron acceleration both to provide synchrotron tracing of the SNR dynamics, and to provide evidence for shock acceleration of ions, firm conclusions about the nature and location of the process are scarce. The observed synchrotron emissivities of older SNRs known to have cooling, high-compression shocks do not require any further electron acceleration beyond the compression of ambient cosmic-ray electrons and magnetic field (van der Laan 1962), though such acceleration may occur (Blandford & Cowie 1982). However, the inferred emissivities of brighter, adiabatic-phase remnants almost certainly require generation of new electrons or amplification (beyond mere compression) of magnetic field. Two mechanisms and locations are available for electron acceleration: in addition to shock acceleration, so-called second-order Fermi (stochastic) acceleration in the remnant interior, wherever levels of MHD turbulence are high, for instance at the unstable contact discontinuity between shocked ejecta and shocked ISM (Cowsik and Sarkar 1984) or just in the region behind the blast wave (Schlickeiser and Fürst 1989). Arguments based on morphology, reviewed in Reynolds (1988), Dickel et al. (1989), and Fulbright and Reynolds 1990, suggest that shock acceleration of electrons provides most or all of the required synchrotron emissivity , and various authors have invoked it, in varying degrees of detail (Bell 1978; Reynolds & Chevalier 1981; Heavens 1984a,b). However, shock acceleration at the blast wave is not unambiguously demanded by the observations, and has never been considered at the reverse shock.

Even if electron acceleration occurs primarily at shock waves, the synchrotron emissivity may still be dominated by magnetic-field effects, such as turbulent amplification at the contact discontinuity. It will be necessary to resolve such issues as the relative importance of electron and magnetic-field effects in SNR morphology in order to make use of radio data to constrain SNR physics. A start in this direction was accomplished by Jenkins, Blondin, and Reynolds (1993) who constructed predicted morphologies of synchrotron radiation from young remnants assuming a fixed fraction of turbulent energy density appearing as magnetic field, and with shock acceleration of electrons. The turbulence profiles were formed from suitable averages over 2-D hydrodynamic simulations (Chevalier, Blondin, and Emmering 1992). For a large range of values of the (unknown) aspect angle of the local magnetic field with respect to the line of sight, models showed a thin rim of emission at the shock and a thicker interior band of emission, resembling Tycho's supernova remnant (e.g., Dickel et al. 1991). More work along these lines should solidify our understanding of the non-thermal processes in SNRs, at least well enough to allow the use of synchrotron radiation as a powerful dynamical diagnostic.

3. Dynamics of SNRs

The dynamical evolution of young SNRs through various stages has been documented by 1D numerical and analytic models. Soon after the expanding shock wave breaks out of the progenitor star a reverse shock is formed where the remains of the star (the ejecta) collide supersonically with the CSM that has been shocked by the leading blast wave. At early times the density profile of the ejecta is well described by a steep power law in radius (e.g., Arnett 1988). If the ambient gas is also fit by a power law, as is the case for a progenitor stellar wind or a constant-density ISM, then the region between the reverse and forward shocks can be modelled with a self-similar solution (Chevalier 1982). This self-similar driven wave (SSDW) phase applies until either the shock runs into something (e.g., a wind-swept shell; Chevalier and Liang 1989) or the ejecta density profile flattens out (Band and Liang 1988). In the latter case the SNR will evolve into a Sedov-Taylor (ST) blast wave. In the former case the forward shock will penetrate the shell and break out into the (presumably) low-density interstellar medium. In either case the expansion velocity will eventually slow to the point where the recently shocked gas is cool enough to radiate a substantial fraction of its energy, and the SNR enters the radiative pressure-driven snow-plow (PDS) phase (Cioffi, McKee, and Bertschinger 1988). Eventually the remnant expansion becomes subsonic, and the SNR merges with the ISM.

This theoretical picture of SNR evolution is lacking in several respects. First, the dynamics of the transition phases (e.g., between the SSDW and the ST phases) have not been sufficiently explored. Secondly, all these results are for 1D, spherically symmetric models, thus ignoring the possibility of non-radial instabilities or evolution into a non-spherical CSM. Multidimensional calculations are needed to understand the role of instabilities in creating the X-ray and radio morphology of SNRs, in forming clumps in young SNRs, in mixing of ejecta with the surrounding medium, in the geometry and amplification of magnetic fields, and possibly in the generation of energetic particles. The ability to model non-spherical SNRs will also allow one to use SNRs as tracers of the ambient density structure, both in the immediate CSM and on the larger scale of several parsecs.

Virtually every phase of evolution, including the transition phases, can be shown to be dynamically or thermally unstable to non-radial perturbations. Gull (1973) has shown that the contact discontinuity separating shocked ejecta and shocked CSM in young SNRs is subject to a convective or Rayleigh-Taylor instability. Goodman (1990) and Ryu and Vishniac (1991) found that Sedov-Taylor blast waves are subject to a convective instability for sufficiently steep radial density profiles in the surrounding medium. The transition to the radiative PDS phase has been shown to be unstable to radial perturbations in numerical simulations (Falle 1981), and analytic work by Bertschinger (1986) suggests that this instability is also present in non-radial modes. The radiative PDS phase is itself subject to an oscillatory over-stability (Vishniac 1983).

These dynamical instabilities may have a profound effect on the observed morphology of SNRs. As an example, the X-ray image of Tycho's remnant apparently shows emission from clumps within a shell interior to the leading shock wave (Seward, Gorenstein, and Tucker 1983). The radio emission also shows a broad region of emission peaking inside the shock front (Dickel, van Bruegel, and Strom 1991), providing at least a suggestion that instabilities of the contact discontinuity are important in this remnant.

Non-radial instabilities may affect the global dynamics as well, changing the results of 1D solutions. If the instabilities can redistribute the entropy within the remnant, the global expansion rate may deviate from the spherically-symmetric solution. If this is the case, many commonly accepted inferences about SNR ages, energetics, and statistics may be inaccurate. A possible example of this situation may arise in the transition from the SSDW phase to the ST phase, where a large fraction of the interior may be unstable as it is swept up by the reverse shock after it reflects off itself near the center of the remnant (a process that itself is likely to be grossly distorted if spherical symmetry is assumed).

The studies of non-radial dynamical stability of SNRs mentioned above all rely on linear perturbation analysis. While such an analysis can provide valuable information on regions of instability and linear growth rates, it cannot address the question of nonlinear evolution and hence cannot explain the impact of the instabilities on the global evolution and structure of SNRs. To answer questions related to the non-linear evolution of dynamical instabilities one must turn to numerical simulation. Two recent examples of numerical studies of dynamical instabilities in SNRs are Chevalier, Blondin, and Emmering (1992; see Figure 1) and MacLow and Norman (1993). Both of these works confirmed the linear perturbation analysis of the relevant instability and went on to quantify the non-linear evolution.

The use of 1D spherically symmetric models is also a limitation in that many SNRs are distinctly non-spherical. In the case of young SNRs with massive progenitors, such objects provide information on the CSM surrounding the supernova. The evidence for an asymmetric distribution of CSM is provided by very young SNRs such as SN 1988Z (Chevalier & Fransson 1993), SN 1986J (Bartel, Rupen, and Shapiro 1989) or SN 1987A (Crotts, Kunkel, & Heathcote 1994). Somewhat older remnants with the asymmetric CSM include Cas A, Kepler's SNR, or N132D in the Large Magellanic Cloud.

In the case of older SNRs, large asymmetries may be tracing density gradients in the ISM. An excellent example is 3C 391 (Reynolds & Moffett 1993), where two levels of brightness gradient may reflect a complex structure in the surrounding medium. The bright shell is only about two-thirds complete, with the brightest region opposite the faintest, while a broad, faint plateau to the SE extends far beyond the shell. The asymmetry in the bright structure can be simply modeled with a picture of shock acceleration and magnetic field varying with external density, so that a density gradient in the NW direction in the external medium can produce a partial shell like that observed (Reynolds & Fulbright 1990). The broad extension may mark a "breakout" of the SNR into a region of considerably lower density. These somewhat speculative conclusions could be placed on a much firmer foundation by incorporating HD simulations into the model imaging.

Finally, scant effort has been invested relating the theoretical models to observations in a critical manner. While the self-similar solutions can be used to derive global parameters such as the age of a remnant, there is no way to check that the assumed evolutionary stage is in fact correct. Robust observational signatures of each stage are not available, making more detailed inferences difficult or impossible. X-ray spectral analysis has had the most success in determining important physical properties of SNRs; radio spectra are difficult to understand and have hardly been analyzed, since a nonlinear theory of electron acceleration is required (but see Reynolds & Ellison 1992). Since most SNRs have complex, confused morphology, imaging observations have been interpreted even less, though some work has been done (see below). Here again, however, basic observational signatures of important features such as the reverse shock remain undetermined. We shall work to correct this major gap between theory and observations, so that the observations can in fact serve as a useful check on the theoretical work.

B. Proposed Methodology

We propose a comprehensive investigation of the structure and evolution of SNRs through the various stages outlined above, emphasizing how dynamical instabilities affect the structure and evolution of SNRs, and how these various stages may be discerned from X-ray and radio observations. At the same time, we shall study the processes by which electrons are accelerated and heated in SNRs, as intrinsically important physics and as essential to relate dynamical calculations to observations. Our project contains two main ingredients: hydrodynamic simulations to study the dynamical evolution of SNRs, and model images and spectra in X-rays and radio to critically compare theoretical results with observations. One of us (SPR) will continue work (funded elsewhere) on radio and X-ray observations of SNRs which will complement this project very well.

i. Hydrodynamic Simulations

The first ingredient of our investigation is numerical hydrodynamic simulations. The primary workhorse for the proposed computations will be VH-1, a hydrodynamics code developed by two of us (JMB and KJB) and colleagues at the University of Virginia. This code has been thoroughly tested and successfully applied to a variety of problems in astrophysics. VH-1 is based on the piece-wise parabolic method of Colella and Woodward (1984) and includes radiative cooling, external forces, and several orthogonal grid geometries. The third-order accuracy in the spatial representation of the fluid variables allows this code to resolve flow patterns on significantly cruder grids than would otherwise be required with other types of numerical methods. The use of a Riemann solver at zone interfaces insures accurate shock jump conditions and alleviates the need for artificial viscosity, which broadens shocks and may dampen the growth of the dynamical instabilities under study. These properties of VH-1 make it an ideal code for the study of dynamical instabilities in supersonic flows, as demonstrated in Chevalier, Blondin, and Emmering (1992).

Computational hydrodynamics will allow us to study the full evolution of dynamical instabilities as well as the evolution of asymmetric SNRs. Because of the minimal amount of dissipation in the PPM algorithm, we can accurately study the growth of instabilities in the linear regime where we can make contact with linear analysis and quantify linear growth rates. We can also study the non-linear evolution of instabilities with computational hydrodynamics, and thereby assess their impact on the structure and evolution of SNRs. Multidimensional simulations can also address the question of mixing between ejecta and the ISM with these techniques.

ii. Model Images and Spectra

We bring a well-developed formalism to bear here in both the X-ray and radio wavebands. To calculate X-ray spectra from hydrodynamical simulations, Borkowski, Sarazin & Blondin (1994) have devised a method for calculating the nonequilibrium ionization and kinetic state for the gas in two-dimensional hydrodynamical flows. This is the only fully developed code that we know of that can do this. Its construction took a considerable amount of effort because of a number of difficult technical problems associated with the complexity and the scale of the problem. Briefly, the idea is to perform hydrodynamical calculations first, storing pressure, density, and Lagrangian "labels" at intermediate times in "history" files. These labels are set to initial coordinates of fluid particles at the start of simulations, and then advected with the fluid across the computational grid with the help of the hydrodynamics code VH-1. The information stored at intermediate times is sufficient to reconstruct how pressure and density varied for each fluid element located at each computational cell at the final time. This reconstruction uses an efficient method of interpolation on irregular grids. It is then possible to find electron temperatures and nonequilibrium ionization for each fluid element, the latter with an efficient algorithm for computing nonequilibrium ionization developed by Hughes & Helfand (1985). In this form, this technique is applicable to nonradiative flows where gas cooling through X-ray emission does not affect the dynamics of the flow, allowing for separate hydrodynamical and ionization calculations. The nonradiative condition is satisfied in young SNRs. The final step involves calculations of X-ray spectra and images, which again must be computationally efficient because of a large number of grid cells required in a typical hydrodynamical simulation of a SNR.

We are hopeful that strong constraints on the structure of the CSM, on ejecta abundances, and on blast-wave properties can emerge from the comparison of these sophisticated model X-ray images and spectra with observations from *Einstein*, ROSAT, ASCA, AXAF, and other future X-ray imaging instruments. We will have the ability to produce predicted images in any instrumental bandpass, including all important spectral effects, at a level far beyond what has been accomplished in the past. This will allow us to make effective use of previously obtained images, as well as to anticipate results of upcoming observations. It will always be possible, of course, to integrate over images to obtain total spectra for comparison with observations of unresolved SNRs.

We have also been successful in developing model radio imaging techniques. Model radio synchrotron images of SNRs have been used to study obliquity-dependence of electron acceleration (Fulbright and Reynolds 1990), appearance of SNRs in a density gradient (Reynolds and Fulbright 1990), and preshock electron diffusion and synchrotron "halos" (Achterberg, Blandford, and Reynolds 1994; Reynolds 1994). In addition, work has begun on the contribution of turbulent amplification of magnetic fields to observed morphology of SNRs (Jenkins, Blondin, and Reynolds 1993). Our fully developed code can be immediately adapted for different dynamical substrates, and for different prescriptions for the sources and evolution of relativistic electrons and magnetic field. Additional near-term projects include allowing for additional electron acceleration at locations of strong turbulence, and examining the degree of order in the magnetic field with model polarization images, including internal Faraday depolarization effects as well as intrinsic disorder in the magnetic field. We hope by these means to set limits on the location, and fractional turbulent energy, in relativistic electron acceleration and magnetic-field amplification, and to constrain the possible extent of electron heat conduction.

iii. *Comparison with Observations*

Model X-ray images and spectra for an individual SNR must be compared with X-ray observations of this remnant. This task can be accomplished with the recently established High Energy Astrophysics Science Archive Center (HEARSAC) Facility at NASA/GSFC. Archived X-ray data can be easily accessed through the HEARSAC On-line Service. Many ROSAT observations are already available to the general astronomical community, and ASCA observations from the performance verification phase, including many bright SNRs, will become available in 1995 January. This date nearly coincides with the anticipated starting date for this project. We will compare our models with observations with the help of the XANADU (XSPEC, XIMAGE) software package supported by the HEARSAC, as we had already done for Kepler's SNR (Borkowski et al. 1994).

In order to ensure the most timely and efficient interpretation of X-ray observations of individual SNRs in terms of our two-dimensional models, we plan on close collaboration with X-ray observers. This would address instrumental and calibration problems associated with X-ray data, would allow us to concentrate our attention on the most important issues, and would keep us in close contact with X-ray observers. For example, existing data on Cas A SNR are quite extensive and the most recent observations with the ASCA satellite are likely to be subject to poorly-understood instrumental problems so early in the satellite's lifetime. We intend to work on this remnant with A. Szymkowiak from the NASA/GSFC, who was himself involved in recent BBXRT observations of this remnant. Moderately frequent visits to the NASA/GSFC are anticipated in order to foster this collaboration.

C. Proposed Work

In spite of a great deal of observational and theoretical effort, the connection between the theoretically calculated hydrodynamic evolution of SNRs, and the spectral and morphological properties of objects observed in X-rays, radio, and other wavebands is far from clear. Without a combined approach, the hydrodynamical modeling cannot be validated (or refuted), much less used to infer secondary properties such as SN energetics and the nature of surrounding material. We propose to use the twofold approach described above to advance the current understanding of SNRs and to provide a more physical model with which to interpret new high-quality X-ray observations. This work will take on two direct lines of attack: modelling individual SNRs in X-ray and radio to infer specific properties of those systems, and a systematic modelling of the continuum of evolutionary stages of SNRs. The first tack allows us to extract as much information as possible from a given set of observations. The second tack will build a new theoretical framework for the future interpretation of X-ray and radio observations. In all this work, the combination of state-of-the-art hydrodynamical modeling and sophisticated X-ray and radio visualization of the dynamical results is essential for success. The best hydrodynamical calculations are useless without the ability to predict the resulting appearance of objects to the X-ray or radio observer.

1. Modelling Individual SNRs

One of the main objectives of this proposal is to simulate X-ray spectra and X-ray images of young, X-ray bright SNRs, and compare them with most recent high-quality X-ray observations, such as imaging spectroscopy with the ASCA satellite. It is best to illustrate our strategy for addressing these issues by discussing a couple of well-known SNRs. The youngest (about 320 yr old) galactic SNR, Cas A, is a good example because of the availability of good-quality data from radio to γ-rays. Another historical SNR, SN 1006, is an ideal test case for studying electron heating and particle acceleration processes in view of its striking bipolar morphology in both X-rays and radio. Still another extremely interesting SNR is SN 1987A, not discussed below, but which will become a showcase at the end of this century. One of us (JMB) has already theoretically investigated the formation of its ring of circumstellar material—the next logical step is to study the interaction of the SN ejecta with the CSM, the outcome of which is of considerable interest to the astronomical community.

i. Cassiopeia A

There is overwhelming evidence that the progenitor of Cas A was a massive star, the most compelling being the detection of the radioactive ^{44}Ti by the Gamma-Ray Observatory (Iyudin et al. 1994). Recent spectroscopic data (Fesen & Becker 1991) indicate that the SN progenitor was a massive stellar core composed mainly of He, with a thin hydrogen layer at the surface. Most of the H-rich envelope was expelled from the star shortly before the SN explosion. This envelope is now seen in the remnant through its interaction with the SN ejecta.

The shell-like X-ray morphology of Cas A consists of a bright clumpy shell and outer, weaker shell-like emission (Fabian et al. 1980; Jansen et al. 1988). The remnant's morphology is strongly asymmetric—optical and X-ray spectroscopy reveal that the clumpy shell is actually a ring inclined to the line of sight (Markert et al. 1983). New data from the ASCA satellite provided images in various emission lines, high-quality spectra in various locations within the remnant, with over a dozen of emission lines detected at each location. Position-dependent Doppler shifts have been found, attesting to the asymmetric nature of the remnant (Mushotzky 1994).

X-ray modeling of Cas A has been limited to very simple models; for example, Jansen et al. (1988) interpreted EXOSAT data in terms of a simple point-blast explosion (Sedov) model. In view of the complex remnant morphology, it is doubtful that such models are adequate to describe even pre-ASCA observations. It is obvious that a major theoretical effort is required to interpret new X-ray data on Cas A, surpassing anything that has been done before.

Our approach is straightforward. We propose to start with a supernova ejecta structure expected for an exploded stellar core with a thin H envelope. We would then impose an appropriate density structure for the ambient medium, which would involve an asymmetric distribution of the circumstellar medium (CSM), possibly in a form of a ring such as seen in SN 1987A. We would then follow the interaction of SN ejecta with the ambient CSM and ISM with the help of the 2-D nonequilibrium-ionization hydrodynamical code described in § IIB. At the current remnant's age, we would calculate X-ray images and X-ray spectra, and compare them with observations. By matching the observed X-ray morphology of the remnant in various spectral features, it should be possible to determine the geometry of the remnant, locate the shocked ejecta, the shocked CSM shell and find the primary shock (blast wave). Chemical abundances can be determined by fitting model spectra to observed spectra at each location within the remnant. The amount of electron heating in collisionless shocks can be found by analyzing the slope and intensity of high-energy X-ray continuum throughout the remnant.

Hydrodynamical instabilities might be very important in determining the remnant's morphology. This is suggested by the work of Chevalier, Blondin & Emmering (1992) who studied the development of a Rayleigh-Taylor instability for the reverse-shock solutions in spherical geometry. They found that the apparent thickness and clumpiness of various large-scale emission features can be modeled in the framework of hydrodynamical instabilities. A similar numerical study can be also done for Cas A SNR with a two-dimensional hydrodynamical code. It could then be possible to interpret the observed morphology (such as the shell thickness) in terms of hydrodynamical instabilities.

ii. SN 1006

SN 1006 was the brightest supernova recorded in history (Clark and Stephenson 1977), and has produced an extremely interesting remnant. Its height above the Galactic plane of almost over 400 pc probably accounts for both its faintness at radio and X-ray wavelengths and its extremely regular, bipolar morphology in both spectral regions (Reynolds and Gilmore 1986; Pye et al. 1981), because of a presumably lower-density, more orderly ambient medium than is true immediately in the Galactic plane. SN 1006 is customarily classified as Type Ia, because of the large z-height and the implied large peak brightness of the SN. Extensive searches have failed to turn up any evidence of a collapsed object at the center, and spectral work (e.g., Hamilton, Sarazin, and Szymkowiak 1986a) seems consistent with a Type Ia origin. Optical emission is confined to a few faint Hα filaments in one region of the periphery, classing SN 1006 with Kepler, Tycho, and several Magellanic Clouds remnants as "Balmer-dominated" SNRs (Smith et al. 1991).

The shell X-ray morphology, with its two bright limbs and fainter regions between, strikingly resembles the radio, which has been explained as due to preferential shock acceleration of electrons in regions of particular obliquity Θ_{Bn}. This acceleration may also provide preferential electron heating to explain the coincidence of the radio and X-ray bipolar structures. The soft X-ray emission is less concentrated in a shell, with significant emission from the remnant interior (Pye et al. 1981). The X-ray spectrum has always been problematic: early work (Becker et al. 1980) suggested a lineless power-law as the best fit, but later observations (summarized in Hamilton, Sarazin, and Szymkowiak 1986a; HSS) disagreed with the slope of this power-law and provided evidence for weak lines below 1 keV. Very recent ASCA data (summarized by J. Hughes, January 1994 AAS Meeting, Washington, DC) indicate the spectra from the fainter regions of the remnant show perfectly normal line spectra, while those from the brightest limbs persist in exhibiting a lineless power-law of energy index about −1.8.

The puzzling X-ray spectrum was modeled by Reynolds and Chevalier (1981) as the extension of the radio synchrotron spectrum, steepened by losses. This picture requires accelerating electrons to 10^{12} eV or higher near the shock, after which they diffuse into a region of higher magnetic field somewhat further behind the shock. The new spectral slope reported by ASCA causes serious problems for this simple picture. An elaborate non-equilibrium ionization thermal model was put forward by HSS, assuming the progenitor was a white dwarf. This picture used a particular 1-D analytic dynamical model of a forward and reverse shock, and explained the lack of lines by a strict segregation of ejecta by radius, so that the X-ray continuum was produced in an outermost carbon-rich layer of dense X-ray emitting gas. The model predicted a rather high expansion rate, marginally at odds with radio observations of the expansion (Moffett, Goss, and Reynolds 1993). In addition, while HSS did not make detailed morphological predictions, they mention the difficulty of explaining the interior soft X-ray emission with their model.

The resolution of these problems seems to be at hand with the new ASCA observations. Because of the anticipated pivotal role of SN 1006 in understanding of electron heating and particle acceleration in SNRs, a thorough interpretation of X-ray observations of this remnant is desirable. Although SN 1006 appears more homogeneous on small scales than other SNRs, radio and X-ray observations still show a fair amount of clumpiness. On theoretical grounds, hydrodynamical instabilities must have been operating in this remnant during its lifetime (Chevalier, Blondin, & Emmering 1992). Our goal is to construct a hydrodynamical model which would include these instabilities, as well as electron heating at strong shocks, possibly dependent on the magnetic field inclination to the shock normal in order to reproduce the observed bipolar morphology. (We are also well equipped to study a hypothetical contribution to X-ray emission from relativistic electrons if necessary.) We will take the structure of expanding SN ejecta appropriate for an exploding white dwarf, and study its interaction with the uniform ambient medium. We will be able to study the development of instabilities, estimate the amount of energy contained in turbulent motions, calculate radio and X-ray maps and position-dependent X-ray spectra. SN 1006 is clearly a prime target for our combined approach.

2. Evolutionary Stages of SNRs

To complement the work described above in which observations of individual remnants are analysed, we propose to study the various evolutionary stages of an SNR using the same techniques. Specifically, we will use hydrodynamic simulations in 1D and 2D to study the SSDW phase, the transition from SSDW to ST, and the transition from ST to a radiative SNR. These hydrodynamic simulations can then be coupled with X-ray emission calculations to investigate differences between 1D and 2D models and to search for important clues in the X-ray spectra that may betray the evolutionary status of a given remnant.

The end product of this line of research will be an X-ray catalog of simulated images and spectra covering an appropriate range in the parameter space, somewhat similar to a catalog constructed by Hamilton, Sarazin, & Chevalier (1983) for the Sedov solution. Because of the availability of standard X-ray software packages such as XANADU, the emphasis will be on a creation of a small computer database compatible with the XSPEC and XIMAGE packages within XANADU. Our results could then be used by X-ray observers to interpret new and archival observations of SNRs. The final form of such computer catalog will be decided upon successful completion of the goals outlined below.

i. Early Evolution: SSDW

The work of Chevalier, Blondin, and Emmering (1992) represents an important step in the study of SNRs dynamics, and we propose to take advantage of this pioneering work by extending it in several directions. The first immediate step will be the calculation of the X-ray emission to study the effects of the dynamical instabilities on the resulting X-ray spectra and morphology. In particular, dynamical instabilities can have the effect of broadening the region of X-ray emission, producing a more extended region of X-ray emission compared to the relatively thin shell of X-rays implied by 1D models.

To complement the X-ray analysis, we will also create model radio images from the 2D hydrodynamic simulations. We will use these models to investigate the possibility that turbulent amplification of CSM magnetic field can enhance the synchrotron emission from SNRs. The dynamical instability in the SSDW phase converts roughly 5% of the post-shock energy into turbulent motions. If one accepted the postulate that this turbulent motion can efficiently amplify an ambient magnetic field, one would expect a fairly large local magnetic energy density in a thick shell interior to the forward shock wave.

We propose to extend these simulations of the SSDW phase to a study of young SNRs for which the reverse shock is radiatively cooling. This situation may be relevant to a class of SNRs that are believed to be propagating into very dense circumstellar material resulting in very high densities in the shocked ejecta, and hence short cooling times (Chevalier 1990). An interesting example of this class of young SNRs is SN 1986J, the brightest SN observed in the radio. This object has been mapped in the radio using VLBI, and shows strong deviations from spherical symmetry (Bartel, Rupen, and Shapiro 1989). These observations suggest either an initially very non-uniform beginning or a very strong non-radial instability. Could dynamical instabilities be sufficiently strong to distort the forward shock wave? While the work on instabilities in the adiabatic SSDW phase suggests that the convective instability is confined to the interior of the forward shock, it may be possible that strong cooling at the reverse shock can enhance the instability by increasing the entropy contrast in the shocked shell. This scenario can be readily tested with the present HD code by including an appropriate energy loss rate for the shock-heated ejecta.

Another investigation of the SSDW phase is the structure of an SNR evolving into an asymmetric CSM. A direct question to be answered by this work is "What magnitude of asymmetry is required in the CSM to produce observed asymmetric SNRs?" Such a question is more directly applicable to "old" supernovae (or very young SNRs) both on an observational basis and on theoretical grounds that the immediate CSM of a SN may be much more asymmetric than the surrounding ISM. As an example, the interpretation of SN 1988Z given by Chevalier and Fransson (1993) requires an extremely large ambient density asymmetry, with the polar region being 10^3 times less dense than the equatorial region. However, if the asymmetry leads to significant non-radial flow, as would be revealed by multidimensional HD simulations, then such a large asymmetry might not be required. This project represents a very straightforward integration of the simulations in Chevalier, Blondin, and Emmering (1992) and the simulations of stellar wind bubble evolution into an asymmetric CSM in Blondin and Lundqvist (1992). This work would apply to many astrophysical systems, including the SN explosion of a Be star with a dense equatorial wind, or the SN explosion of a red supergiant in a moderately close binary system.

ii. Transition from SSDW to ST

The transition from the SSDW phase to the ST phase has not been investigated in detail, and yet it is expected to be relevant to many observed SNRs because of the length of time a given SNR spends in this transitional phase. Band and Liang (1988) used 1D numerical HD to study this phase, but their computational techniques produced considerable numerical noise, they considered only 3 specific models, and they did not evolve their models long enough to see some of the more interesting effects. We have begun preliminary work using 1D HD simulations to quantify the behavior of SNRs in this phase. We have found that for SNRs propagating into a preexisting stellar wind this transition is very slow, lasting as much as a thousand times longer than the SSDW phase. For SNRs propagating into a constant density ambient medium, this transition is much quicker and more violent. The reverse shock is driven into the center by the high pressure in the shocked shell, reaching the center roughly 7 times the age at which the turn-over in the ejecta density profile reaches the reverse shock (defining the beginning of the transitional phase). After bouncing off itself at the center of the SNR, this reverse shock then travels outward, re-accelerating the shocked ejecta to a velocity such that it expands with the self-similar rate of the expanding forward shock. From then on the ejecta occupy a region of roughly one-tenth the volume of the entire SNR. This 1D result in and of itself is very exciting, but the evolution will be even more interesting in 2D when the effects of this reflected reverse shock on the hydrodynamic instabilities are included, and when the "bounce" of the reverse shock is treated realistically. Perhaps the interaction of the reflected shock and the clumpy ejecta could enhance the local X-ray emission leading to a center-brightened X-ray SNRs.

We propose to expand this 1D investigation of the transitional phase to include predictions of the X-ray and optical morphology of such a SNR, and to extend the HD calculations to 2D. The study of this transitional phase in 2D will be a straight forward extension of the work described in Chevalier, Blondin, and Emmering (1992), with the exception that all of the interior of the blast wave must be included in the evolution. This will necessitate the use of a cylindrical grid in order to avoid prohibitively small time steps in the time-explicit HD calculations. For the constant ambient density case the instability properties may be substantially different from the SSDW phase. First of all, the rapid deceleration of the shocked shell may enhance the growth rate of the convective instability studied in the SSDW phase. Secondly, the passage of the reflected reverse shock may re-invigorate the instability growth through the action of the Richtmyer-Meshkov instability.

iii. Transition from ST to Radiative PDS

The onset of radiative cooling in an expanding SNR has been shown to induce large-amplitude radial perturbations (Falle 1981), which is likely to effect the global morphology of older SNRs. This is an example of the global thermal overstability of radiative shocks first studied analytically by Chevalier and Imamura (1982) and later examined with the aid of 1D numerical simulations by, among others, Imamura, Wolff, and Durisen (1984). It is possible that this overstability is enhanced with the addition of non-radial wave modes.

Bertschinger (1986) used a 3D linear stability analysis to investigate the role of non-radial perturbations in the evolution of a radiative spherical shock wave. He found slight differences in 3D compared to the 1D results, but again, the linear analysis does not provide information on the end state of the overstability. Recently, Strickland & Blondin (1994) have applied hydrodynamic simulations to study this overstability in 1D and 2D. This work has shown that the back of the radiative shock—where the shock-heated gas has cooled down to the preshock temperature—develops strong non-radial structure with a wavelength of order the length of the radiative shock.

We propose to continue this multidimensional study of the thermal overstability of radiative shock waves, extending the recent non-linear simulations to spherical geometry with specific application to SNRs. In particular, we will study the effect of this overstability in multidimensions during the transition to the radiative phase, when the wavelength of instability is of order the radius of the remnant. As an example of the kinds of investigations that we can accomplish with relatively little effort, we will examine the evolution of a SNR entering this transition phase in a slightly non-uniform environment. If one side of the SNR begins to cool slightly ahead of the other side, a strong asymmetry—much larger than the asymmetry imposed by the surrounding medium—could develop in the SNR. One of the goals of this work will be to find observational signatures that can distinguish between the different possible sources of structure in older SNRs. For example, the filamentary structure observed in many SNRs has been attributed to shocked clouds in the ISM, local thermal instabilities in the radiative shock of an SNR (Blondin & Cioffi 1989), the global thermal instability to be studied here, and the dynamical thin-shell instability discovered by Vishniac (1983) and studied numerically by Mac Low & Norman (1993).

While this work is less relevant to X-ray emission (the bulk of the X-ray flux from these older SNRs will come from the interior of the remnant, away from the interesting hydrodynamics), it is directly applicable to high-resolution optical observations using the *Hubble Space Telescope*. Such observations will provide detailed information on the scale and structure of the shock front, relating directly to dynamical instabilities we intend to investigate.

D. Institutional Facilities

Crucial to the success of the computational projects described in this proposal is continuing access to large allocations of supercomputer time. One of us (JMB) has maintained access to such large allocations through the North Carolina Supercomputing Center (NCSC—conveniently located only twenty miles from campus). The current annual allocation to JMB is over 900 cpu hours on a Cray YMP. This allotment is extended some 40% by running large jobs during non-prime hours. At the time of this writing NCSC is planning on upgrading to a next-generation supercomputer, possibly the new Cray Triton. Whatever the upgrade, the state of North Carolina appears committed to providing state-of-the-art supercomputing to its residents. NCSC is also a vital resource in the field of scientific visualization. This ready access to supercomputing and visualization at NCSC is an indispensable resource supporting the PI, co-I's and students in the activities here proposed.

Another essential element is local workstations that possess intermediate-level processing power and substantial amounts of disk space for down-loading and post-processing data from supercomputer simulations. The astrophysics group at NCSU maintains a cluster of 5 UNIX workstations, 4 of which provide better than 15 Mflops of computing speed each. One of these workstations also possesses advanced 3D graphics hardware for visualization and analysis of 3D data. This will be particularly advantageous for the proposed research, in that we will be able to rapidly produce volume rendered images of the X-ray emission predicted from multidimensional simulations. Substantial disk space (over 9Gb) and archiving (Exabyte and DAT) are also available, although significantly more disk space will be added for the research proposed herein due to the extremely large datasets generated by such work. These local facilities represent an institutional commitment toward the development of a computational astrophysics laboratory at NCSU.

III. BIBLIOGRAPHY

Achterberg, A., Blandford, R. D., & Reynolds, R. P. 1993, AA, submitted

Arnett, W. D. 1988, ApJ, 331, 337

Band, D. L., & Liang, E. P. 1988, ApJ, 334, 266

Bartel, N., Rupen, M., & Shapiro, I. I. 1989, ApJ, 337, L85

Becker, R. H., Szymkowiak, A. E., Boldt, E. A., Holt, S. S., & Serlemitsos, P. J. 1980, ApJ, 240, L33

Bell, A.R. 1978, MNRAS, 182, 443

Bertschinger, E. 1986, ApJ, 304, 154

Blandford, R. D., & Cowie, L. L. 1982, ApJ, 260, 625

Blandford, R. D., & Eichler, D. 1987, Physics Reports, 154, 1

Blondin, J. M., & Lundqvist, P. 1993, ApJ, 405, 337

Borkowski, K. J., Blondin, J. M., & Sarazin, C. L. 1992, ApJ, 400, 222

Borkowski, K. J., Sarazin, C. L., & Blondin, J. M. 1994, ApJ, 429, in press

Canizares, C. R. 1990, in Physical Processes in Hot Cosmic plasmas, ed. W. Brinkmann, A.C. Fabian & F. Giovanelli (Dordrecht: Kluwer), 17

Cargill, P. J., & Papadopoulos, K. 1988, ApJ, 329, L29

Chevalier, R. A., 1982, ApJ, 258, 790

Chevalier, R. A. 1988, in Supernova Remnants and the Interstellar Medium, ed. R.S. Roger & T.L. Landecker (Cambridge University Press: Cambridge), 31

Chevalier, R. A. 1990, in Supernovae, ed. A. G. Petschek (Berlin: Springer), p.91

Chevalier, R. A., Blondin, J. M., & Emmering, R. T. 1992, ApJ, 392, 118

Chevalier, R. A., & Fransson, C. 1993, preprint

Chevalier, R. A. & Imamura, J. N. 1982, ApJ, 261, 543

Chevalier, R. A., & Liang, E. P. 1989, ApJ, 344, 332

Cioffi, D. F., McKee, C. F., & Bertschinger, E. 1988, ApJ, 334, 252

Colella, P., & Woodward, P. R. 1984, J. Comp. Phys., 54, 174

Cowsik, R., & Sarkar, S. 1984, MNRAS, 207, 745

Cox, D.P., and Anderson, P.R. 1982, ApJ, 253, 268

Crotts, A. P. S., Kunkel, W. E., & Heathcote, S. R. 1994, ApJ, in press

Dickel, J. R., Eilek, J. A., Jones, E. M., & Reynolds, S.P. 1989, ApJS, 70, 497

Dickel, J. R., van Bruegel, W. J. M., & Strom, R. G. 1991, AJ, 101, 2151

Fabian, A. C., Willingale, R., Pye, J. P., Murray, S. S., & Fabbiano, G. 1980, MNRAS, 193, 175

Falle, S. A. E. G. 1981, MNRAS 195, 1011

Fesen, R. A., & Becker, P. 1991, ApJ, 371, 621

Fulbright, M.S., & Reynolds, S.P. 1990, ApJ, 357, 591

Goodman, J. 1990, ApJ, 358, 214

Gronenschild, E.H.B.M., and Mewe, R. 1982, A&AS, 48, 305

Gull, S. F. 1973, MNRAS, 161, 47

Hamilton, A. J. S., Sarazin, C. L., & Chevalier, R. A. 1983, ApJS, 51, 115

Hamilton, A. J. S., Sarazin, C. L., & Szymkowiak, A. E. 1986a, ApJ, 300, 698

Hamilton, A. J. S, Sarazin, C. L., & Szymkowiak, A. E. 1986b, ApJ, 300, 713

Heavens, A. F. 1984a, MNRAS, 210, 813

Heavens, A.F., 1984b, MNRAS, 211, 195

Hughes, J. P., & Helfand, D. J. 1985, ApJ, 291, 544

Hughes, J.P. 1987, ApJ, 314, 103

Imamura, J. N., Wolff, M. T., & Durisen, R. H. 1984, ApJ, 276, 667

Itoh, H. 1977, PASJ, 29, 813

Itoh, H., Masai, K., and Nomoto, K. 1988, ApJ, 334, 279

Iyudin, A. F., Diehl, R., Bloemen, H., Hermsen, W., Lichti, G. G., Morris, D., Ryan, J., Schönfelder, V., Steinle, H., Varen-dorff, M., deVries, C., & Winkler, C. 1994, A&A, 248, L1

Janesen, E., Smith, A., Bleeker, J. A., de Korte, P. A. J., Peacock, A., & White, N. E. 1988, ApJ, 331, 949

Jenkins, G. C., Blondin, J. M., & Reynolds, S. P. 1993, BAAS, 25, 1421

Jones, F.C., & Ellison, D.C. 1991, Ap&SS, 58, 259

Kaastra, J. S., & Jansen, F. A. 1993, AAS, 97, 873

Kesteven, M.J., & Caswell, J.L., 1987, A&A, 183, 118

Mac Low, M.-M., & Norman, M. L. 1993, ApJ, 407, 207

Markert, T. H., Canizares, C. R., Clark, G. W., & Winkler, P. F. 1983, ApJ, 268, 134

Masai, K., Hayakawa, S., Inoue, H., Itoh, H., & Nomoto, K. 1988, Nature, 335, 804

Matsui, Y., Long, K. S., Dickel, J. R., and Greisen, E. W. 1984, ApJ, 287, 295

Matsui, Y., Long, K.S., and Tuohy, I.R. 1988, ApJ, 329, 838

Moffett, D. A., Goss, W. M., & Reynolds, S. P. 1993, AJ, 106, 1566

Moffett, D. A., & Reynolds, S. P. 1994, ApJ, in press

Mushotzky, R. 1994, in the proceedings of a symposium held at STScI in 1994 May, The Analysis of Emission Lines, in press

Nugent, J.J., Pravdo, S.H., Garmire, G.P., Becker, R.H., Tuohy, I.R., & Winkler, P.F. 1984, ApJ, 284, 612

Petre, R., Canizares, C. R., Kriss, G. A., and Winkler, P. F. 1982, ApJ, 258, 22

Pye, J. P., et al. 1981, MNRAS, 194, 569

Reid, P. B., Becker, R. H., and Long, K. S. 1982, ApJ, 261, 485

Reynolds, S. P.: 1988, in Galactic and Extragalactic Radio Astronomy, eds. G. L. Verschuur and K. I. Kellermann, Springer-Verlag, New York, p. 439.

Reynolds, S. P. 1994, ApJS, 90, 845

Reynolds, S.P., & Chevalier, R.A. 1981, ApJ, 245, 912

Reynolds, S.P., & Ellison, D.C. 1992, ApJ, 399, L75

Reynolds, S.P., & Fulbright, M.S. 1990, in Proc. 21st Int. Cosmic Ray Conf. (Adelaide), 4, 72

Reynolds, S.P., & Gilmore, D.M. 1986, AJ, 92, 1138

Reynolds, S.P., & Moffett, D.A. 1993, AJ, 105, 2226

Ryu, D. S., & Vishniac, E. J. 1991, ApJ, 368, 411

Schlickeiser, R., & Fürst, E. 1989, AA 219, 192

Seward, F. D., Gorenstein, P., & Tucker, W. 1983, ApJ, 266, 287

Shull, J.M. 1982, ApJ, 262, 308

Smith, R. C., Kirshner, R. P., Blair, W. P., & Winkler, P. F. 1991, ApJ, 375, 652

van der Laan, H. 1962, MNRAS, 124, 125

Vishniac, E. T. 1983, ApJ, 274, 152

White, R. L., & Long, K. S. 1991, ApJ, 373, 543

Bibliography

· ·

· · · · · · · · · · · · · ·

Works Cited

Abbott A. 1996. Scientists lose cold fusion libel case. Nature 380:369.

Adams JL. 1976. Conceptual blockbusting: a pleasurable guide to better problem solving. New York: W. W. Norton.

[AIP] American Institute of Physics. 1990. AIP style manual. 4th ed. New York: American Institute of Physics.

Altman LK. 1994, February 10. Antimicrobial drugs endorsed for ulcers in a major U.S. shift. New York Times; Sect A:1, D:21.

Amato I. 1993. Pons and Fleischmann redux? Science 260:895.

Angier N. 1992 May 10. Women still scarce on membership rolls of science academy. News and Observer (Raleigh, NC); Sect A:18.

Anholt RRH. 1994. Dazzle 'em with style: the art of oral scientific presentation. New York: Freeman and Company.

Aristotle. 1954. The rhetoric and the poetics of Aristotle. Rhys Roberts W, Ingram Bywater, translators. New York: Random House.

Associated Press. 1992 Sept. 5. Funding withheld for conference on genetics-crime link. News and Observer (Raleigh, NC); Sect A:7.

Associated Press. 1996 May 24. Quiz of adults discovers unscientific Americans. News and Observer (Raleigh NC); Sect A:9.

Bacon F. 1605, 1859. The advancement of learning. The works of Francis Bacon. Volume 3. Spedding J, Ellis RL, and Heath DD, editors. New York: Longman.

Bazerman C. 1983. Scientific writing as a social act: a review of the literature of the sociology of science. In: Anderson P, Brochmann R, Miller C, editors. New essays in technical and scientific communication: research, theory, and practice. Farmingdale, NY: Baywood. p 156–84.

Bazerman C. 1985. Physicists reading physics: Schema-laden purposes and purpose-laden schema. Written Communication 2(1):3–23.

Benditt J. 1995. Conduct in science. Science 268:1705.

Berkenkotter C, Huckin TN. 1995. Genre knowledge in disciplinary communication: cognition/culture/power. Hillsdale, NJ: Lawrence Erlbaum.

Biddle AW, Bean DJ. 1987. Writer's guide: life sciences. Lexington, MA: D. C. Heath and Company.

Blakeslee AM. 1994. The rhetorical construction of novelty: presenting claims in a letters forum. Science, Technology, and Human Values 19:88–100.

Boyd R. 1979. Metaphor and theory change: what is "metaphor" a metaphor for? In: Ortony A, editor. Metaphor and thought. London: Cambridge University Press. p 356–408.

Bradford A, Whitburn M. 1962 Analysis of the same subject in diverse periodicals: one method for teaching audience adaptation. Technical Writing Teacher 9: 58–62.

Brody H. 1996 September. Wired science. Technology Review [Online]. Available: http://web.mit.edu/afs/athena/org/t/techreview/www/tr.html.

Bronowski J. 1965. Science and human values. Rev. ed. New York: Harper and Row.

Brummett B. 1976. Some implications of "process" or "intersubjectivity": postmodern rhetoric. Philosophy and Rhetoric 9:21–51.

Burke, Kenneth. 1968. Counter-Statement. Berkeley: University of California Press.

Campbell JA. 1975. The polemical Mr. Darwin. Quarterly Journal of Speech 61:365–90.

Campbell KK. 1982. The rhetorical act. Bellmont, CA: Wadsworth Publishing Co.

Carey J. 1992 10 Aug. What Barry Marshall knew in his gut. Business Week 68–69.

Carnap R. 1950. Logical foundations of probability. Chicago: University of Chicago Press.

[CBE] Council of Biology Editors. 1994. Scientific style and format: the CBE manual for authors, editors, and publishers. 6th ed. New York: Cambridge University Press.

[CBE] Council of Biology Editors. 1995. Report from the annual meeting—1994. In [CBE] Views 18(1):14–16.

Chazin S. 1993 Oct. The doctor who wouldn't accept no. Reader's Digest 119–24.

Cohen J. 1995. The culture of credit. Science 268: 1706–11.

Cohen J. 1996 Jan 5. AIDS trials take on peer review. Science 271:20–21.

Cole KC. 1990 Jan. 7. Science under scrutiny. New York Times; Sect EDUC:18.

Crease RP, Samios NP. 1989 Oct. 1. The science of things that aren't so. News and Observer (Raleigh NC); Sect D:1, 3.

Dagani R. 1993, June 14. Latest cold fusion results fail to win over skeptics. Chemical and Engineering News 71:38–41.

Davis RM. 1985. Publication in professional journals: a survey of editors. IEEE Transactions on Professional Communication 28(2):34–42.

Day RA. 1994. How to write and publish a scientific paper. Phoenix, AZ: Oryx Press.

[DOC] U.S. Department of Commerce. 1994. Statistical abstract of the United States 1994. Washington, DC: Government Printing Office.

Dondis DA. 1973. A primer of visual literacy. Cambridge, MA: MIT Press.

Ehninger D, Gronbeck BE, Monroe AH. 1984. Principles of speech communication. 9th brief ed. Glenview, IL: Scott, Foresman and Co.

Fahnestock J. 1986. Accommodating science: the rhetorical life of scientific facts. Written Communication 3:275–96.

Farrell TB, Goodnight GT. 1981. Accidental rhetoric: The root metaphors of Three Mile Island. Communication Monographs 48:271–300.

Fisher WR. 1987. Human communication as narration: toward a philosophy of reason, value, and action. Columbia: University of South Carolina Press.

Flower L. 1993. Problem-solving strategies for writing. 4th ed. New York: Harcourt Brace Jovanovich.

Garfield E. 1995. Giving credit only where it is due: The problem of defining authorship. The Scientist 9:13.

Gilbert GN, Mulkay M. 1984. Opening Pandora's box: A sociological analysis of scientists' discourse. Cambridge: Cambridge University Press.

Greenland PT. 1994. Essay review: the story of cold fusion. Contemporary Physics 35:209–11.

Gross AG. 1985. The form of the experimental paper: A realization of the myth of induction. Journal of Technical Writing and Communication 15:15–26.

Hall J. 1971. Decisions, decisions, decisions. Psychology Today 5(6):51–54, 86, 88.

Halloran SM. 1984. The birth of molecular biology: an essay in the rhetorical criticism of scientific discourse. Rhetoric Review 3:70–83.

Harmon JE. 1992. Current contents of theoretical scientific papers. Journal of Technical Writing and Communication 22:357–75.

Havelock EA. 1986. Orality, literacy, and Star Wars. Written Communication 3:411–20.

Herndl CG, Fennell BA, Miller CR. 1991. The accidents at Three Mile Island and the shuttle Challenger. In: Bazerman C, Paradis J, editors. Textual dynamics of the professions. Madison: University of Wisconsin Press. p 279–305.

Herrington AJ. 1985. Writing in academic settings: A study of the contexts for writing in two college chemical engineering courses. Research in the Teaching of English 19 (4):331–61.

Holton G. 1973. Thematic origins of scientific thought: Kepler to Einstein. Cambridge, MA: Harvard University Press.

Hoshiko T. 1991 Oct. 28. Facing ethical dilemmas: scientists must lead the charge. The Scientist, 11, 13.

Hsia HJ. 1977. Redundancy: is it the lost key to better communication? Audiovisual Communication Review 25:63–85.

Huizenga JR. 1992. Cold fusion: the scientific fiasco of the century. Rochester, NY: University of Rochester Press.

Huler S. 1990 Nov 12. How to get your research published: editors' thoughts. The Scientist 23.

Jones SE, Palmer EP, Czirr JB, Decker DL, Jensen GL, Thorne JM, Taylor SF, Rafelski J. 1989. Observation of cold nuclear fusion in condensed matter. Nature 338:737.

Katz MJ. 1985. Elements of the scientific paper: a step-by-step guide for students and professionals. New Haven, CT: Yale University Press.

Katz SB. 1992a. Narration, technical communication, and culture: *The soul of a new machine* as narrative romance. In: Secor M, Charney C, editors. Constructing rhetorical education. Carbondale, IL: Southern Illinois Press. p 382–402.

Katz SB. 1992b. The ethic of expediency: classical rhetoric, technology, and the Holocaust. College English 54:255–75.

Katz SB. 1993. Aristotle's rhetoric, Hitler's program, and the ideological problem of praxis, power, and professional discourse. Journal of Business and Technical Communication 7:37–62.

Katz SB, Miller CR. 1996. The low-level radioactive waste siting controversy in North Carolina: toward a rhetorical model of risk communication. In: Herndl CG, Brown SC, editors. Green culture: environmental rhetoric in contemporary America. Madison: University of Wisconsin Press. p 111–40.

Koshland D Jr. 1989 Aug. 25. Minimizing fraud without stifling science. News and Observer (Raleigh, NC); Sect A:16.

Kovacs D., Directory Team, and Kent State University Libraries. 1994. The directory of scholarly electronic conferences [Online]. Available: listserv@kentvm.bitnet

Kuhn TS. 1970. The structure of scientific revolutions. 2nd ed. Chicago: University of Chicago Press.

Kuhn TS. 1979. Metaphor in science. In: Ortony A, editor. Metaphor and thought. London: Cambridge University Press. p 409–19.

Lakoff G, Johnson M. 1980. Metaphors we live by. Chicago: University of Chicago Press.

Lancet. 1984. Spirals and ulcers. Lancet 1:1336.

Latour B, Woolgar S. 1979. Laboratory life: the social construction of scientific facts. Beverly Hills, CA: Sage.

Leary DE. 1990. Psyche's muse: the role of metaphor in the history of psychology. In: Leary DE, editor. Metaphors in the history of psychology. New York: Cambridge University Press. p 1–78.

Maddox J. 1989. End of cold fusion in sight. Nature 340:15.

Marshall E. 1995. Dispute slows paper on "remarkable" vaccine. Science 268:1712–15.

McDonald KA. 1995 April 28. Too many co-authors? The Chronicle of Higher Education 42:A35.

McMillan VE. 1997. Writing papers in the biological sciences. 2nd ed. Boston: Bedford Books.

Medawar PB. 1964 August 1. Is the scientific paper fraudulent? Saturday Review 47:42–43.

Mehlenbacher B. 1992. Rhetorical moves in scientific proposal writing: a case study from biochemical engineering [dissertation]. Pittsburgh, PA: Carnegie Mellon. Unpublished data.

Mehlenbacher B. 1994. The rhetorical nature of academic research funding. IEEE Transactions on Professional Communication 37:157–62.

Memory JD, Arnold JF, Stewart DW, Fornes RE. 1985 March. Physics as a team sport. American Journal of Physics 53:270–71.

Miller CR. 1992. Kairos in the rhetoric of science. In: Witte SP, Nakadate N, Cherry RD, editors. A rhetoric of doing. Carbondale, IL: Southern Illinois UP. p 310–327.

Miller CR, Halloran SM. 1993. Reading Darwin, reading nature; or, on the ethos of historical science. In: Selzer J, editor. Understanding scientific prose. Madison: University of Wisconsin Press. p 106–26.

Monmaney T. 1993 Sept 20. Marshall's hunch. New Yorker 64–72.

Monroe J, Meredith C, Fisher K. 1977. The science of scientific writing. Dubuque, IA: Kendall/Hunt.

Myers G. 1985. The social construction of two biologists' proposals. Written Communication 2:219–45.

Myers G. 1991. Stories and styles in two molecular biology review articles. In: Bazerman C, Paradis J, editors. Textual dynamics of the professions. Madison: University of Wisconsin Press. p 45–75.

[NAS] National Academy of Sciences. 1989. On being a scientist. Washington, DC: National Academy Press.

[NAS] National Academy of Sciences. 1995. On being a scientist: responsible conduct in research. 2nd ed. Washington, DC: National Academy Press.

Nelkin D. 1975. The political impact of technical expertise. Social Studies of Science 5:35–54.

Nelkin, D. 1979. Controversy: Politics of technical decisions. 2nd ed. Beverly Hills: Sage Publications.

[NIH] National Institutes of Health, Consensus Development Panel on Helicobacter pylori in Peptic Ulcer Disease. 1994. Helicobacter pylori in peptic ulcer disease. Journal of the American Medical Association 272:65–69.

North Carolina Low-Level Radioactive Waste Management Authority. 1996. Spectrum [newsletter]. 8(2):104.

Olsen LA, Huckin TN. 1983. Principles of communication for science and technology. New York: McGraw-Hill.

Olsen LA, Huckin TN. 1991. Technical writing and professional communication. 2nd ed. New York: McGraw-Hill.

Paul D, Charney D. 1995. Introducing chaos (theory) into science and engineering: effects of rhetorical strategies on scientific readers. Written Communication 12:396–438.

Pechenik JA. 1987. A short guide to writing about biology. Boston: Little, Brown & Co.

Penrose AM, Fennell BA. 1993 April 1–3. Discourse conventions and the making of knowledge: Linguistic socialization in academic disciplines. Conference on College Composition and Communication, San Diego, CA.

Petrasso RD, Chen X, Wenzel KW, Parker RR, Li CK, Fiore C. 1989a. Problems with the g-ray spectrum in the Fleischmann et al. experiments. Nature 339:183–85.

Petrasso RD, Chen X, Wenzel KW, Parker RR, Li CK, Fiore C. 1989b. Response to Fleischmann et al. Nature 339:667–69.

Polanyi M. 1958. Personal knowledge: towards a post-critical philosophy. Chicago: University of Chicago Press.

Polkinghorne DE. 1988. Narrative knowing and the human sciences. Albany: State University of New York Press.

Pons S, Fleischmann M. 1993. Calorimetry of the Pd-D_2O system: from simplicity via complications to simplicity. Physics Letters A 176:118–29.

Popper K. 1959. The logic of scientific discovery. London: Hutchinson.

Regalado A. 1995 April 7. Multiauthor papers on the rise. Science 268:25.

Reif-Lehrer L. 1990 May 14. For today's scientist, skill in public speaking is essential. The Scientist 25.

Roy R. 1993. Science publishing is urgently in need of major reform [opinion]. The Scientist 7:11.

Rymer J. 1988. Scientific composing processes: how eminent scientists write journal articles. In Jolliffe DA, editor. Advances in writing research. Norwood, NJ: Ablex. Vol 2, Writing in academic disciplines; p 211–50.

Schwitters RF. 1996 February 16. The substance and style of 'big science.' The Chronicle of Higher Education 43:B2.

Science. 1995. Conduct in science [series]. Science 268:1705–19.

Seiken J. 1992 Mar 2. A reviewer's-eye view of evaluation processes at NIH, NSF. The Scientist 19.

SerVaas C. 1994 May/June. The ulcer cure from down under. Saturday Evening Post 62–65.

Shabecoff P. 1989 May 8. Scientist's words on global warming changed by OMB to soften conclusions. News and Observer (Raleigh, NC); Sect A: 1, 4.

Sullivan D. 1991. The epideictic rhetoric of science. Journal of Business and Technical Communication 5:229–245.

Swales J. 1984. Research into the structure of introductions to journal articles and its application to the teaching of academic writing. In: Williams R, Swales J, Kirkman J, editors. Common ground: shared interests in ESP and communication studies. New York: Pergamon. p 77–86.

Swales J. 1990. Genre analysis: English in academic and research settings. Cambridge: Cambridge UP.

Taubes G. 1995. McGill: Analyzing the data. Science 268:1714.

Thompson DK. 1993. Arguing for experimental "facts" in science: a study of research article results sections in biochemistry. Written Communication 8:106–28.

Toulmin S, Rieke R, Janik A. 1984. An introduction to reasoning. 2nd ed. New York: Macmillan. Chap 14, Argumentation in science; p 229–64.

Turbayne CM. 1970. The myth of metaphor. Columbia, SC: University of South Carolina Press.

Watson JD. 1968. The double helix. New York: New American Library.

Wilkes J. 1990 Jan 8. Scientists should spend more time communicating with the public. The Scientist 15.

Winsor DA. 1993. Constructing scientific knowledge in Gould and Lewontin's "The Spandrels of San Marco." In: Selzer J, editor. Understanding scientific prose. Madison: University of Wisconsin Press. p 127–44.

Wright P, Reid F. 1973. Written information: some alternatives to prose for expressing the outcomes of contingencies. Journal of Applied Psychology 57:160–66.

Young RE, Becker AL, Pike KL. 1970. Rhetoric: discovery and change. New York: Harcourt Brace Jovanovich.

• • • • • • • • • • • • • •

Samples Cited

American Physical Society. 1989. Program of the 1989 Annual Meeting for the Division of Nuclear Physics. Asilomar, CA: American Physical Society.

American Society for Microbiology. 1990. Call for abstracts. Anaheim, CA: American Society for Microbiology.

Annals of Internal Medicine. 1992. Information for authors. Annals of Internal Medicine 117(1):I-9–I-12.

Astrophysical Journal. 1990. Instructions to authors. Astrophysical Journal 357(1):i–v.

Atmospheric Environment. 1996. The preparation of papers for Atmospheric Environment. Atmospheric Environment 30:ix–x.

Bell GD. 1991. Anti-*Helicobacter pylori* therapy: clearance, elimination, or eradication? Lancet 337:310–11.

Blackman CF. 1994 January 27–29. Interaction of state and time varying magnetic fields with biological systems. Ohlendorf Foundation Meeting, Margarita Island, Venezuela.

Blackman CF, Benane SG, House DH. 1985, June 16–20. A pattern of frequency dependent responses in brain tissue to ELF electromagnetic fields. The Annual Meeting of the Bio-electromagnetics Society, San Francisco, CA.

Blaser MJ. 1987. Gastric *Campylobacter*-like organisms, gastritis, and peptic ulcer disease. Gastro-enterology 93:371–83.

Blaser MJ. 1996 Feb. The bacteria behind ulcers. Scientific American 104–7.

Brody AJ, Pelton MR. 1988. Seasonal changes in digestion in black bears. Canadian Journal of Zoology 66:1482–84.

Brown FA. 1954 Apr. Biological clocks and the fiddler crab. Scientific American 190:34–37.

Brown FA. 1959. Living clocks. Science 130:1535–44.

Brown FA. 1960. Life's mysterious clocks. Saturday Evening Post 233:18–19, 43–44.

Brown FA. 1962. Responses of the planarian, Dugesia, and the protozoan, Paramecium, to very weak horizontal magnetic fields. Biological Bulletin 123:264–81.

Burkholder JM, Lewitus AJ. 1994. Trophic interactions of ambush predator dinoflagellates in estuarine microbial food webs. Proposal submitted to the National Science Foundation. North Carolina State University, Department of Botany.

Burkholder JM, Noga EJ, Hobbs CH, Glasgow HB Jr, Smith SA. 1992. New "phantom" dinofla-gellate is the causative agent of major estuarine fish kills. Nature 358:407–10; Nature 360:768.

Burkholder JM, Rublee PA. 1994. Improved detection of an ichthyotoxic dinoflagellate in estu-aries and aquaculture facilities. Proposal submitted to the (U.S.) National Sea Grant Col-lege Program. North Carolina State University, Department of Botany.

Byrne T, Hibbard J. 1987. Landward vergence in accretionary prisms: The role of the backstop and thermal history. Geology 15:1163-1167.

Clay DE, Clapp CE, Linden DR, Molina JAE. 1989. Nitrogen-tillage-residue management: 3. observed and simulated interactions among soil depth, nitrogen mineralization, and corn yield. Soil Science 147:319–25.

[DOE] US Department of Energy. 1991. Guide for the submission of unsolicited proposals [Brochure DOE/MA-0095]. Washington, DC: Department of Energy, Procurement and Assistance Management Directorate.

Fabrizio MC. 1987. Growth-invariant discrimination and classification of striped bass stocks by morphometric and electrophoretic methods. Transactions of the American Fisheries Soci-ety 116:728–36.

Fleischmann M, Pons S. 1989. Electrochemically induced nuclear fusion of deuterium. Journal of Electroanalytical Chemistry 261:301–8.

Fulbright MA, Reynolds SP. 1990. Bipolar supernova remnants and the obliquity dependence of shock acceleration. Astrophysical Journal 357:591–601.

Goodbred SL Jr, Hine AC. 1995. Coastal storm deposition: salt-marsh response to a severe extra-tropical storm, March 1993, west-central Florida. Geology 23:679–82.

Graham DY, Evans DJ Jr, Evans DG. 1989. Detection of *Campylobacter pylori* infection. Lancet 1:569.

Graham DY, Lew GM, Klein PD, Evans DG, Evans DJ Jr, Saeed ZA, Malaty HM. 1992. Effect of treatment of *Helicobacter pylori* infection on the long-term recurrence of gastric or duodenal ulcer. Annals of Internal Medicine 116:705–8.

Hamilton MJ, Kennedy ML. 1987. Genic variability in the racoon procyon lotor. The American Midland Naturalist: 118:266–74.

Huyghe P. 1993 Apr. Killer Algae. Discover 14(4):71–75.

Journal of Heredity. 1996. Information for contributors. Journal of Heredity 87:iii–iv.

Lam SK. 1989. Duodenal ulcer relapse after eradication of *Campylobacter pylori*. Lancet 1:384.

Lancet. 1990. Information for authors. Lancet 336:1117.

Lim EM, Rauzier J, Timm J, Torrea G, Murray A, Gicquel B, Portnoï D. 1995. Identification of *Mycobacterium tuberculosis* DNA sequences encoding exported proteins by using *phoA* gene fusions. Journal of Bacteriology 177:59–65.

Loffeld RJLF, Stobberingh E, Arends JW. 1989. Response to David. Y. Graham et al. Lancet 1:569–70.

Lorimer CG, Chapman JW, Lambert WD. 1994. Tall understorey vegetation as a factor in the poor development of oak seedlings beneath mature stands. Journal of Ecology 82:227–37.

Mallin MA, Burkholder JM, Larsen LM, Glasgow HB Jr. 1995. Response of two zooplankton grazers to an ichthyotoxic estuarine dinoflagellate. Journal of Plankton Research 17:351–63.

Marshall BJ, Warren JR. 1984. Unidentified curved bacilli in the stomach of patients with gastritis and peptic ulceration. Lancet 1:1311–15.

Marshall B, Armstrong J, McGechie D, Glancy R. 1985. Attempt to fulfil Koch's postulates for pyloric campylobacter. Medical Journal of Australia 142:436–39.

Marshall BJ, McCallum RW, Prakash C. 1987. *Campylobacter pyloridis* and gastritis. Gastroenterology 92:2051.

Marshall BJ, Warren JR, Goodwin CS. 1989. Duodenal ulcer relapse after eradication of *Campylobacter pylori*. Lancet 1:836–37.

[MEPS] Marine Ecology Progress Series. 1996. Editorial statement. Marine Ecology Progress Series 134.

Morris AJ, Nicholson GI, Perez-Perez GI, Blaser MJ. 1991. Long-term follow-up of voluntary ingestion of *Helicobacter pylori*. Annals of Internal Medicine 114:662–63.

[NASA] National Aeronautics and Space Administration. 1993. NASA research announcement soliciting proposals for theory in space astrophysics. NRA 93-OSSA-06. Washington, DC: National Aeronautics and Space Administration, Office of Space Science.

[NASA] National Aeronautics and Space Administration. 1995. Instructions for responding to NASA research announcements (NFSD 89-19). Washington, DC: National Aeronautics and Space Administration.

National Sea Grant College Program. 1994. Statement of opportunity for funding: marine biotechnology. Raleigh, NC: North Carolina State University, UNC Sea Grant College Program.

Nature. 1995. Guide to authors. Nature 378:6552.

[Neuroscience] Society for Neuroscience. 1994. Preliminary program for 24th annual meeting, Nov 13–14; Miami Beach, FL.

[NIH] National Institutes of Health. 1993. Helpful hints: preparing an NIH research grant application. Bethesda, MD: National Institutes of Health, Division of Research Grants.

[NIH] National Institutes of Health. N.D. Peer review of NIH research grant applications. Washington, DC: National Institutes of Health, Division of Research Grants.

North Carolina Low-Level Radioactive Waste Management Authority. 1996. Authority begins comprehensive site assessment. Spectrum [newsletter] 8:1,4.

[NSF] National Science Foundation. 1990. Grants for research and education in science and engineering: an application guide [NSF 90-77]. Washington, DC: National Science Foundation.

[NSF] National Science Foundation. 1995. Grant proposal guide [NSF 95-27]. Arlington, VA: National Science Foundation.

Perez-Perez GI, Dworkin BM, Chodos JE, Blaser MJ. 1988. *Campylobacter pylori* antibodies in humans. Annals of Internal Medicine 109:11–17.

Peterson WL. 1989. Antimicrobial therapy of duodenal ulcer? Hold off for now! Gastroenterology 97:508–10.

Phytopathology. 1988. Author's guide for manuscript preparation. Phytopathology 78:4–6.

Quigley MF, Slater HH. 1994. Mapping forest plots: A fast triangulation method for one person working alone. Southern Journal of Applied Forestry 18:133–36.

Rauws EAJ, Langenberg W, Houthoff HJ, Zanen HC, Tytgat GNJ. 1988. *Campylobacter pyloridis*-associated chronic active antral gastritis: a prospective study of its prevalence and the effects of antibacterial and antiulcer treatment. Gastroenterology 94:33–40.

Reynolds SP, Borkowski KJ, Blondin JM. 1994. X-ray emission and dynamics of supernova remnants. Proposal submitted to NASA's Astrophysics Theory Program. North Carolina State University, Department of Physics.

Reynolds SP, Lyutikov M, Blandford RD, Seward FD. 1994. X-ray evidence for the assocation of G11.2-0.3 with the supernova of 386 AD. Monthly Notices of the Royal Astronomical Society 271:L1–L4.

Ross PI, Jalkotzy MG. 1992. Characteristics of a hunted population of cougars in southwestern Alberta. Journal of Wildlife Management 56:417–26.

Schwab AP, Kulyingyong S. 1989 March. Changes in phosphate activities and availability indexes with depth after 40 years of fertilization. Soil Science 147:179–85.

Sud VK, Sekhon GS. 1989. Blood flow through the human arterial system in the presence of a steady magnetic field. Physics in Medicine and Biology 34:795–805.

Sylvester, RJ. 1988. A bayesian approach to the design of phase II clinical trials. Biometrics 44:823–36.

Terracini ED, Brown FA. 1962. Periodisms in mouse "spontaneous" activity synchronized with major geophysical cycles. Physiological Zoology 35:27–37.

Veldhuyzen van Zanten SJO, Goldie J, Riddell RH, Hunt RH. 1988. *Campylobacter pylori* infection. Annals of Internal Medicine 109:925.

Warren JR, Marshall B. 1983. Unidentified curved bacilli on gastric epithelium in active chronic gastritis. Lancet 1:1273–75.

Watson JD, Crick FHC. 1953. A structure for deoxyribose nucleic acid. Nature 171:737–38.

Zukav G. 1979. The dancing Wu Li masters: an overview of the new physics. New York: William Morrow.

Acknowledgments (*continued from copyright page*)

Chapter 9

Barry J. Marshall, J. Robin Warren. Letters to *The Lancet* (June 4, 1983); "Unidentified Curved Bacilli in the Stomach of Patients with Gastritis and Peptic Ulceration." *The Lancet,* Saturday 16 June 1984. Letters from B. J. Marshall, J. R. Warren, and C. S. Goodwin (April 15, 1989); R. J. L. F. Loffeld, E. Stobberingh, and J. W. Arends (September 2, 1989); S. K. Lam (1989); G. D. Bell (February 2, 1991).Reprinted by permission of The Lancet Limited, London, England.

David Y. Graham, MD et al. "Effect of Treatment of Helicobacter pylori Infection on the Long-term Recurrence of Gastric or Duodenal Ulcer." *Annals of Internal Medicine,* 1992: 116, pp. 705–8. Letter to *Annals of Internal Medicine* (1988) from S. J. O. Veldhuyzen van Zanten, MD, MPH, J. Goldie, R. H. Riddell, MD, Richard H. Hunt, MD. Reprinted by permission of the American College of Physicians.

Martin J. Blaser, MD. "Gastric Camphylobacter-like Organisms, Gastritis, and Peptic Ulcer Disease." *Gastroenterology* 93: 371–83 (1987). Reprinted by permission of the author and W. B. Saunders Company, Orlando, FL.

Martin J. Blaser, MD. "The Bacteria Behind the Ulcers." *Scientific American* (February 1996), pp. 104–7. Copyright © 1996 by Scientific American, Inc. All rights reserved. Reprinted with permission of the author and Scientific American, Inc.

Chapter 10

JoAnn M. Burkholder et al. "New 'phantom' dinoflagellate is the causative agent of major estuarine fish kills." *Nature* 358: 407–10, July 30, 1992. Copyright © 1992 Macmillan Magazines Limited. Reprinted with permission from *Nature* and the author.

Michael A. Mallin, JoAnn M. Burkholder, L. Michael Larsen, and Howard B. Glasgow, Jr. "Response of two zooplankton grazers to an ishthyotoxic estuarine dinoflagellate." *Journal of Plankton Research* 17: 351–63 (1995). Reprinted by permission.

Patrick Huyghe. "Algae." *Discovery* magazine 14 (4), 1993. Copyright © 1993 by Patrick Huyghe. Reprinted by permission of the Walt Disney Publishing Group.

Chapter 11

Stephen P. Reynolds et al. "X-ray evidence for the association of G11.2-0.3 with the supernova of 386 AD." *Monthly Notices of the Royal Astronomical Society* 271, No.3, L1–L4 (December 1, 1994). Reprinted by permission of Blackwell Science Ltd., Oxford, England.

Michael S. Fulbright and Stephen P. Reynolds. "Bipolar supernova remnants and the obliquity dependence of shock acceleration." *The Astrophysical Journal* 357, 591–601 (July 10, 1990). Published by The University of Chicago Press for The American Astronomical Society. Reprinted by permission.

Index of Names and Titles

Subject Index